π-Stacked Polymers and Molecules

Tamaki Nakano
Editor

π-Stacked Polymers and Molecules

Theory, Synthesis, and Properties

Springer

Editor
Tamaki Nakano
Catalysis Research Center (CRC)
Hokkaido University
Sapporo, Japan

ISBN 978-4-431-56112-5 ISBN 978-4-431-54129-5 (eBook)
DOI 10.1007/978-4-431-54129-5
Springer Tokyo Heidelberg New York Dordrecht London

Printed on acid-free paper

Springer is part of Springer Science+Business Media (www.springer.com)

Preface

π-Electronic polymer and molecular systems exhibit photo-electronic properties such as conduction, emission, optical non-linearity, and even photo-catalytic activity. The properties and functions are based on the mobility of a hole or an electron in the π-conjugated systems, within a molecule and between molecules. It is hence important to design a polymer or a molecular system so that charges can be transported smoothly and transport can be controlled.

The most successful and practical examples in this discipline are main-chain conjugated polymers represented by polyacetylene, whose conduction has been ascribed to mobility of charges injected by doping through a conjugated chain and among chains. Since the finding of conduction of polyacetylene by Shirakawa, Heeger, and MacDiarmid (awarded the Nobel Prize in Chemistry 2000), various main-chain conjugated polymers have been designed and synthesized, among which poly(3,4-ethylenedioxythiophene) (PEDOT) in combination of poly(styrenesulfonate) (PSS) is one of the most industrially applied examples. Along with the excellent conduction behaviors of main-chain conjugated polymers, they have some drawbacks including deep color, poor solubility, and relatively high HOMO levels. These properties are inherent in the long main-chain conjugation: a narrow gap in HOMO–LUMO levels due to the long conjugation allows photo-excitation by visible light, and rigidity of the polymer chain arising from limited rotation freedom of bonds that constitute the long main-chain conjugation system promotes interchain interactions. Although facile doping is feasible due to the rather high HOMO level, the other side of this nature is oxidation by O_2 in the air which deteriorates electronic properties.

An alternative π-electronic macromolecular system is a π-stacked polymer. A stack of aromatic groups is an ideal assembly where mobility of charges is realized. Piling up multiple aromatic groups without forming covalent bonds among them in an orderly stack in a polymer system may create novel photo-electronic polymeric materials free from the shortcomings noted above and even with improved properties. Based on this idea, π-stacked polymeric and molecular systems have been designed. The first four chapters of this book were written by four scientists/groups who synthesize and characterize such systems, and the last chapter by a group who sheds light on the relation between π-stacked structure and their photo-electronic properties from a theoretical point of view.

Prior to development of synthetic π-stacked polymers, characteristic features of π-stacked molecules had been studied for small molecule systems. A representative example is the tetrathiafulvalene (TTF)-tetracyanoquinodimethane (TCNQ) complex in crystal prepared by the Johns Hopkins group [Cowan et al., *Journal of the American Chemical Society*, vol. 95, p. 948 (1973)]. In the TTC–TCNQ complex, charges generated by the donor–acceptor interaction between TTF and TCNQ are transported/delocalized through independent columns of TTF and TCNQ stacks in crystal, leading to a molecular conductor. In order to find parallels of the TTF–TCNQ crystal, various combinations of donor and acceptor molecules were studied.

In addition, for biological macromolecular systems, DNA duplexes have stacked aromatic moieties within their helices: base pairs form a long stacking system. Charge transport through a DNA duplex was pointed out by Barton in 1993 [Barton et al., *Science*, vol. 262, p. 1025 (1993)]. Barton's group synthesized DNA double helices having an emitting chromophore and a quencher group that were separated by stacked base pairs, and charge transport was established based on emission measurements where quenching requires charge transport. The DNA π-stack was named "π-way" by the group.

Although the stacked crystal systems and DNA π-stack are intriguing systems from the point of view of molecular electronic materials, TTC–TCNQ and its family, being crystalline materials, are difficult to fabricate into desired forms such as thin film; and DNAs, with alternating copolymers of sugar units and phosphonate units, may not be durable-enough organics under practical conditions. Artificial π-stacked polymers are expected to overcome these drawbacks and find wide application in organic electronics complementing main-chain conjugated macromolecular systems.

Sapporo, Japan Tamaki Nakano

The editor thanks Dr. Yasuhito Koyama, Associate Professor at Catalysis Research Center, Hokkaido University, Japan, who kindly pointed out corrections and suggested improvements.

Contents

Contributors

Antonella Caterina Boccia Istituto per lo Studio delle Macromolecole (CNR), Milano, Italy

Andrea Cappelli Dipartimento Farmaco Chimico Tecnologico and European Research Centre for Drug Discovery and Development, Università degli Studi di Siena, Siena, Italy

Yoshiki Chujo Department of Polymer Chemistry, Graduate School of Engineering, Kyoto University, Kyoto, Japan

David M. Collard School of Chemistry and Biochemistry, Georgia Institute of Technology, Atlanta, GA, USA

Alessandro Donati Dipartimento Farmaco Chimico Tecnologico and European Research Centre for Drug Discovery and Development, Università degli Studi di Siena, Siena, Italy

Germano Giuliani Dipartimento Farmaco Chimico Tecnologico and European Research Centre for Drug Discovery and Development, Università degli Studi di Siena, Siena, Italy

Giorgio Grisci Dipartimento Farmaco Chimico Tecnologico and European Research Centre for Drug Discovery and Development, Università degli Studi di Siena, Siena, Italy

Raniero Mendichi Istituto per lo Studio delle Macromolecole (CNR), Milano, Italy

Yasuhiro Morisaki Department of Polymer Chemistry, Graduate School of Engineering, Kyoto University, Kyoto, Japan

Tamaki Nakano Catalysis Research Center (CRC), Hokkaido University, Sapporo, Japan

Marco Paolino Dipartimento Farmaco Chimico Tecnologico and European Research Centre for Drug Discovery and Development, Università degli Studi di Siena, Siena, Italy

Filippo Samperi Istituto di Chimica e Tecnologia dei Polimeri (CNR), Catania, Italy

Tetsuya Taketsugu Department of Chemistry, Faculty of Science, Hokkaido University, Sapporo, Japan

Takao Tsuneda Fuel Cell Nanomaterials Center, University of Yamanashi, Kofu, Japan

Salvatore Vomero Dipartimento Farmaco Chimico Tecnologico and European Research Centre for Drug Discovery and Development, Università degli Studi di Siena, Siena, Italy

Synthesis and Properties of π-Stacked Vinyl Polymers

Tamaki Nakano

Introduction

Control of spatial arrangements of constituent molecules is crucial in designing organic materials with desired functions because molecules in a regulated assembly often indicate properties that are different from those in solution or in gas phase due to specific intermolecular interactions. Hence, ultimate fabrication of molecule-based organic material can be achieved by arranging molecules exactly in the positions where they should be. Although this is not a simple task, various methods have been developed for molecular assembly control of organic molecules ranging from the relatively well-established methods such as LB membrane formation, crystal engineering, and molecular complex formation based on metal–ligand coordination or hydrogen bonding, in which accumulation of relatively weak intermolecular forces enables construction of a long-range molecular arrangements [1–9], to instrument-assisted methods such as molecular manipulation using optical tweezers or microscopes [10–12].

Another way to organize functional molecules is to align them as a part of a polymer chain with a regulated conformation and configuration where the polymer backbone acts as a structural template or scaffold. This is feasible either by polymerizing functional molecules as monomers or by attaching functional molecules through chemical reaction with a polymer chain. In the former case, functional molecules to be aligned have a polymerizable moiety such as vinyl group; they become a part of monomeric unit incorporated into a chain, typically as side-chain groups. In this case, monomer structure itself may significantly contribute to the formation of regulated polymer chain structure through intermolecular repulsion or attraction. In the latter case, the polymer chain and the functional molecules to be

T. Nakano (✉)
Section of Catalytic Assemblies, Catalysis Research Center (CRC), Hokkaido University, N 21, W 10, Kita-ku, Sapporo 001-0021, Japan
e-mail: tamaki.nakano@cat.hokudai.ac.jp; tamaki.nakano@gmail.com

T. Nakano (ed.), *π-Stacked Polymers and Molecules: Theory, Synthesis, and Properties*, DOI 10.1007/978-4-431-54129-5_1,

attached have corresponding reactive or interacting groups. An advantage of these polymer-based methods over the other methodologies is that molecular orders formed along a polymer chain are generally more stable due to covalent bonds. Also, a very tight or close packing of molecules (monomeric units), which is not plausible for the method based on weak intermolecular attractions, is possible when the enthalpy gain in covalent bond formation by polymerization or by polymer reaction outweighs the loss in free energy in packing molecules tightly into a small space around the polymer chain. Although these polymer-based methods generally require more synthetic efforts than the intermolecular-force-based methods which are mainly applied for small molecules, a wider variety of molecular orders may be possible based on the recent advances in polymerization chemistry which have realized various macromolecular architectures. In addition, a clear advantage of polymers over small molecule systems is that polymeric materials may be readily fabricated into desired forms by solution processing such as spin coating or drop casting to form a thin film.

Molecules can assemble in various different fashions. Among them, a relatively simple order known for aromatic molecules is a π-stacked assembly where planar aromatic molecules form an orderly pile or heap in a co-facial manner. π-Stacked molecules often show characteristic physical properties such as reduced photo absorption intensity (hypochromism, hypochromic effect), excimer or exciplex formation on photo excitation, and charge transport. The last property has practical importance and has been demonstrated for organic molecules and metallic complexes, such as the conducting tetrathiafulvalene (TTF)-7,7,8,8,-tetracyanoquinodimethane (TCNQ) complex where TTF and TCNQ form independent columnar stacks [13] and the tetramethyltetraselenafulvalene (TMTSF)-PF_6 complex where TMTSF molecules are in a stacked arrangement [14]. In either case, the stacked molecules play an important role in effectively transport a hole or an electron [15–20].

Although the molecular stacks of TTF-TCNQ and (TMTSF)-PF_6 complexes were achieved in crystal, π-stacked structure has been also constructed using polymer chains. Stable and regulated π-stacked structures in polymers are observed in DNAs [21], poly(phenyleneethylnylene)s [22, 23], oligomers consisting of alternating donor and acceptor monomeric units [24, 25], oligomers having perylene moieties in the main chain [26–28], and polymers of dibenzofulvene (**DBF**) and its derivatives [29–40]. In a DNA duplex, paired bases are regularly stacked inside the helical strand. In the poly(phenyleneethylnylene)s, phenylene groups in the main chain closely overlap due to helical folding of the polymer chain. In oligomers consisting of aromatic donor and acceptor units, the chain folds into a pleated conformation due to a donor–acceptor. In poly(**DBF**), the side-chain aromatic groups are tightly stacked on top of each other along the almost *all-trans* main chain.

A practical property expected for π-stacked polymers is charge transport. This was first proposed for DNA duplexes and was examined using double helices having an emitting organic group at one end and a quenching group [41–43]. Charge transport through π-stacked base pairs was established based on the fact that quenching requires transfer of charges between the emitting and quenching groups. Through this work, DNA double helix was given a name, "π-way." However, DNA double helix is not a durable chemical from a view of material sciences as it can be readily hydrolyzed, metabolized by DNases, and unfolded into single, unstacked chains at an elevated temperature.

Poly(**DBF**) is a vinyl polymer and is kin of polystyrene, a versatile and durable material unlike the other π-stacked polymers. However, poly(**DBF**) and its derivatives have characteristic structural and functional features that are not known for polystyrene and its analogues and can be recognized, in the context of this chapter, as a polyethylene derivative which creates an ordered alignment of fluorene, a chromophore molecule. Hereafter, the synthesis, structure, and properties of π-stacked poly(**DBF**) and its derivatives are described in detail.

Synthesis of π-Stacked Poly(dibenzofulvene)

Structure and Reactivity of Dibenzofulvene

Dibenzofulvene (**DBF**), the monomer consisting of fluorene backbone and vinyl (vinylidene, 1,1-ethenediyl) group, can be readily synthesized from 9-hydroxymethylfluorene (Aldrich) by the reaction in a methanol solution containing KOH [44, 45] and purified by recrystallization from hexane. Polymerization of this compound was first described in 1937; no details of the reaction or the structure of the products have been reported [46].

The structure of **DBF** is closely related to that of 1,1-diphenylethylene (**DPE**) which is known as a representative vinyl compound that does not produce a homopolymer through any kind of catalysis. **DPE**'s low reactivity has been ascribed to steric reasons [47–50]. The only difference between **DBF** and **DPE** in chemical structure is the presence/absence of a single bond connecting two benzene rings to form a fused ring system. In spite of this tiny difference, **DBF** readily affords a polymer by anionic, radical, and cationic catalyses while **DPE** does not (Fig. 1).

To obtain information on why **DBF** gives a polymer, while **DPE** does not, the electron density of the vinyl group was first considered. The ^{13}C chemical shift of the β-vinyl carbon of **DBF** was 107.7 ppm, while those of **DPE** and styrene were 114.3 and 113.6 ppm, respectively, as determined in $CDCl_3$ at 23 °C, suggesting that

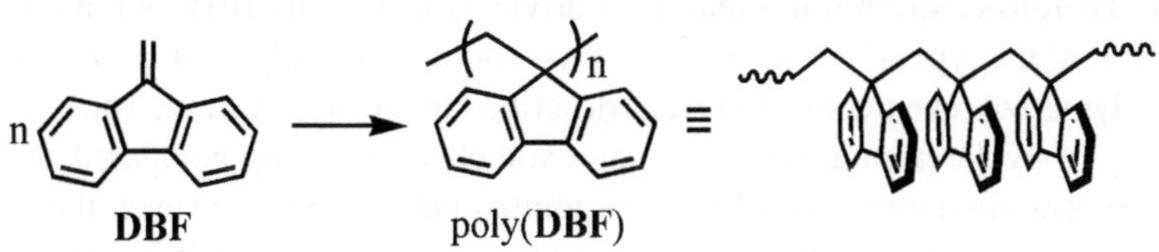

Fig. 1 Polymerization of **DBF**

Table 1 Mulliken charges of vinyl group of DBF, DPE, and styrene[a]

	DBF		**DPE**		Styrene	
Basis set	α-Carbon	β-Carbon	α-Carbon	β-Carbon	α-Carbon	β-Carbon
6-31G*	0.154	−0.463	0.057	−0.423	−0.138	−0.416
6-31G**	0.121	−0.326	0.036	−0.298	−0.084	−0.300

[a]Estimated by ab initio calculations (HF//MP2/6-31G*, HF//MP2/6-31G**) for PM3-optimized structures

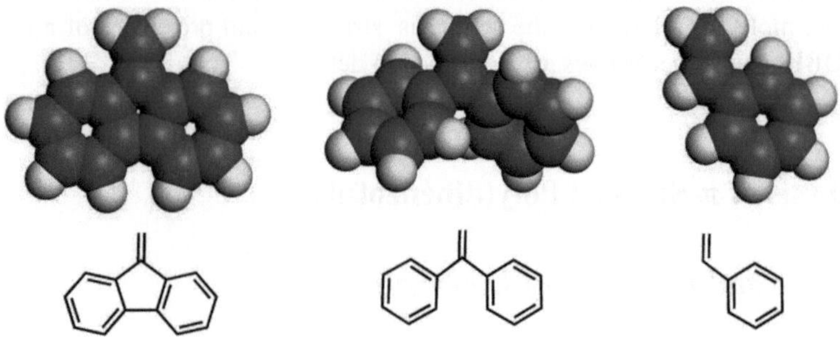

DBF **DPE** styrene

Fig. 2 Structures of **DBF**, **DPE**, and styrene optimized by AM1 calculation

the β-carbon of **DBF** has a higher electron density compared with those of **DPE** and styrene. This was supported by the Mulliken charges obtained for **DBF**, **DPE**, and styrene by ab initio calculations (HF//MP2/6-31G* and HF//MP2/6-31G**) (Table 1). The monomer structures were optimized by PM3 calculation [51] prior to the ab initio analysis. The results support the finding that β-carbon of **DBF** has a higher electron density; however, these results only indicate the reactivity of **DBF** during cationic polymerization.

Next, the structure of **DBF** was examined using semiempirical molecular orbital calculations. The AM1 calculation [51] afforded a nearly planar structure for **DBF** which is not possible for **DPE** (Fig. 2). The nearly planar structure should not only reduce steric hindrance but also effectively stabilize the anionic, cationic, or radical species at the α-carbon through electron conjugation. In addition, the bond angle between the two vinyl-to-phenyl bonds in **DBF** was 105.6°, which is significantly deviated from 120°, implying that the vinyl group of **DBF** contains significant strain energy. Therefore, strain relief may be a driving force of the **DBF** polymerization regardless of the type of active species. This would reasonably explain the fact that **DBF** polymerizes even with *t*-BuOK as described in the next section. The enhanced reactivity of cyclic monomers having an exomethylene group compared with the acyclic parent vinyl compound has been pointed out [52–55]; however, the acyclic counterparts do slowly produce homopolymers in the examples in [52–55], whereas **DPE** does not produce a homopolymer.

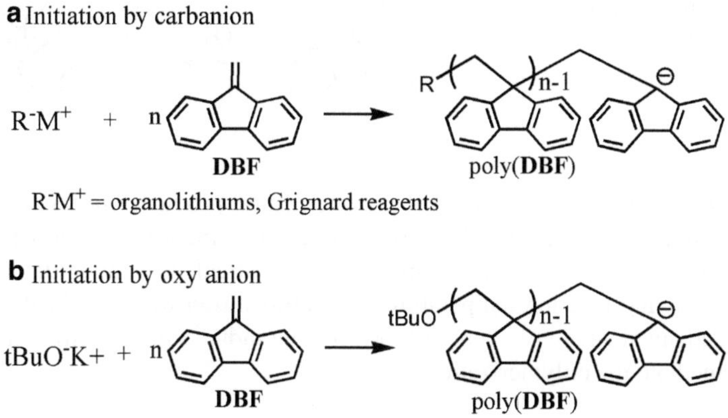

Fig. 3 Anionic polymerization of **DBF** using carbanion species (**a**) and *t*-BuOK (**b**)

Polymerization

DBF can be polymerized by anionic, cationic, and radical modes similarly to styrene [29]. In the anionic polymerization, the reactions using *n*-butyllithium (*n*-BuLi) and 9-fluorenyllithium (FlLi) led to nearly complete monomer consumption at −78 °C, while that using a Grignard reagent (CH_3MgBr) required room temperature to achieve a reasonable monomer conversion (Fig. 3a). It is interesting that *t*-BuOK, a weak nucleophile, very effectively polymerized **DBF** at −78 °C (Fig. 3b). In this case, a 9-fluorenyl anion, a carbanion, is generated from the oxy anion (*t*-BuO⁻) by the reaction with **DBF**. Although the generation of a carbanion from an oxy anion has been accomplished using Si-containing cyclic compounds, the method using a hydrocarbon compound is only possible with **DBF** so far [56, 57]. The anionic methods were used to synthesize uniform oligomers of **DBF** having completely controlled chemical and stereochemical structures used for the detailed structure analyses as described in the following section [30].

Although radical polymerization is more versatile and convenient than anionic or coordination polymerization, it is generally less selective in stereochemistry except for limited examples [58–60]. As exceptional examples, bulky acrylic monomers such as 1-phenyldibenzosubery methacrylate and triphenylmethyl methacrylate can be polymerized in a highly configuration- and conformation-specific manner, giving a highly isotactic, helical polymer [61–63]. As shown by these examples, it is of significant benefit if a polymer with a regulated conformation can be obtained by facile radical polymerization rather than the anionic polymerization which requires a strict control of reaction conditions.

The radical polymerization of **DBF** has been studied in detail [31]. No significant difference in chemical structure was found between poly(DBF)s synthesized by radical and anionic polymerizations. The conformation-specificity (stereospecificity) of the **DBF** radical polymerization was also similar to that in the anionic

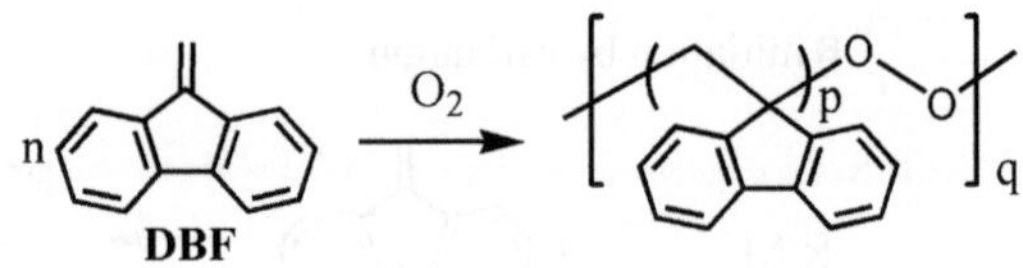

Fig. 4 Copolymerization of **DBF** with O_2

polymerization. However, the radical polymerization products appeared to have a very small amount of irregular, defective conformation incorporated into the mostly π-stacked chain. The conformation-specificity was affected by the initial monomer concentration and reaction temperature. This observation was explained by a proposed mechanism where two types of the growing radicals having different conformations mediate the polymerization.

Copolymerization DBF with O_2

Due to its high reactivity, **DBF** reacts even with oxygen contained in the air in the solid state and affords a copolymer consisting of **DBF** and –O-O– units (Fig. 4). Solid-state polymerization is an efficient method in controlling polymer structure. Several diyne and diene monomers undergo solid-state polymerization, leading to polymers with controlled chemical and stereostructures [64–70]. However, solid-state vinyl polymerization is rare ([71] and references therein; [72–75]).

Upon standing in the flask under air or pure O_2 atmosphere, **DBF** crystals gave a polymer. The monomer crystals became slightly yellow during the reaction, but there was no change in the apparent state of the solid material, indicating that the reaction took place in the solid state. The reaction under pure O_2 resulted in higher polymer yield than that under air while no reaction was confirmed under N_2 atmosphere. These results indicate that oxygen, as a part of air or as a pure purge gas, is involved in the polymerization.

The polymerization occurred in the dark as well as in the presence of ambient light. The ambient light seems to have little effect on the polymerization behavior. The observed polymerization activity in the dark is in contrast to the fact that sunlight or UV irradiation is necessary for the octadecyl sorbate-O_2 copolymerization system [76]. Pyrolysis experiments gave a support to the structure: molecular weight of the polymer significantly decreased in a heat treatment at 100–175 °C due to thermal scission of the –O-O– bonding taking place on heating.

*An Alternative, Non-polymerization Method to Synthesize Oligo(*DBF*)*

Short **DBF** oligomers have been synthesized by small-molecule-oriented synthetic method (Fig. 5) [39, 40]. In this case, reaction of fluorene with paraformaldehyde in the presence of a base afforded a dimer having hydrogen atoms at the terminals.

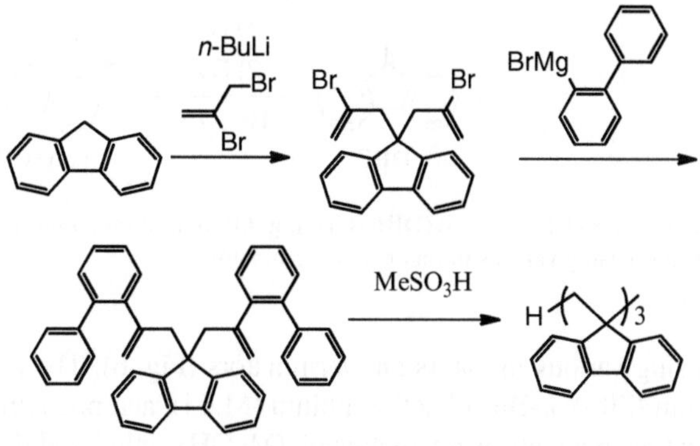

Fig. 5 An alternative way to synthesize π-stacked **DBF** oligomers

Palladium-catalyzed coupling of the 2-biphenylmagnesium bromide to 9,9-bis(2-bromopropenyl)fluorene which was prepared from fluorene and 2,3-dibromopropene followed by an acid catalyzed (intramolecular) Friedel–Crafts alkylation led to a trimer. However, this method has a limitation that it can only produce a short chain, not polymer but up to tetramer.

Proofs of π-Stacked Structure

Synthesis and Isolation of Uniform Oligomers

Determining polymer conformations without any ambiguity is still not a straightforward task even using state-of-the-art analytical techniques. Principally, no conclusive methods have been devised to learn conformation or intermolecular structures of amorphous polymers. For a polymer chain, the following way may be most convincing as far as feasible: (1) oligomers with defined chain length (degree of polymerization) without molar mass distribution are isolated, (2) their single crystal structures are elucidated by X-ray analyses (solid-state structures), (3) their solution conformation is proposed based on the solid-state conformation where theoretical simulations comparing the solid-state structure with other possible structures are useful, (4) credibility of the conformation proposed in (3) is examined by solution property measurements, and (5) dependence of solution and/or solid-state properties on chain length is evaluated. Through step (5), a conformation of an oligomer whose degree of polymerization (DP) is beyond the point where changes in properties depending on DP saturate can be taken as a polymer conformation. This tactics worked out well for poly(**DBF**) though it may not be applicable for a conformation that is stable only for a rather long chain such as helix for poly(meth) acrylates [61].

The π-stacked structure of poly(**DBF**) has been firmly established both in the solid state (crystal) and in solution [30]. The structural analyses were performed using pure oligomers with various terminal groups prepared by anionic polymerization

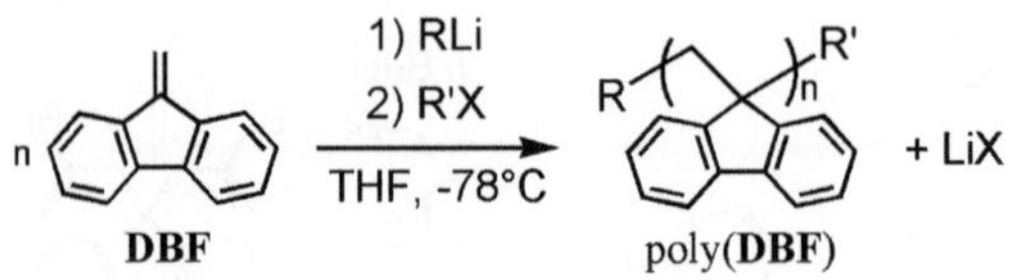

Fig. 6 Synthesis of poly(**DBF**)s having different terminal groups (R and R′) by anionic polymerization using various initiators and terminators

using various initiators and terminators (Fig. 6). The initiators were 9-fluorenyllithium (FlLi), *n*-BuLi, methyllithium (MeLi), and phenyllithium (PhLi), and the terminating reagents were methanol (MeOH), ethyl iodide (EtI), and benzyl bromide ($PhCH_2Br$). Oligomers having uniform degree of polymerization (DP) were isolated by preparative size-exclusion chromatography (SEC). The conditions and results of the reactions with the expected polymer structures are summarized in Table 2.

In all cases, the **DBF** monomer was consumed in a high yield; however, reaction products were partially insoluble in THF or chloroform at a higher [**DBF**]/[initiator] ratios. The THF-soluble and THF-insoluble parts indicated very similar IR signals, suggesting that chemical structures of the soluble and insoluble parts are identical. Higher-molecular-weight fractions may tend to become insoluble probably due to interchain interactions of poly(**DBF**) chain.

The reactions afforded the products (oligomers and polymers) with five terminal-group combinations, **1**–**5**. In the structure of **1**, $n-1$ instead of n has been used to describe the number of monomeric units incorporated into the chain through polymerization (DP) so that n corresponds to the number of fluorene units in a chain. This is because the number of fluorene units in a chain is more important in discussing the formation of π-stacked conformation of **1** than DP. For **2**–**5**, the number of fluorene units corresponds to DP.

The efficiency of introducing terminal groups was confirmed by MALDI mass spectrum. The observed mass numbers supported the expected polymer structures. The spectra for **1, 2,** and **5** are shown in Fig. 7 as examples. Although unintended protonation of the growing anion with protonic contamination could lead to a polymer having hydrogen at the termination end in entries 3, 5, and 6 aiming at structures **2, 4,** and **5**, respectively, the corresponding spectra in Fig. 7 showed no clear peaks based on oligomers having hydrogen at the termination end of the chain.

In order to obtain detailed information on the π-stacked conformation of the **DBF** polymer and its effect on photophysical and electrochemical properties, THF-soluble products were separated into oligomers that are uniform in terms of the number of fluorene units in a chain (**1**–**5** in Table 1; $n=2$–8) and polymers ($n \geq 9$, mixture) using a preparative-scale SEC apparatus (Fig. 8). Chain lengths of the isolated oligomers were determined by MALDI mass spectra. The purity of the separated oligomers **1** and **2**, which were used for the solution structure and property analyses, was checked by analytical scale SEC. The isolated

Table 2 Anionic polymerization of dibenzofulvene (DBF) in THF

Entry	Polymer structure[a]	DBF (mmol)	Initiator (mmol)	Terminator (mmol)	Temp. (°C)	Time (h)	Conv.[b] (%)	THF-solublec part yield (%)	M_n[d]
1	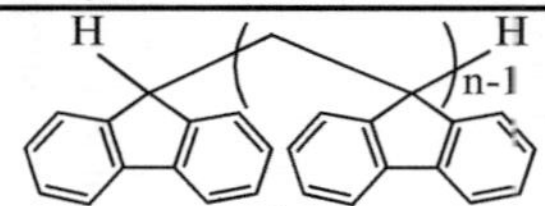 **1**	5.6	9-FlLi (1.1)	MeOH (4.0)	−78	24	>99	>99	860
		4.6	9-FlLi (9.3)	MeOH (12.1)	0	1	>99	>99	350
2	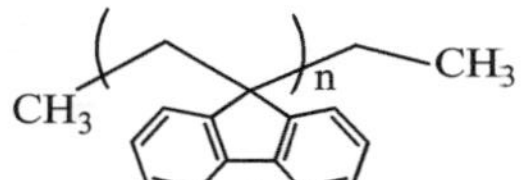 **2**	10.0	MeLi (2.0)	CH_3CH_2I (25.0)	−78	48	84	98	380
3	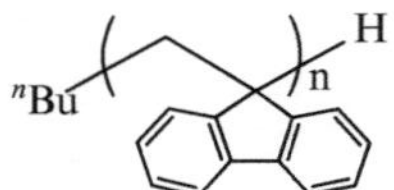 **3**	11.2	*n*-BuLi (2.2)	MeOH (6.2)	−78	24	>99	81	790
3	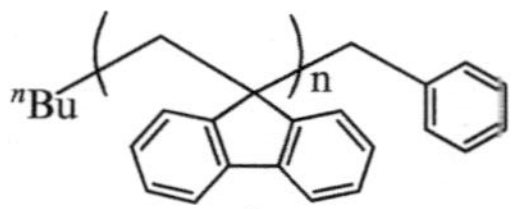 **4**	2.8	*n*-BuLi (0.93)	$PhCH_2Br$ (8.4)	−78	24	>99	99	990[e]
5	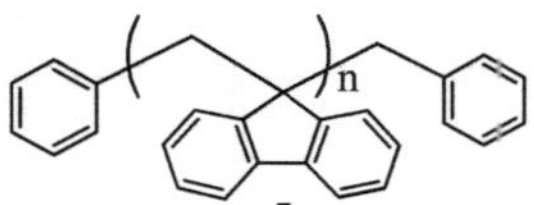 **5**	10.0	PhLi (2.0)	$PhCH_2Br$ (16.8)	−78	48	84	63	950

[a] *n*−1 and *n* denote of polymerization for entries 1 and 2–5, respectively
[b] Determined by ^{1}H NMR analysis of the reaction mixture
[c] THF-sol. part was a mixture of oligomers and unreacted monomer when the monomer conversion was not quantitative. The yield of this part was calculated excluding the weight of the unreacted monomer
[d] Determined by GPC using two OligoPore columns connected in series with oligo(DBF)s as the standard (eluent THF)
[e] Determined by GPC using two TSKgel G1000HHR columns connected in series with oligo(DBF)s as the standard (eluent THF)

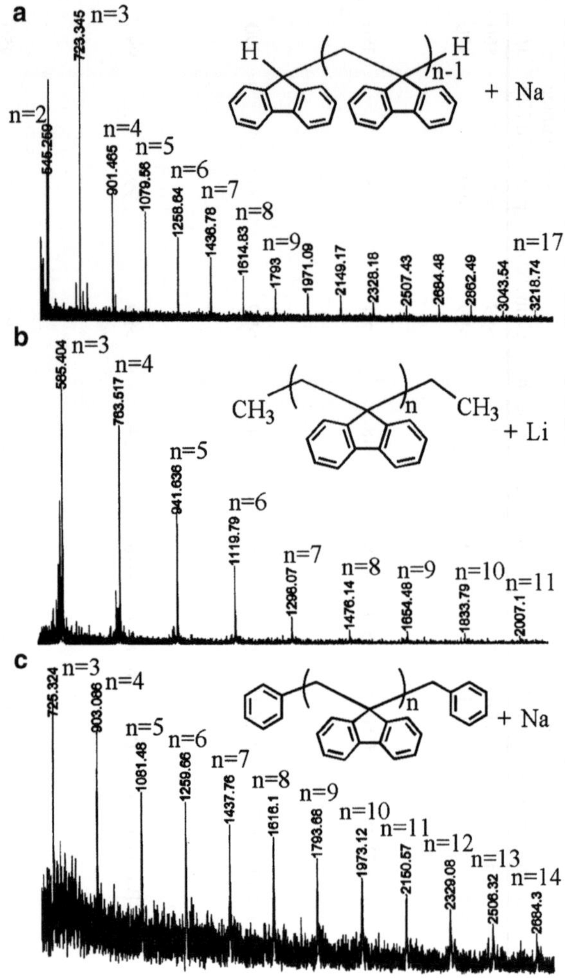

Fig. 7 MALDI-TOF mass spectra of oligo(**DBF**)s **1** (Table 2, entry 1) (**a**), **2** (Table 2, entry 2) (**b**), and **5** (Table 2, entry 6) (**c**). Reprinted with permission from Nakano et al. [30]. Copyright 2003, American Chemical Society

oligomers contained a small amount of higher- and lower-molecular-weight fractions. The purity of oligomers **1** of $n = 2$–8 was in the range of 95–99 % and that of oligomers **2** of $n = 2$–8 in the range of 92–99 %.

X-ray Crystal Analyses

Some of the oligomers isolated by SEC led to single crystals suitable for X-ray analyses by recrystallization from a chloroform solution, and they were subjected to crystallization for X-ray crystal structure analysis [30, 35]. Crystal structures were solved for **1** ($n = 2$ and 4), **2** ($n = 2$ and 6), **3** ($n = 3$), **4** ($n = 4$), and **5** ($n = 2$ and 5).

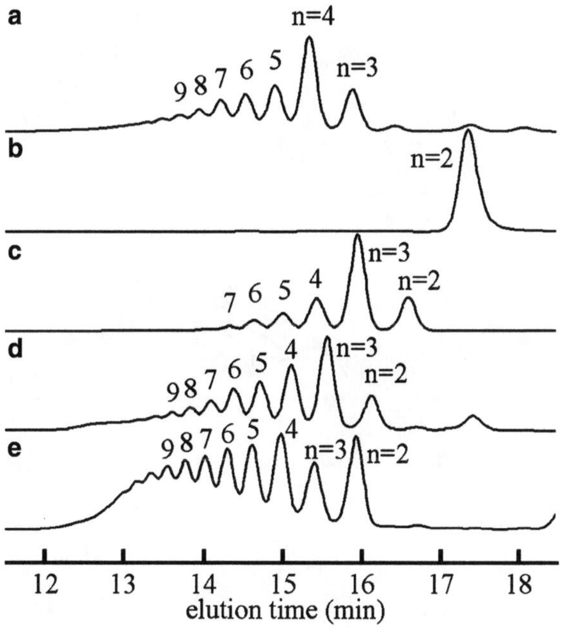

Fig. 8 SEC curves (II) of oligo(**DBF**)s **1** (Table 1, entry 1) (**a**), **1** (Table 2, entry 2) (**b**), **2** (Table 2, entry 3) (**c**), **3** (Table 2, entry 4) (**d**), **5** (Table 2, entry 6) (**e**). Reprinted with permission from Nakano et al. [30]. Copyright 2003, American Chemical Society

The crystal structures are shown in Fig. 9. The single crystal structures indicated two conformational characteristics: (1) the in-chain fluorene moieties of oligomers of $n = 4$–6 have π-stacked conformation with the corresponding main-chain carbon–carbon bonds having a slightly twisted *trans–trans* structure regardless of the terminal groups; and (2) the chain-terminal fluorene moiety is π-stacked when the terminal group is ethyl or bulkier, while they are flipped when the terminal group is hydrogen. It is also noteworthy that the π-stacking is not in a perfectly face-to-face manner, but the fluorene groups are slightly twisted (slipped); the twist takes place in one direction in a chain leading to a helical structure. In addition, the stacked fluorene groups are not completely parallel to each other, which is probably due to steric repulsion. The twisted π-stacked structure is most clearly confirmed in **2** of $n = 6$, the longest molecule for which the crystal structure was solved. This oligomer has a π-stacked conformation throughout the chain and forms a long-pitched, single-handed helical structure where ca. 9 monomeric units form one turn as a result of the twisted alignment of the fluorene groups. The conformationally enantiomeric right- and left-handed helices are paired in the solid state, i.e., the crystal formed as racemic crystal. Induction of single-handed helicity (optical activity) is described in a later section.

Experimental and Theoretical NMR Analyses

^{1}H NMR spectra of the oligomers were measured to confirm the chemical structure and to obtain information on the solution conformation of oligomers **1** and **2**

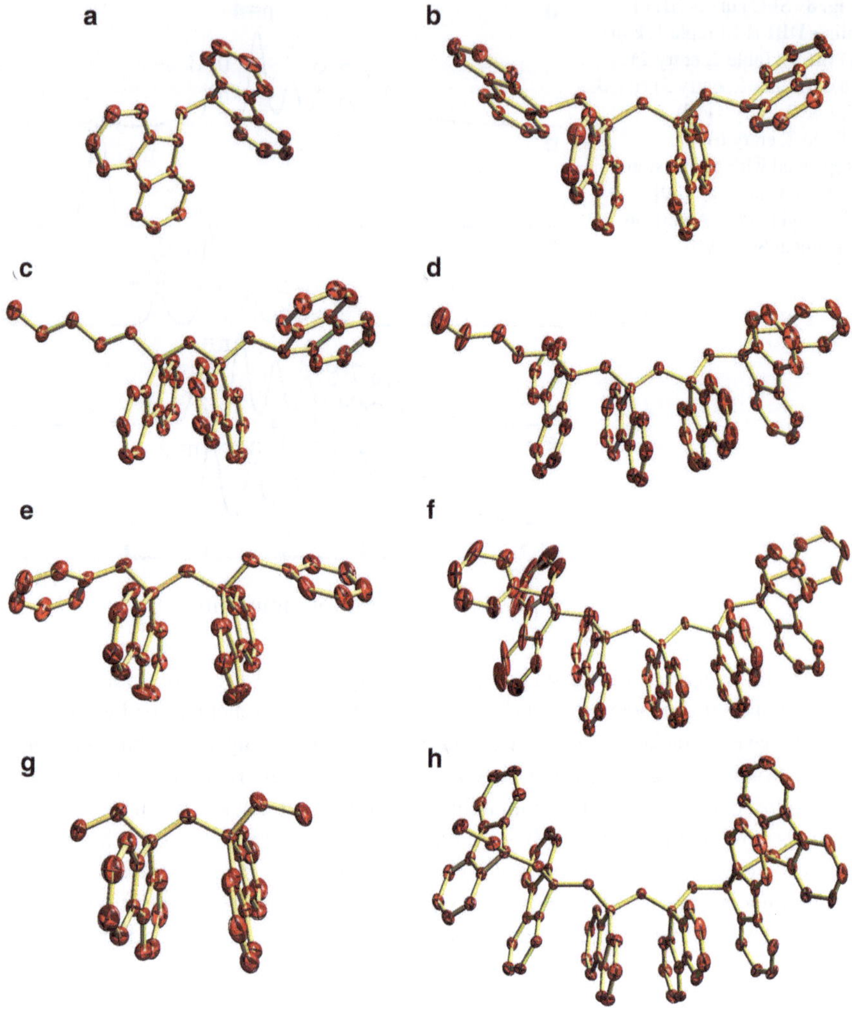

Fig. 9 Crystal structures of oligo(**DBF**)s **1** (n=2 (**a**) and n=4 (**b**)), **3** (n=3) (**c**), **4** (n=4) (**d**), **5** (n=2 (**e**) and n=5 (**f**)), and **2** (n=2 (**g**) and n=6 (**h**)). Hydrogen atoms are omitted for clarity. Reprinted with permission from Nakano et al. [30]. Copyright 2003, American Chemical Society

(n=2–8) and polymers **1** (M_n=1,700, $\bar{n}$=9.6, mixture of $n \geq 9$) and **2** (M_n=1,890, n=10.4, mixture of $n \geq 9$). Because poly(**DBF**) has no stereocenter in the main chain, the chemical shift information of a proton signal should only reflect conformation of the oligomer in its vicinity.

The 1D spectra are shown in Figs. 10 and 11, and the chemical shift values are summarized in Tables 3 and 4. The peak assignments were performed on the basis of H–H COSY and NOESY spectra. The structures of **1** and **2** of n=8 with the proton numbering system are shown in Fig. 12 as generic specimens.

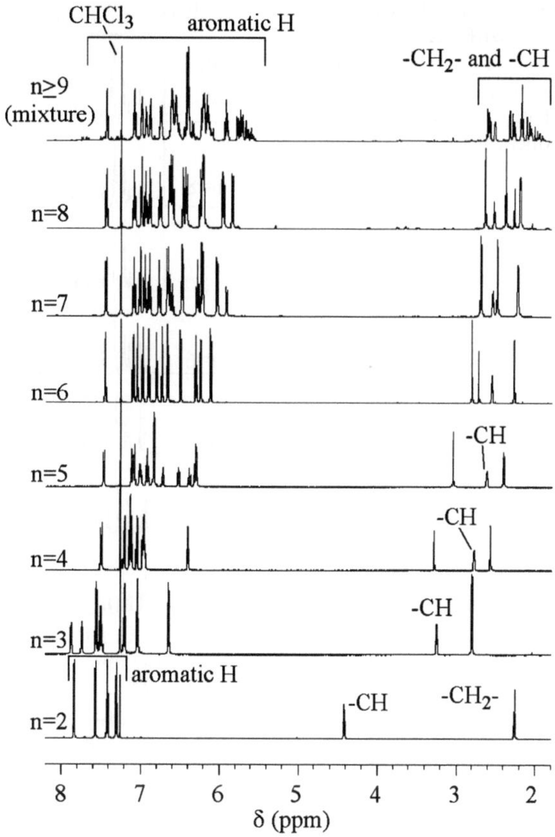

Fig. 10 ^{1}H NMR spectra of oligo(**DBF**)s (**1**) (600 MHz, $CDCl_3$, r.t.). Reprinted with permission from Nakano et al. [30]. Copyright 2003, American Chemical Society

The protons showing COSY correlation peaks are marked by blue arrows and those showing NOESY correlations are marked by red arrows in Fig. 12. The same proton–proton correlations were also observed for oligomers of $n=2$–7. As examples of the 2D NMR spectra, H–H COSY and NOESY spectra of **1** of $n=8$ are shown in Figs. 13 and 14, respectively, and those of 2 of $n=8$ in Figs. 15 and 16.

As for oligomers **1**, the triplet peaks having an intensity of two protons were first assigned to the terminal methine protons, and the signals due to the methylene protons neighboring the methines were then found by the COSY experiment. In the case of oligomers **2**, the triplet peaks in the highest magnetic-field region were assigned to the terminal methyl protons, and the neighboring methylene signals were identified by the COSY experiment.

The peak assignments of the remaining main-chain methylene and aromatic protons were performed based on the 1D spectral profiles and the NOESY spectra. In the 1D spectra, all oligomers indicated only one set of terminal peaks: triplet methine signals for **1** and ethyl signals consisting of a methyl triplet and methylene quartet for **2**. In addition, the number of observed peak sets for the methylene protons was only half the number of existing methylene groups for the oligomers of

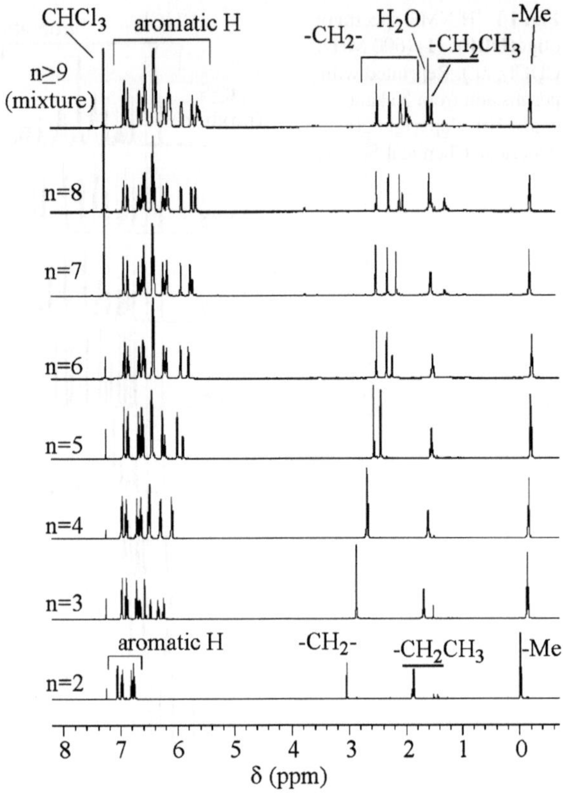

Fig. 11 ^{1}H NMR spectra of oligo(**DBF**)s (**2**) (500 MHz, $CDCl_3$, r.t.). Reprinted with permission from Nakano et al. [30]. Copyright 2003, American Chemical Society

$n=3$, 5, and 7 or [(number of existing methylenes)-1]/2$+1$ for the oligomers of $n=2$, 4, 6, and 8. This profile may be explained in terms of a conformation symmetrical with respect to the central methylene (oligomers of $n=2$, 4, 6, and 8) or the central fluorene group (oligomers of $n=3$, 5, and 7), such as the crystal structures of **2** of $n=6$ and **5** of $n=5$ shown in Fig. 9, if the conformation is stable in solution. A fast exchange between different conformations may also lead to a simple NMR profile. However, fast conformational dynamics in solution seem unreasonable because clear NOE correlations were observed that support a stable all-trans conformation as follows.

In the NOESY spectra of all oligomers analyzed here, all in-chain methylene protons indicated correlation peaks with one or two different methylene groups in addition to aromatic protons. This finding is best interpreted assuming an all-trans conformation as illustrated in Fig. 12; only the neighboring groups in such a conformation will show correlation peaks in the NOESY spectra. Hence, the methylene protons were assigned as summarized in Tables 3 and 4 assuming an all-trans conformation, which is consistent with the regular numerical relations between the number of observed methylene signals and the existing methylene groups noted above.

Table 3 ^{1}H NMR chemical shifts (**δ**, ppm) of oligo(DBF)s (**1**)[a]

n	2	3	4	5	6	7	8	≥9[b]
H_A	4.42 (t)	3.25 (t)	2.78 (t)	2.64 (t)	2.56 (t)	2.55 (t)	2.53 (t)	2.62–1.88
H_B	2.26 (d)	2.80 (d)	2.57 (d)	2.42 (d)	2.28 (d)	2.23 (d)	2.20 (d)	(methylene
H_C			3.28 (s)	3.06 (s)	2.82 (s)	2.70 (s)	2.64 (s)	protons)
H_D					2.73 (s)	2.49 (s)	2.38 (s)	
H_E							2.27 (s)	
H_{A1}	7.57 (d)	6.64 (d)	6.41 (d)	6.31 (d)[c]	6.25 (d)	6.23 (d)	6.22 (d)[c]	5.55–7.44
H_{A2}	7.30 (dd)	7.04 (dd)	6.96 (dd)[c]	6.93 (dd)	6.91 (dd)	6.91 (dd)	6.90 (dd)	(aromatic
H_{A3}	7.42 (dd)	7.20 (dd)	7.14 (dd)[c]	7.13 (dd)	7.10 (dd)	7.10 (dd)	7.10 (dd)	protons)
H_{A4}	7.84 (d)	7.56 (d)	7.50 (d)	7.48 (d)	7.46 (d)	7.45 (d)	7.44 (d)	
H_{B1}		7.74 (d)	7.05 (d)	6.85 (d)[c]	6.74 (d)	6.68 (d)	6.64 (d)	
H_{B2}		7.49 (dd)[c]	6.97 (dd)[c]	6.85 (m)[c]	6.81 (dd)	6.79 (dd)	6.77 (dd)	
H_{B3}		7.51 (dd)[c]	7.13 (dd)[c]	7.02 (m)	6.98 (dd)	6.97 (dd)	6.95 (dd)	
H_{B4}		7.88 (d)	7.22 (d)	7.10 (d)	7.06 (d)	7.03 (d)	7.01 (d)	
H_{C1}				6.33 (d)[c]	6.12 (d)	6.04 (d)	5.96 (d)	
H_{C2}				6.40 (dd)	6.31 (dd)	6.29 (dd)	6.25 (dd)[c]	
H_{C3}				6.74 (dd)	6.67 (dd)	6.66 (dd)	6.61 (dd)[c]	
H_{C4}				6.54 (d)	6.50 (d)	6.49 (d)[c]	6.43 (d)	
H_{D1}						5.93 (d)	5.85 (d)	
H_{D2}						6.24 (dd)	6.22 (dd)[c]	
H_{D3}						6.62 (dd)	6.61 (dd)[c]	
H_{D4}						6.50 (d)[c]	6.47 (d)	

[a]See Fig. 12 for proton numbering. Spectra were recorded in $CDCl_3$ at room temperature (600 MHz). Residual $CHCl_3$ signal (7.26 ppm) was used as an internal reference
[b]A polymer (M_n = 1,700 (vs. oligo(DBF)), *n* = 9.6, *n* > 9 mixture)
[c]Indicate interchangeable assignments

The aromatic signals were assigned also using the 2D NMR information. The observed NOE signals were interpreted to correlate the protons at the 1- and 8-positions of the fluorene ring, denoted by $H_{A\text{-}D1}$ in Fig. 12, and the methylene groups in the vicinity. The position of each fluorene ring was assigned in this way. The peak assignments of the remaining aromatic protons were completed using the COSY information. NOE correlations between protons of neighboring fluorene units in stacked conformation were also observed although they are not shown in Fig. 12 for clarity.

It is noteworthy that aromatic proton signals of all samples of **2** were within the range of 5.6–7.1 ppm. This is in stark contrast to the fact that poly(9,9-dimethyl-2-vinylfluorene) [77] and poly(2-vinylfluorene) (Nakano T, unpublished results), which probably have a flexible conformation, showed their aromatic proton signals in a much lower magnetic-field range (6–8 ppm). This supports the stable π-stacked conformation of **2** throughout the chain in solution.

The aromatic proton signals of **1** appeared in a wider chemical shift range compared with those of **2**, suggesting that **1** may have a less regular π-stacked conformation. However, the samples of **1** of *n* = 4 and larger showed all aromatic signals, except for

Table 4 ^{1}H NMR chemical shifts (**δ**, ppm) of oligo(DBF)s (**2**)[a]

n	2	3	4	5	6	7	8	>9[b]
Me–	−0.01 (t)	−0.14 (t)	−0.16 (t)	−0.20 (t)	−0.22 (t)	−0.23 (t)	−0.24 (t)	−0.25 (t)
H_A	1.88 (q)	1.69 (q)	1.61 (q)	1.55 (q)	1.53 (q)	1.51 (q)	1.50 (q)	1.49 (q)
H_B	3.05 (s)	2.87 (s)	2.68 (s)	2.56 (s)	2.51 (s)	2.48 (s)	2.47 (s)	2.46–1.86
H_c			2.67 (s)	2.45 (s)	2.34 (s)	2.28 (s)	2.25 (s)	(methylene
H_D					2.24 (s)	2.12 (s)	2.07 (s)	protons)
H_E							2.00 (s)	
H_{A1}	6.77 (d)	6.58 (d)	6.50 (d)[c]	6.45 (d)[c]	6.42 (d)[c]	6.40 (d)[c]	6.39 (d)[c]	5.52–6.91
H_{A2}	6.83 (dd)	6.72 (dd)	6.71 (dd)	6.68 (dd)	6.66 (dd)	6.65 (d)	6.64 (dd)	(aromatic
H_{A3}	6.98 (dd)	6.90 (dd)	6.90 (dd)	6.87 (dd)	6.86 (dd)	6.85 (d)	6.85 (dd)	protons)
H_{A4}	7.07 (d)	6.98 (d)	6.98 (d)	6.94 (d)	6.93 (d)	6.92 (d)	6.92 (d)	
H_{B1}		6.25 (d)	6.10 (d)	6.00 (d)	5.94 (d)	5.91 (d)	5.90 (d)	
H_{B2}		6.34 (dd)	6.30 (dd)	6.26 (dd)	6.24 (dd)	6.22 (dd)	6.21 (dd)	
H_{B3}		6.66 (dd)	6.65 (dd)	6.62 (dd)[c]	6.60 (dd)[c]	6.59 (dd)	6.58 (dd)	
H_{B4}		6.48 (d)	6.48 (d)[c]	6.45 (d)[c]	6.42 (d)[c]	6.41 (d)[c]	6.40 (d)[c]	
H_{C1}				5.90 (d)	5.81 (d)	5.75 (d)	5.72 (d)	
H_{C2}				6.22 (dd)	6.19 (dd)	6.16 (dd)[c]	6.15 (dd)[c]	
H_{C3}				6.60 (dd)[c]	6.58 (dd)[c]	6.56 (dd)[c]	6.55 (dd)[c]	
H_{C4}				6.45 (d)[c]	6.42 (d)[c]	6.40 (d)[c]	6.39 (d)[c]	
H_{D1}						5.70 (d)	5.65 (d)	
H_{D2}						6.15 (dd)[c]	6.13 (dd)[c]	
H_{D3}						6.56 (dd)[c]	6.54 (dd)[c]	
H_{D4}						6.38 (d)[c]	6.36 (d)	

[a]See Fig. 12 for proton numbering. Spectra were recorded in $CDCl_3$ at room temperature (500 MHz). Residual $CHCl_3$ signal (7.26 ppm) was used as an internal reference

[b]A polymer (M_n = 1,890 (vs. oligo(DBF), $\bar{n}$ = 10.4, $n \geq 9$ mixture)

[c]Indicate interchangeable assignments

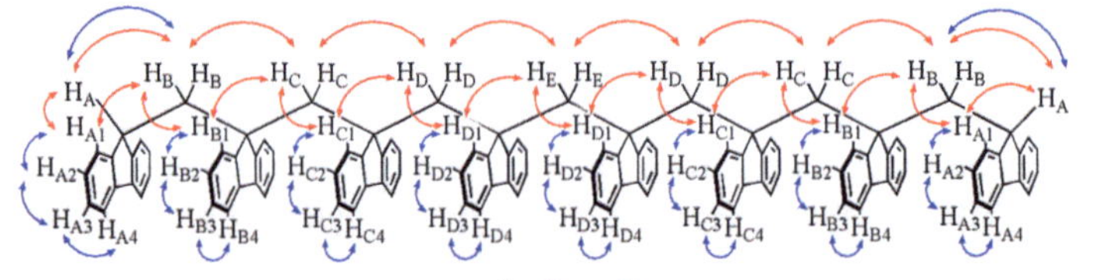

1 of n = 8

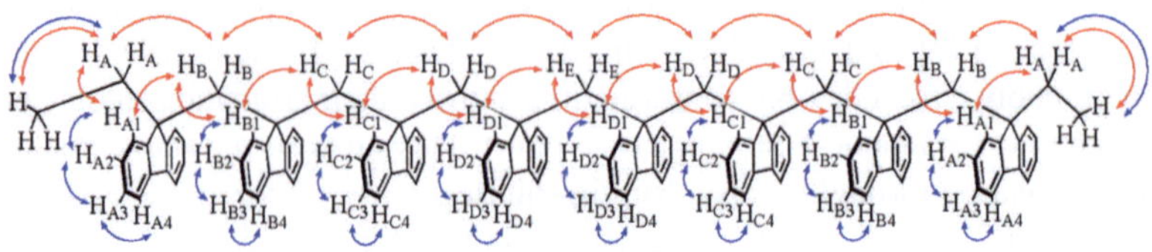

2 of n = 8

Fig. 12 Structures of **1** and **2** of n=8 with the proton numbering systems COSY and NOESY correlations were marked by *blue* and *red arrows*, respectively. NOESY correlations between aromatic protons are omitted. Reprinted with permission from Nakano et al. [30]. Copyright 2003, American Chemical Society

Fig. 13 H–H COSY spectra of oligo(**DBF**) **1** of $n=8$: alkyl region (**a**) and aromatic region (**b**) (500 MHz, $CDCl_3$, r.t.). Reprinted with permission from Nakano et al. [30]. Copyright 2003, American Chemical Society

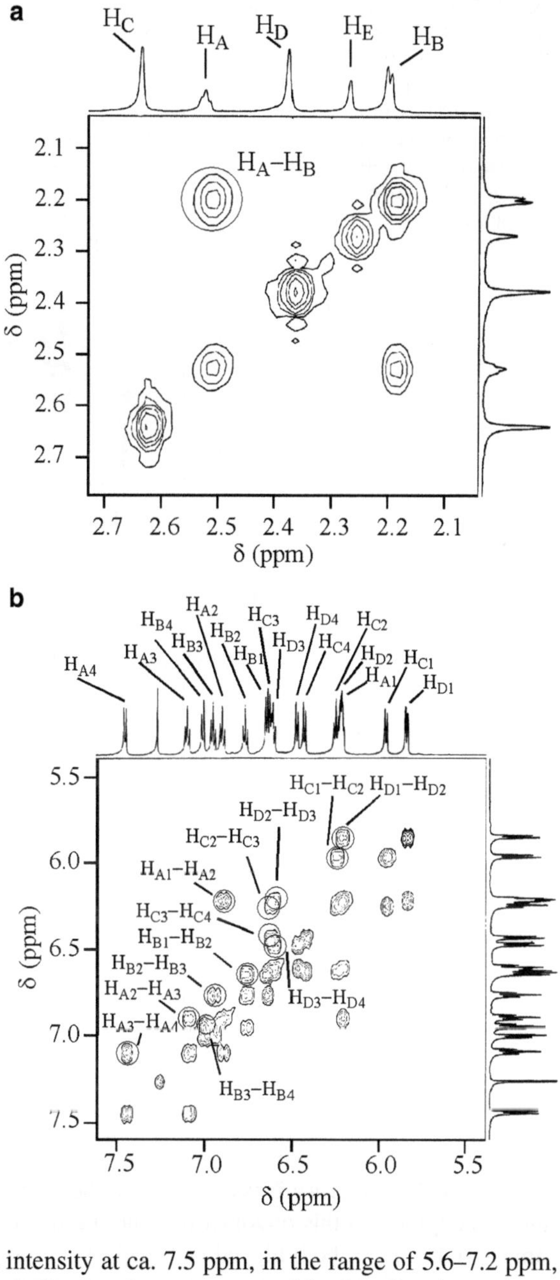

a doublet with a four-proton intensity at ca. 7.5 ppm, in the range of 5.6–7.2 ppm, similarly to all samples of **2**, indicating that most part of **1** of $n=4$ or larger has a regular π-stacked conformation similar to that of **2**. The exceptional four-proton intensity signals are due to the 4- and 5-positions of the terminal fluorene rings according to the peak assignments described above. This may indicate that the terminal fluorene rings of **1** are flipped as observed in the crystal structures of **1**.

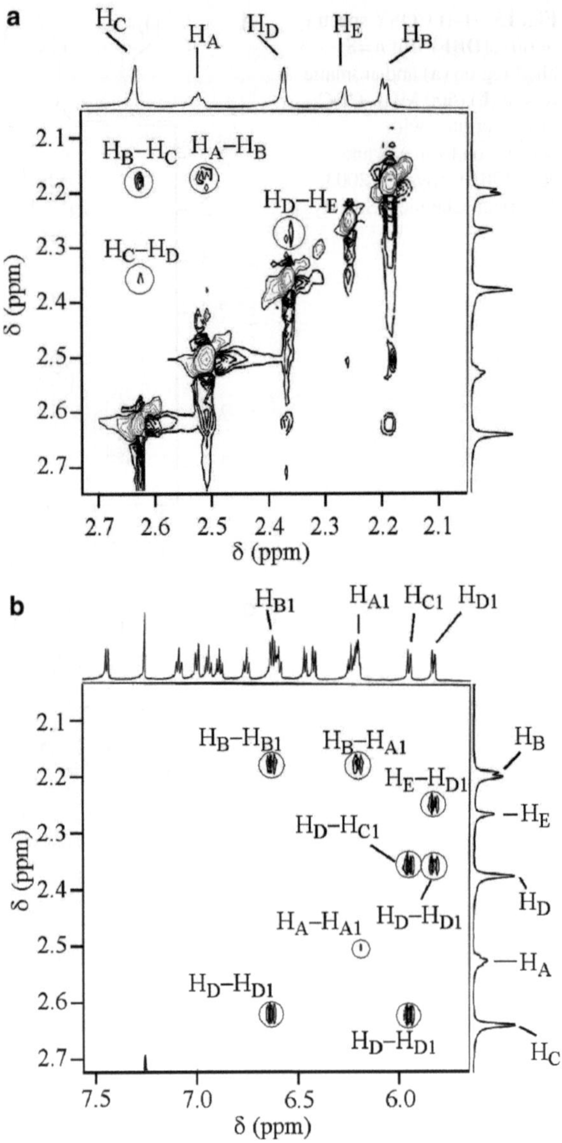

Fig. 14 NOESY spectra of oligo(**DBF**) **1** of $n = 8$: alkyl–alkyl region (**a**) and alkyl-aromatic region (**b**) (500 MHz, $CDCl_3$, r.t.). Reprinted with permission from Nakano et al. [30]. Copyright 2003, American Chemical Society

The signals of **1** of $n = 2$ were similar to those of fluorene in the chemical shift range, implying that this oligomer does not have a π-stacked conformation. In the sample of **1** of $n = 3$, both flipped and stacked conformation may exist because the spectral profile in the aromatic region is just between those of the longer oligomer and the shorter one.

In order to obtain information on the stability of the conformation of the oligomers, temperature effects on the NMR spectra were examined for all oligomers in the range of −50 °C to +60 °C. Most peaks only monotonously shifted upfield or

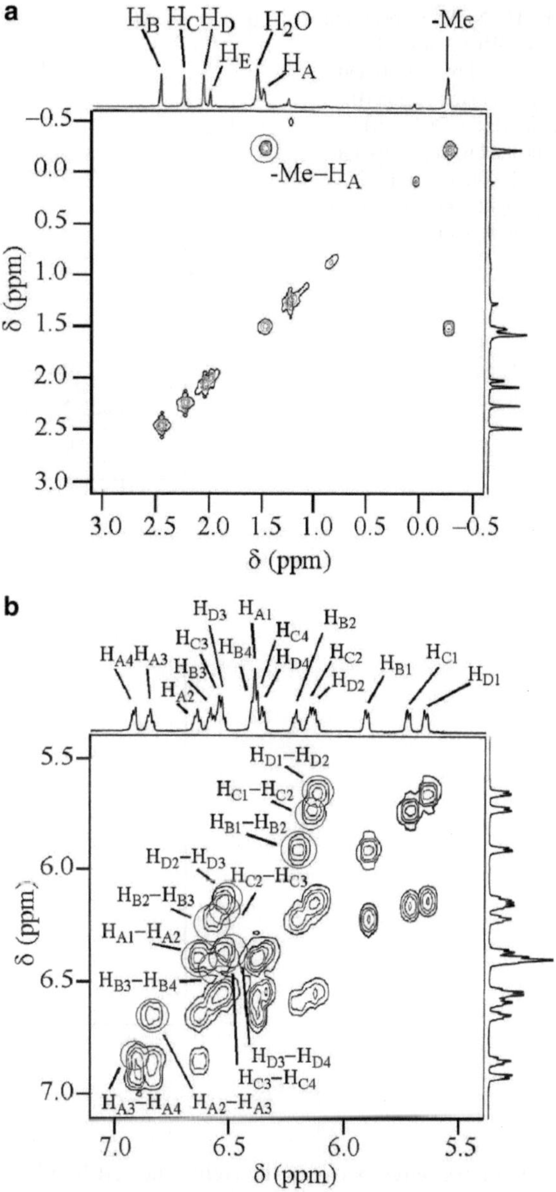

Fig. 15 H–H COSY spectra of oligo(**DBF**) **2** of $n = 8$: alkyl region (**a**) and aromatic region (**b**) (500 MHz, $CDCl_3$, r.t.). Reprinted with permission from Nakano et al. [30]. Copyright 2003, American Chemical Society

downfield with temperature, and the shift amount was within 0.2 ppm. The peak shift was reversible with temperature. These results suggest that the π-stacked conformation in solution proposed so far is stable and does not undergo a drastic conformational mutation in this temperature range.

It should be noted here that no clear NMR information indicating the helical twist in solution was obtained even at lower temperatures. The helical structure may

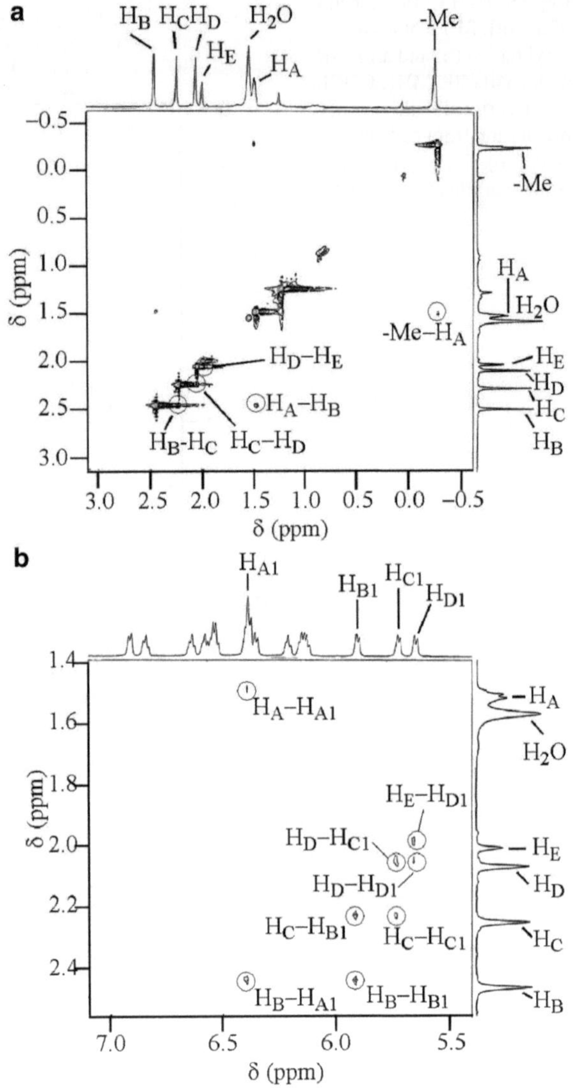

Fig. 16 NOESY spectra of oligo(**DBF**) **2** of $n = 8$: alkyl–alkyl region (**a**) and alkyl-aromatic region (**b**) (500 MHz, $CDCl_3$, r.t.). Reprinted with permission from Nakano et al. [30]. Copyright 2003, American Chemical Society

be in an exchange between the right- and left-handed local twists that is faster than the NMR time scale under the experimental conditions of this study.

In order to rationally connect the NMR spectral information to the solution conformation of the oligomers, the chemical shifts of the main-chain methylene and terminal methine or ethyl protons were computationally estimated by density functional theory (DFT) [78] calculation for some optimized conformers of **1** of $n = 2$ and **2** of $n = 2$ and 4. The conformer structures optimized by the semiempirical PM5 method [51] are shown in Fig. 17 with their heats of formation (H_f) (PM5) and theoretical chemical shifts obtained by the LORG method [79] using B88 [80]

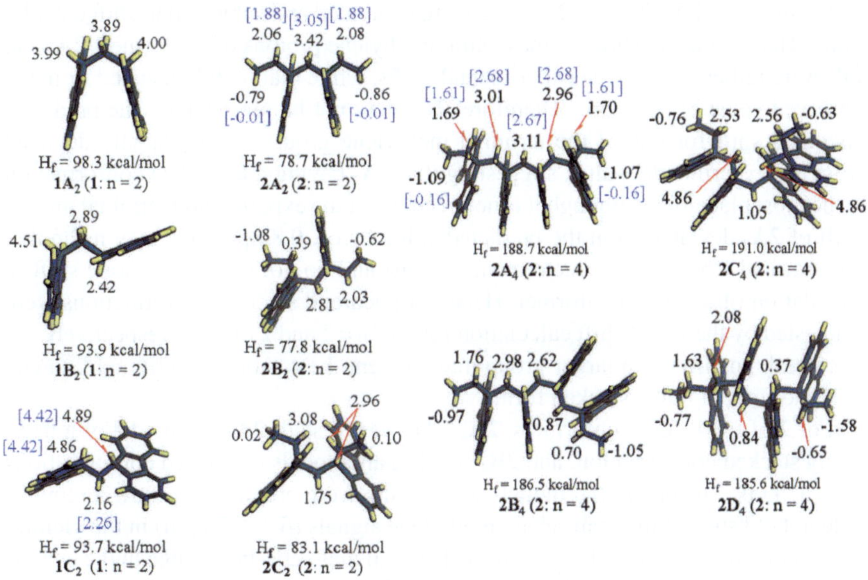

Fig. 17 Conformers of **1** (n=2) and **2** (n=2 and n=4) and calculated chemical shifts of methylene protons. The values in *square brackets* are the experimental chemical shifts. Reprinted with permission from Nakano et al. [30]. Copyright 2003, American Chemical Society

exchange and LYP [81] correlation functionals. The H_f values of the conformers having identical chemical structures were similar, indicating that all conformers shown in the figures could contribute to the experimental NMR spectra at similar probabilities.

Prior to application of the DFT method to the oligomers, the accuracy of the DFT chemical shift calculation was tested using fluorene, 9-methylfluorene, and superphane, which consist of methylene, methine, and an aromatic ring and are expected to have little conformational freedom. The experimental $-CH_2-$ shift for fluorene was 3.95 ppm ($CDCl_3$, 500 MHz, Me_4Si) and that for superphane was 2.98 ppm [82], while the calculated shifts were 3.87 and 3.29 ppm for fluorene and superphane, respectively. The experimental $-CH_3$ and methine shifts for 9-methylfluorene were 1.53 and 3.95 ppm, respectively ($CDCl_3$, 500 MHz, Me_4Si), and the corresponding calculated values were 1.43 and 4.03 ppm. These results mean that the theoretical calculation in this study may contain an error of 0.1–0.3 ppm.

Conformers **1A$_2$–1C$_2$** were evaluated for **1** of n=2. The calculated methylene shift of **1A$_2$** deviated from the experimental value by 1.6 ppm, and one of the calculated methine shifts of **1B$_2$** deviated from the experimental value by 2.0 ppm, suggesting that these are not the most plausible conformers. In contrast, the deviations of the calculated methylene and methine shifts of **1C$_2$** from the experimental values were only 0.4 and 0.1 ppm, respectively. These results suggest that **1C$_2$**, in which the two fluorenyl groups are flipped, is the most probable conformer though other conformers could also exist in conformational dynamics.

Conformers **$2A_2$–$2C_2$** for **2** of $n=2$ were generated and subjected to shift calculations. The theoretical shifts of the central methylene protons of conformers **$2A_2$** and **$2B_2$** were rather close to the experimental shifts, while that of **$2C_2$** deviated from the observed value by 1.3 ppm. Therefore, **$2C_2$** may not be reasonable. The predicted chemical shift for one of the terminal methylene groups of **$2B_2$** largely deviated from the experimental value, suggesting that **$2A_2$** having a π-stacked conformation is the most plausible. Although it is noticeable that the experimental terminal methyl shift of **$2A_2$** deviated from the predicted value by ca. 0.8 ppm, this may mean that the terminal ethyl groups rotate relatively fast and do not give an accurate shift in calculation of a single conformer. Hence, flipped and stacked conformations were suggested by the NMR shift calculation for **1** of $n=2$ and **2** of $n=2$, respectively. As in crystal, an ethyl or a larger group may prevent the rotation of terminal fluorene unit and force it into a stacked form.

For **2** of $n=4$, four conformers, **$2A_4$–$2D_4$** were tested: **$2A_4$** is a model of a flawlessly stacked conformation, and **$2B_4$** and **$2C_4$** are models of stacked conformations with a "kink." Although the models with a kink were predicted to show a remarkable upfield shift of the main-chain methylene signals (δ 1–0.7 ppm) in the vicinity of the kink, no peak was observed in such a high magnetic-field range in the experimental spectra. In addition, the calculated methylene shifts of **$2A_4$** were rather close to the experimental data. Therefore, **$2A_4$** is considered to be the most plausible conformation. In this case also, calculated shifts of the terminal methyl protons significantly deviated from the observed values, probably due to the relatively fast motion of the terminal ethyl group as noted above.

As described here, the chemical shift calculation appears to be an effective method of obtaining information on the oligomer conformation. Although longer oligomers and polymers were not subjected to chemical shift calculation, their experimental methylene shifts seem to converge into a value around 2 ppm, and no peaks of the methylene group were observed in a range higher than 1 ppm in the spectra. This suggests that the longer oligomers and polymers have a stable π-stacked conformation without a kink in the chain.

Conformational Simulations

Oligomer conformation was evaluated by force field and semiempirical molecular orbital calculations using MM3* [83], MMFF94s [84], and COMPASS [85] force fields and a semiempirical PM5 [51] method.

First, Monte Carlo conformational searches with MM3* and MMFF94s force fields were performed on a **DBF** hexamer having ethyl groups at the chain terminals (**2**, $n=6$) [86]. The relative arrangements of the fluorene moieties were fixed, and the dihedral angles around the 12 main-chain bonds were changed at an interval of 60° in generating initial geometries (Fig. 18). Each conformer generated in the Monte Carlo process was fully minimized using MM3* or MMFF94s force field until the RMS residue went below 0.05 kcal/mol/Å or 1,000 iterations were

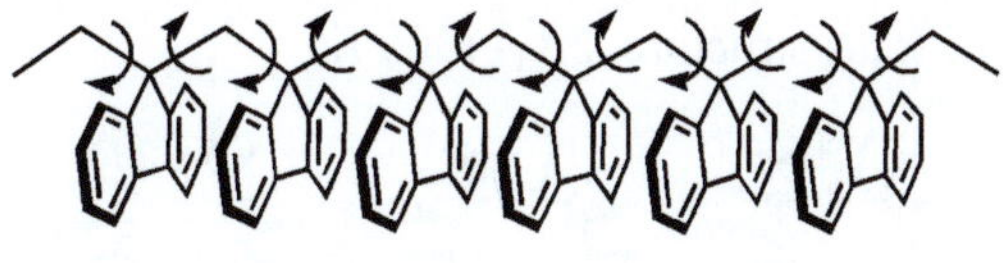

Fig. 18 Dihedral angle sets for the Monte Carlo simulation of **2** of $n=6$. Reprinted with permission from Nakano et al. [30]. Copyright 2003, American Chemical Society

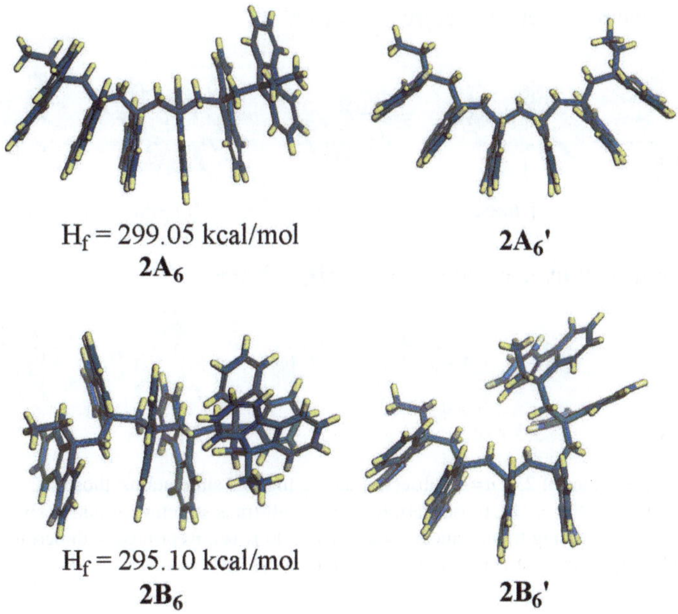

Fig. 19 Conformers of **2** of $n=6$ found most stable in the Monte Carlo simulation and optimized by the PM5 method, $\mathbf{2A_6}$ (force filed, MM3*) and $\mathbf{2B_6}$ (force field, MMFF94s), and those after MD simulation, $\mathbf{2A_6}'$ (300 K, 10 ns) and $\mathbf{2B_6}'$ (300 K, 8.2 ns), starting from $\mathbf{2A_6}$ and $\mathbf{2B_6}$, respectively. Reprinted with permission from Nakano et al. [30]. Copyright 2003, American Chemical Society

achieved. Conformers most frequently found within 15 kJ/mol energy range from the global minimum were considered to be the most stable ones.

In this simulation, two distinctive structures were found to be the most stable conformations using MM3* and MMFF94s force fields out of 20,000 (MM3*) and 19,550 (MMFF94s) conformers. These two most stable conformers were further minimized by the PM5 method to afford structures $\mathbf{2A_6}$ and $\mathbf{2B_6}$ shown in Fig. 19. Structure $\mathbf{2A_6}$ was similar to the crystal structure (Fig. 9) while structure $\mathbf{2B_6}$ had a partially trans–gauche-like structure. The two conformers indicated similar values of H_f by the PM5 method.

Next, the stability of conformations $\mathbf{2A_6}$ and $\mathbf{2B_6}$ were tested by molecular dynamics (MD) simulations using the COMPASS force field under NVT conditions with Berendsen's thermostat [87] implemented in the Discover module of the

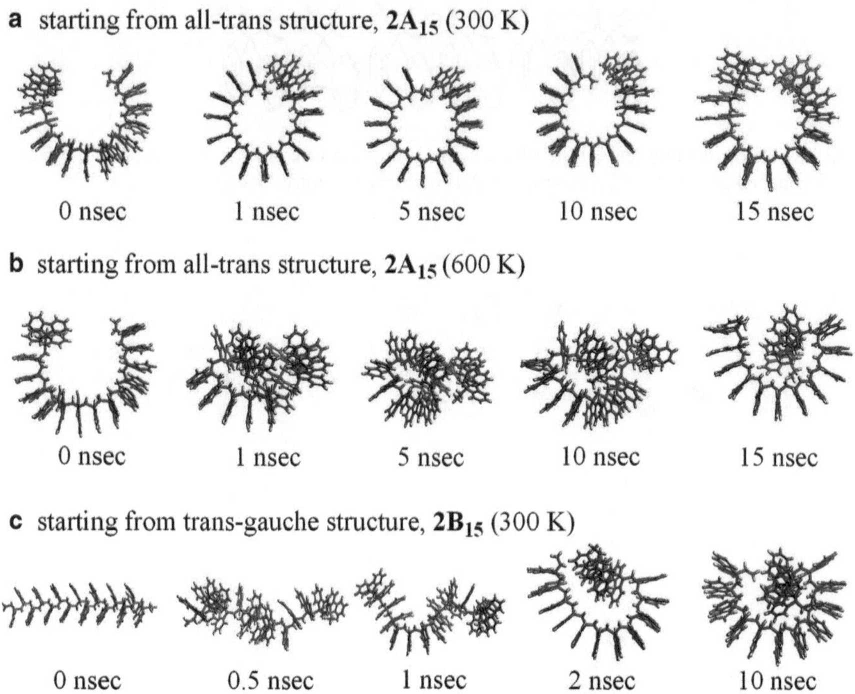

Fig. 20 Conformations of **2** of $n = 15$ observed during the MD simulations: those starting from an all-trans structure at 300 K (**a**), those starting from an all-trans structure at 600 K (**b**), and those starting from an alternating trans–gauche structure at 300 K (**c**). Reprinted with permission from Nakano et al. [30]. Copyright 2003, American Chemical Society

Material Studio software package(Accelrys). The simulations were performed under a constant NVT condition in which the numbers of atoms, volume, and thermodynamic temperature were held constant. The step time was 1 fs and the decay constant was 0.1 ps. Conformations obtained through MD simulation were saved in trajectory files at every 5 or 10 ps and were optimized by MM simulation. The results are shown in Fig. 19 (right-hand side). While conformation $\mathbf{2A_6}$ did not largely change in the MD simulation at 300 K for 10 ns to give structure $\mathbf{2A_6}'$, conformation $\mathbf{2B_6}$ quickly changed into a mostly all-trans, π-stacked conformation with a kink (gauche–gauche main-chain conformation) ($\mathbf{2B_6}'$) in the middle of the chain at 300 K within 0.1 ns, and structure $\mathbf{2B_6}'$ did not significantly change within 8 ns. These results suggest that a π-stacked conformation is preferred under dynamic conditions while several different conformers, including the π-stacked structure, may have similar steric energies under static conditions.

The conformational stability test was extended to a model of the polymer (**2**, $n = 15$). In this case, an all-trans 15-mer and an alternating trans–gauche 15-mer were generated as initial conformations and optimized using the COMPASS force field with the Fletcher–Reeves [88] conjugate gradient algorithm until the RMS residue went below 0.01 kcal/mol/Å ($\mathbf{2A_{15}}$ and $\mathbf{2B_{15}}$ in Fig. 20). The molecular mechanically optimized structures were then subjected to MD simulations. The all-trans model, $\mathbf{2A}_{15}$, remained in the π-stacked structure throughout the simulation at 300 K for 15 ns. Even in the simulation at 600 K for 15 ns, the π-stacked structure of $\mathbf{2A}_{15}$ was only partially distorted; the π-stacked sequence remained at a shorter length than the initial structure, and they were connected through gauche–gauche kink structures in the chain. In contrast, structure $\mathbf{2B}_{15}$ promptly changed into a preferentially all-trans, π-stacked conformation with a kink at 300 K within 2 ns of the simulation time and the structure was found unchanged at 10 ns. These results suggest that the π-stacked conformation is most preferable for poly(**DBF**). Although some conformers with a kink in the main chain were observed in the calculations, most of the real molecules are considered to have a π-stacked structure throughout the chain, based on the discussions about the NMR spectra in the preceding section.

These results suggest that the π-stacked conformation is most preferable for poly(**DBF**). Although some conformers with a kink in the main chain were observed in the calculations, most of the real molecules are considered to have a π-stacked structure throughout the chain, based on the discussions about the NMR spectra in the preceding sections.

Properties of π-Stacked Poly(dibenzofulvene)

Absorbance and Emission Spectra

The absorption and emission spectra of the oligomers ($n = 2$–8) and polymers ($n \geq 9$) of **1** and **2** indicated characteristic profiles based on their structures. The absorption spectra are shown in Fig. 21. Oligomer **1** of $n = 2$ bearing two fluorene groups indicated slight hyperchromicity relative to 9-methylfluorene, the model for the monomeric unit, which is obvious in the 240–280 nm region, while oligomers **1** of $n = 3$–8 and the polymer ($M_n = 1{,}700$, $\bar{n} = 9.6$, mixture of $n \geq 9$) exhibited significant hypochromicity. In addition, all the samples of **1** of $n = 3$ and larger showed a red shift of the absorption bands. As for oligomers **2**, all the samples, including the one of $n = 2$, exhibited hypochromicity and a red shift.

Hypochromic effects have been reported for stacked base pairs in a double helical strand in DNAs [89, 90]. Therefore, the hypochromicity described above is consistent with the above discussion that the oligomers and polymers studied here have π-stacked conformation in solution. The degree of hypochromicity found for the oligomers and polymers in this study is larger than that for DNAs. This may indicate that the π-electron systems are more densely packed in the present systems than those in DNAs. The fact that **1** of $n = 2$ was hyperchromic [89] while **2** of $n = 2$ was hypochromic supports the conclusion from the ^{1}H NMR studies that the former compound does not have a stacked conformation while the latter does.

The observed red shift of the absorption bands means that electronic interaction between neighboring fluorene units in the ground state lowered the excitation energy of the oligomers or polymers. In main-chain conjugated polymers, a longer chain

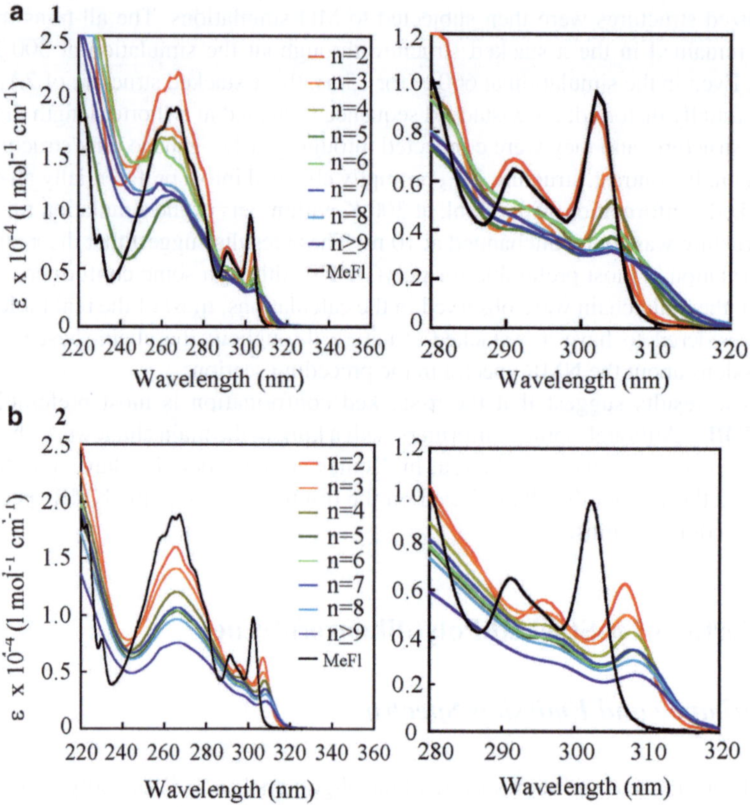

Fig. 21 Absorption spectra of oligo(**DBF**)s **1** (**a**), **2** (**b**) and 9-methylfluorene (MeFl) in THF at room temperature. Reprinted with permission from Nakano et al. [30]. Copyright 2003, American Chemical Society

length results in a significantly reduced band gap [91, 92]. Electronic interaction between closely stacked but not covalently bonded aromatic groups may also make the band gap narrower from the present results although the effect seems to be smaller compared with that for main-chain conjugated polymers [92].

In order to more quantitatively evaluate the red shift and the hypochromicity, the wavelength and molar extinction coefficient (ε) of the lowest-energy absorption peak were plotted against the chain length (n) of oligomers **2** (Fig. 22). The peak-top wavelength increased with the chain length and levelled off around $n=5$ (Fig. 22, left). In addition, ε at the peak top decreased with n and levelled off also around $n=5$ (Fig. 22, right). These results suggest that in the ground state, the electronic interaction between the stacked fluorene groups may extend over as long as ca. 5 units. An exciton may be formed over ca. 5 successive monomeric units in a chain upon photoexcitation.

This may be interpreted as follows. The **DBF** oligomers and polymers have a π-stacked structure in solution as we concluded on the basis of the NMR data and

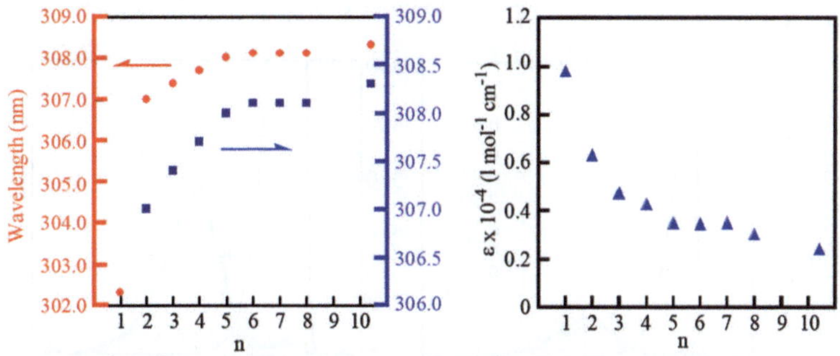

Fig. 22 Wavelength-vs.-n and molar extinction coefficient (ε)-vs.-n plots for the longest wavelength absorption maxima of oligo(**DBF**)s **2**. Data at $n = 1$ are for 9-methylfluorene, and those at $n = 10.4$ are for a polymer ($M_n = 1{,}890$ (vs. oligo(**DBF**)), $\bar{n} = 10.4$). Reprinted with permission from Nakano et al. [30]. Copyright 2003, American Chemical Society

theoretical simulations. However, the π-stacked molecules should experience thermal vibration that does not largely distort the conformation; this may lead to slight heterogeneity of the stacked conformation, i.e., the distance between the fluorene groups and main-chain torsion angle can slightly vary part by part in a chain, especially in a relatively short time range within which photoabsorption takes place though exact time scale of the excitation event is not yet known. The roughly coincidental critical number of fluorene units suggests that ca. five fluorene units are always in one block of very uniform stacked sequence even at a short time scale. In other words, the dynamic "persistence length" of a very uniformly π-stacked sequence from the view of photoexcitation covers five monomeric units under the experimental conditions of this study.

An alternative interpretation may be that the electronic interaction of the stacked fluorene groups in the polymer side chain extends inherently up to ca. five units even for a long, completely uniform "frozen" π-stacked polymer. Absorption measurements at an extremely low temperature or in a solid-state matrix that may significantly retard molecular vibration might provide an answer. Computational studies on the electronic states of poly(**DBF**) may also be necessary.

Figure 23 shows the emission spectra of **1** and **2**. In the spectra of **1** (Fig. 23a), those of $n \geq 4$ indicated predominant emission bands around 400 nm, while that of $n = 2$ showed a shorter-wavelength emission band around 310 nm, which was similar in shape and wavelength to that of monomeric fluorene, and that of $n = 3$ indicated both bands around 310 and 400 nm. The bands around 400 nm are considered to arise from an excited dimer or an excimer in which excited energy delocalizes over two neighboring, stacked fluorene units because the band wavelength is rather close to those of the fluorene excimer [93] (367 nm, in toluene) and the excited side-chain dimer of poly(2-vinylfluorene) [94] (380 nm, in THF). In contrast to **1**, all samples of **2** indicated predominant dimer emission.

As discussed so far, the terminal fluorene moieties are flipped and are not stacked in samples of **1**. The terminal-flipped conformation explains the monomeric fluorene emission from **1** of $n = 2$ and 3. However, the sample of **1** $n \geq 4$ showed only the

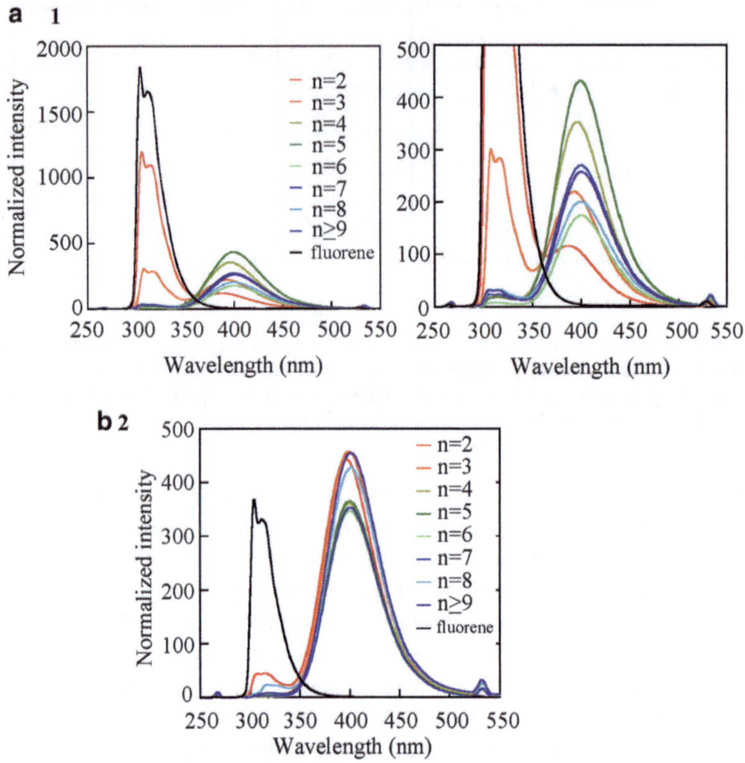

Fig. 23 Emission spectra of oligo(**DBF**)s **1** (**a**), **2** (**b**) and fluorene in THF at room temperature (λ_{ex} = 267 nm). In (**a**), spectra in the *left panel* are expanded in the *right panel*. Intensity was normalized to a constant absorbance at 267 nm. Reprinted with permission from Nakano et al. [30]. Copyright 2003, American Chemical Society

dimer emission. This suggests that the energy transfer from the terminal, flipped fluorene moieties to dimer forming, in-chain stacked fluorene moieties quenches monomeric fluorene in the longer oligomers of **1**. The predominant dimer emission from all **2** samples is consistent with the fact that all oligomers of **2** have a π-stacked structure throughout the chain.

The emission wavelength of the dimer emission of **1** and **2** was slightly red-shifted compared with that of the fluorene excimer [93] and the excited side-chain dimer in poly(2-vinylfluorene) [94], which may mean that the arrangement of two fluorene molecules in excimer is different from that of two fluorene units in the oligomers. As can be seen in the crystal structure, the side-chain fluorene moieties are so densely packed along the chain that the main-chain bends into a slightly arched shape. Such a close packing of two fluorene units probably enhances the electronic interaction between stacked π-electrons and, hence, reduces the delocalized excited energy level.

It is noteworthy that the emission profile of oligomer **1** of $n = 2$ is in sharp contrast to what is expected from "Hirayama's rule" [95], which predicts that a molecular

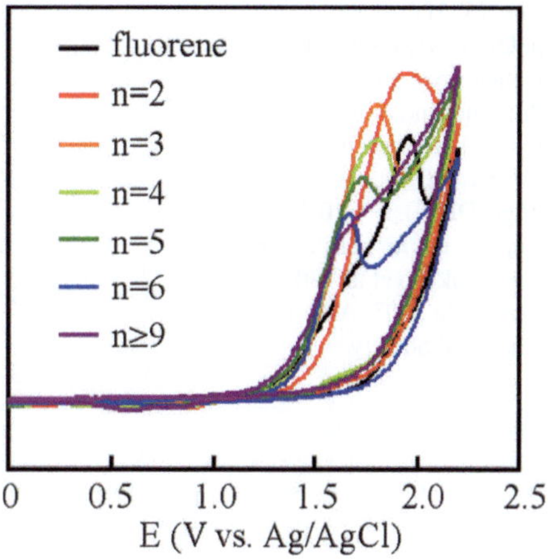

Fig. 24 Cyclic voltammograms of oligo(**DBF**)s (**2**) and fluorene. Reprinted with permission from Nakano et al. [30]. Copyright 2003, American Chemical Society

array with two chromophores connected through three CH_2 groups favors excimer formation more than others with more or less CH_2 groups between the chromophores. Although in the structure of **1** of $n=2$, two fluorenes are connected through three carbons, this oligomer showed only monomer emission. The conformational dynamics of this sample seems to be slower than that of the compounds tested by Hirayama that form a stacked excimer before emission.

Electrochemical Properties

Stacked π-electron systems may facilitate charge delocalization. In order to obtain information on how the stacked fluorene units may affect the electronic properties, cyclic voltammetry measurements were performed for oligomers **2**. The experiments were performed in a THF solution containing tetrabutylammonium perchlorate as a supporting electrolyte (0.1 M) under N_2 flow at room temperature with a conventional three-electrode configuration consisting of a platinum working electrode, an auxiliary platinum electrode, and a Ag/AgCl reference electrode at a sweep rate of 100 mV/s. The half-wave potential ($E_{p/2}$) values were determined as the potential at the current that was half of diffusion limiting current.

The samples showed clear oxidation peaks around 1.5–1.7 V but no clear corresponding reduction peak in the reverse scans from 2 to 0 V (Fig. 24). In addition, the intensity of the oxidation peak gradually decreased in repeated runs although the peak position was unchanged. These results suggest that the oxidized species may undergo some irreversible reaction on the electrode surface.

Figure 25 shows the dependence of oxidation potential (half-wave potential, $E_{p/2}$) on the chain length. The oxidation potential decreased with chain length and was almost constant when an oligomer contains ca. five or more fluorene units. This may mean that a hole is stabilized by delocalization over as many as ca. five

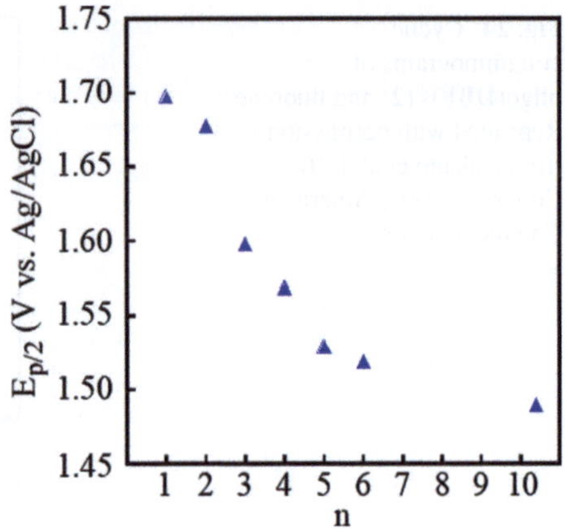

Fig. 25 Half-wave potential-vs.-*n* plot for **2**. Data at $n=1$ are for fluorene, and those at $n=10.4$ are for a polymer ($M_n=1{,}890$ (vs. oligo(**DBF**)), $\bar{n}=10.4$). Reprinted with permission from Nakano et al. [30]. Copyright 2003, American Chemical Society

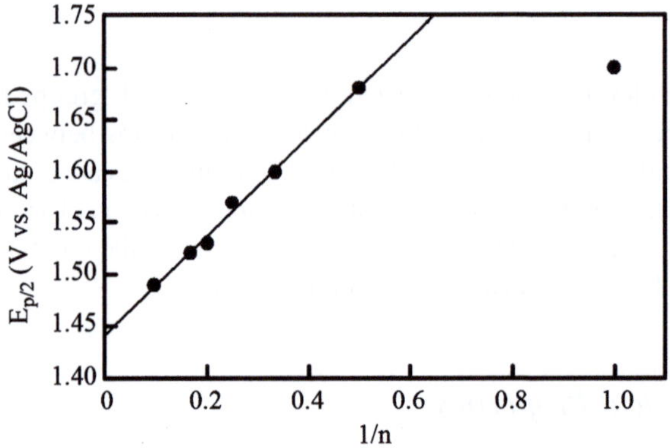

Fig. 26 Half-wave potential-vs.-1/*n* plots for the longest wavelength absorption maxima of oligo(**DBF**)s **2**. Extrapolation of the plot to $1/n=0$ gave 1.44 V vs. Ag/AgCl (1.40 V vs. SCE) for a polymer with an infinite number of fluorene moiety. Data at $1/n=1$ are for fluorene, and those at $1/n=0.0962$ are for a polymer ($M_n=1{,}890$ (vs. oligo(**DBF**)), $\bar{n}=10.4$). Reprinted with permission from Nakano et al. [30]. Copyright 2003, American Chemical Society

stacked fluorene units. This critical number of fluorene units coincides with those for the changes in the absorption intensity and edge wavelength depending on the chain length. As discussed in the preceding section, this may be ascribed to the proposed dynamic "persistence length" of a uniformly π-stacked sequence consisting of five fluorene units or to the nature of charge delocalization through the π-stacked structure of poly(**DBF**) which may inherently be limited to five fluorene units even if the chain has a longer, uniform "frozen" π-stacked structure. A low-temperature or solid-state study minimizing thermal vibration may provide an answer.

Plots of $E_{p/2}$ against 1/*n* led to a potential of 1.44 V (vs. Ag/AgCl) at an infinite n by extrapolation (Fig. 26). This value may correspond to the oxidation potential (half-wave potential) of a long chain of poly(**DBF**).

Fig. 27 A sandwich-type cell for TOF measurements

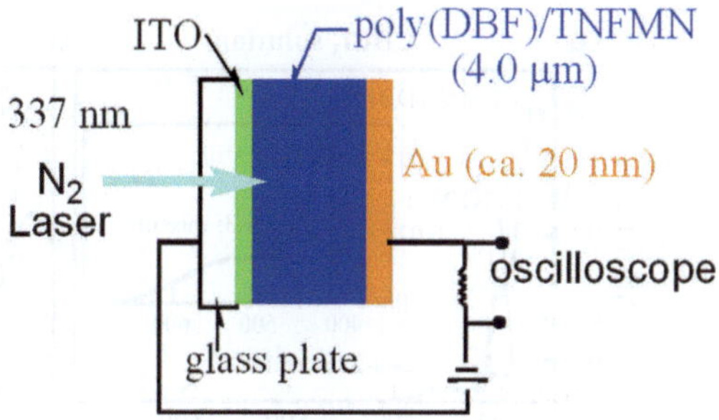

Charge Transport

Conducting polymers are an important class of material required for organic electronics and optoelectronics [96]. While most conducting polymers so far studied have a long electronic conjugation system in the main chain, a few vinyl polymers with no main-chain conjugation indicate intriguing electronic properties as represented by poly(*N*-vinylcarbazole) (poly(NVCz)), which finds practical applications based on its photoconductivity [97, 98]. In this section, the electronic properties of poly(**DBF**) are discussed. Hole drift mobility was measured by a time-of-flight (TOF) method [91, 99, 100] using a cast film of the polymer.

Poly(**DBF**) prepared by the anionic polymerization of using 9-fluorenyllithium in tetrahydrofuran (THF) at −78 °C (M_n=1,070, M_w/M_n=1.38, SEC using oligo(**DBF**)s as the standard sample) (20 mg) and 2,4,7-trinitrofluorenylidene-9-malononitrile (TNFMN) (1.0 mg) as an acceptor were dissolved in chloroform (1.0 mL), and the solution was cast on an indium tin oxide (ITO) glass plate and slowly dried under air to afford a thin film (thickness 5 μm). A circular Au electrode (thickness 20 nm, diameter 5 mm) was vacuum deposited onto the film to produce a sandwich-type cell for the drift mobility measurement (Fig. 27).

In order to investigate the interaction between poly(**DBF**) and the acceptor molecule, absorption spectra were taken in a chloroform solution and in film (Fig. 28). Both in the solution and in the film, a weak band, probably due to donor–acceptor interaction, appeared around 520 nm in the mixture of poly(**DBF**) and TNFMN in addition to the bands due to poly(**DBF**) and TNFMN. These results indicate that TNFMN is an effective electron accepting dopant toward poly(**DBF**). In addition, the spectra indicate that pure poly(**DBF**) absorbs much less than TNFNM and the mixture at 337 nm. This means that N_2 laser irradiation in TOF experiments will mainly excite the acceptor band or the donor–acceptor band to induce charge separation.

Charge drift mobility (μ) was estimated by TOF transient photocurrent measurement under vacuum under N_2 pulse laser irradiation (337 nm, pulse duration 3 ns, 50 μJ). The value of μ was calculated according to $\mu = d/(t_\tau \bullet F)$ (cm^2 V^{-1} s^{-1}), where d is the film thickness, t_τ is the transit time determined by the TOF experiment, and F is the field strength. By switching polarity in the measurement, it was confirmed that holes mediate the charge drift in the present systems. Figure 29 indicates the μ values at different F's and different temperatures. Overall, a higher field strength led

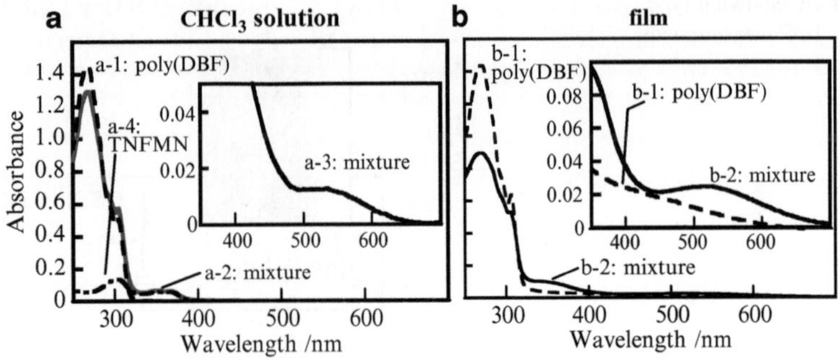

Fig. 28 Absorption spectra of poly(**DBF**) in $CHCl_3$ solution (cell length 0.1 mm) (**a**) and poly(**DBF**) film (0.5 μm) (**b**): (a–1) poly(**DBF**) in $CHCl_3$ ([**DBF** unit] 1.1×10^{-2} M), (a–2) poly(**DBF**)-TNFMN mixture in $CHCl_3$ ([**DBF** unit] 1.1×10^{-2} M, [TNFMN] 2.8×10^{-4} M), (a–3) poly(**DBF**)-TNFMN mixture in $CHCl_3$ ([**DBF** unit] 1.1×10^{-1} M, [TNFMN] 2.8×10^{-3} M), (a–4) TNFMN in $CHCl_3$ (2.8×10^{-4} M), (b–1) poly(**DBF**) film, and (b–2) poly(**DBF**)-TNFMN film ([**DBF** unit]/[TNFMN] = 20). Reprinted with permission from Nakano et al. [33]. Copyright 2004, The Chemical Society of Japan

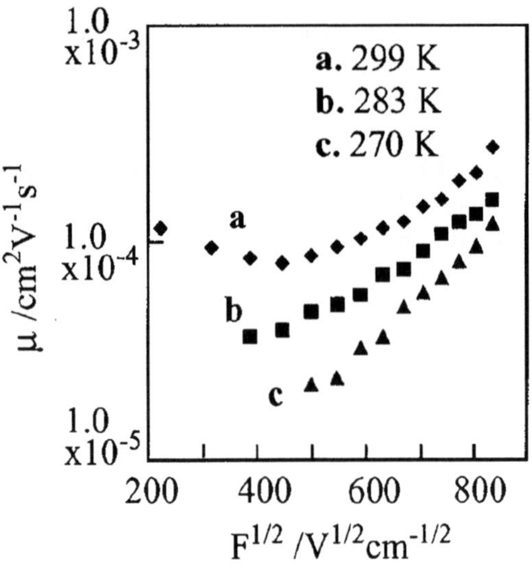

Fig. 29 Hole drift mobility of poly(**DBF**) film (4 μm): (*a*) at 299 K, (*b*) at 283 K, and (*c*) at 270 K. Reprinted with permission from Nakano et al. [33]. Copyright 2004, The Chemical Society of Japan

to a higher drift mobility; however, a reversed relationship was observed in the field strength range below 2×10^5 V/cm. In this range, holes may transport in a different path from that predominantly used at the higher field strength, suggesting that the polymer sample is somewhat heterogeneous in terms of molecular structure and/or intermolecular arrangement. The hole drift mobility was higher at a higher temperature at all field strengths, indicating that the charge drift is thermally activated.

The observed charge drift mobility was as high as 2.7×10^{-4} cm^2V^{-1} s^{-1} at 299 K at $F=7\times10^5$ V/cm. This value is higher by the order of 10^3 than those for other side-chain aromatic vinyl polymers including poly(NVCz) and poly(1-vinylpyrene) (10^{-7} cm^2 V^{-1} s^{-1} order) [101] and is the highest so far reported for a vinyl polymer although mobility values higher than 10^{-2} cm^2 V^{-1} s^{-1} have been reported for some other types of organic polymers [102–104]. The hole drift mobility is even higher than that of main-chain conjugating poly(*p*-phenylenevinylene) (1×10^{-5} cm^2 V^{-1} s^{-1}) [105], is comparable to that of poly(methylphenylsilane) (1×10^{-4} cm^2 V^{-1} s^{-1}) [105], and is slightly lower than Se (10^{-4} cm^2 V^{-1} s^{-1} order) [106], an inorganic semiconductor.

In the film sample, both intramolecular and intermolecular charge transfers may take place, and the observed charge mobility will mainly reflect the slower process. Because the poly(**DBF**) chains used in this study appear to have disordered intermolecular alignments on the basis of the almost linear field dependences of mobility in the higher field ranges, the major limiting step of the drift mobility will be the intermolecular charge transfer. The disordered structure was supported by the fact that the film did not show birefringence in the polarizing microscopy observation. In addition, the poly(**DBF**) sample used here did not show a clear melting point in DSC analysis, further supporting that the chains do not have an ordered alignment.

From the conclusion of the disordered system, it may be assumed that intramolecular charge transfer is faster than the bulk mobility characterized by $\mu=2.7\times10^{-4}$ cm^2 V^{-1} s^{-1}. This idea is supported by cyclic voltammetry measurements of poly(**DBF**) where the oxidation potential for the poly(**DBF**) was significantly lower than that for fluorene, a model of monomeric side chain as discussed in the preceding section. This indicates that a hole may be readily delocalized over the π-stacked fluorene groups in the side chain, leading to fast intramolecular charge transfer. The ordered molecular structure of the π-stacked poly(**DBF**) may act as a charge transport pathway.

From the data in Fig. 29, field-dependent activation energies for the charge transport were calculated to be 0.22 eV at $F=7.0\times10^5$ Vcm^{-1}. This value is much smaller than that of poly(NVCz) (0.4–0.7 eV) [97] and is comparable to that of poly(methylphenylsilane) (0.24 eV) [106]. For poly(NVCz), locally π-stacked chromophores are trap sites for charge drift [107, 108], and the activation energy corresponds to the thermal hopping-up of charges from the π-stacked trap sites [97]. The smaller activation energies of the poly(**DBF**) support that the hole transport mechanism in poly(**DBF**) is completely different from that in poly(NVCz). Although comparisons between poly(**DBF**) and other vinyl polymers with π-stacked structures should be more appropriate, no such polymer is known other than poly(**DBF**). Poly(NVCz) was chosen to be a reference compound in this text as a vinyl polymer showing a relatively high charge mobility.

In summary, the charge mobility of poly(**DBF**) film was as high as some main-chain conjugating polymers. In addition, fast intramolecular charge transport was suggested to take place through the stacked π-electron systems. Charge transport through regularly stacked heteroaromatic systems have been proposed for DNAs where the stacked systems serve as a "*π-way*" [41–43, 109–111] although there is controversy as to whether they really conduct [112–114]. Poly(**DBF**) may be recognized as a new, synthetic "*π-way*" molecule. Further studies are under way to establish the charge transport through single molecule of π-stacked poly(**DBF**) and to expand the π-stacked molecular design to other versatile synthetic polymers.

Charge mobility behavior can be described by a bandlike regime or a charge hopping scheme. In the former model, the total conduction [valence] bandwidth (W_b) and the effective mass of the electron [hole] of a poly(**DBF**) chain are defined by the electronic coupling (transfer integral, t) between adjacent fluorene units as $W_b = 4t$ and $m_{eff} = h^2/2at$, respectively (here h is the reduced Planck's constant and a is the spacing between the fluorene units) [115]. In the latter model, at the microscopic level, a hopping event can be described as a self-exchange electron-transfer (ET) reaction from a charged, relaxed unit to an adjacent neutral unit. The carrier mobility is then controlled by the self-exchange ET rate. At high temperature, when semiclassical ET theory usually applies, the self-exchange electron-transfer (hopping) rate is given by $k_{ET} = A \exp(-\lambda/4k_BT)$ where the reorganization energy, λ, measures the strength of hole–vibration or electron–vibration interactions and prefactor A depends on the strength of the electronic coupling: in the case of weak coupling (nonadiabatic ET regime), $A \sim t^2$. Thus, in both bandlike and hopping regimes, there are two major parameters that impact the charge mobility: (1) the intermolecular transfer integral t, which should be maximized, and (2) the reorganization energy λ (electron–vibration coupling), which needs to be small for efficient transport. For a dimer of **DBF** having terminal ethyl group, t for hole was estimated to be 0.2–0.3 eV from photoelectron spectral measurements, and DFT calculations suggested that λ values for hole and electron transfers were 0.32 and 0.26 eV, respectively [115]. For comparison, t for pentacene crystals estimated by INDO level calculations do not exceed 0.1 eV which is less than half the value for poly(**DBF**). In addition, λ values for hole and electron transfers in pentacene crystals have been computed to be 0.32 and 0.26 eV, respectively [115]. These results indicate that holes in **DBF** chains are characterized by a large transfer integral (bandwidth) and smaller reorganization energy than electrons; charge transport along a poly(**DBF**) chain is therefore expected to be more efficient for holes than electrons. Hole and electron mobilities of poly(**DBF**) and derivatives might tend toward those found in oligoacene systems.

Recently, intramolecular charge mobility of poly(DBF) was estimated to be, at least, of the order of 10^{-1} cm^2 V^{-1} s^{-1} s by time-resolved microwave conductivity measurements (Seki S, Nakano T, unpublished results). This means that performance of poly(DBF) as a charge transport mediator or a conductor would be largely improved by enhancing intermolecular charge transport through regulation of chain alignment or formation of controlled chain assemblies.

Poly(dibenzofulvene) Derivative and Analogues

Derivatives Bearing Alkyl Side Chains

In spite of the charge mobility and colorless (transparent) optical properties, a drawback of poly(**DBF**) as a practical material is its low solubility in solvents: the highest M_n (vs. polystyrene) of soluble fractions was ca. 2,500 [31]. Fractions of higher

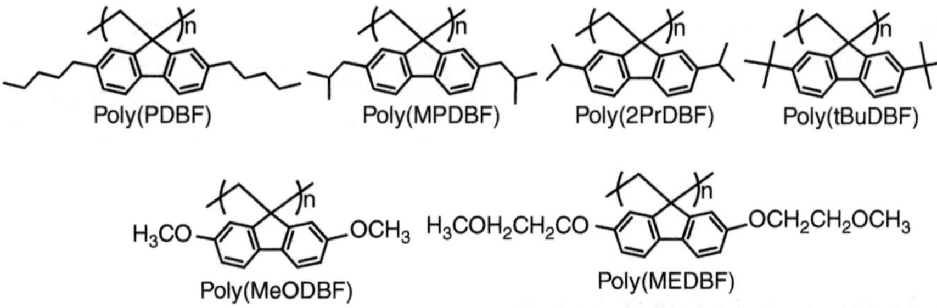

Fig. 30 Poly(**DBF**) derivative having aliphatic and polar side chains at the 2- and 7-positions of the fluorene moiety

molar masses are insoluble. The low solubility hampers fabrications from a solution that is essential for organic polymer materials and often results in brittleness of films of the polymer. In order to overcome this shortcoming, various derivatives of poly(**DBF**) having substituents at the 2- and 7-positions of fluorene moiety have been designed and synthesized. The basic idea was to block interchain interactions by the side-chain groups.

Polymers obtained from 2,7-di-*n*-pentyldibenzofulvene (**PDBF**), 2,7-bis(2-methylpropyl)dibenzofulvene (**MPDBF**), 2,7-di-2-propyldibenzofulvene (2**PrDBF**), and 2,7-bis(*t*-butyl)dibenzofulvene (**tBuDBF**) are the examples in this category (Fig. 30) [116, 117]. **PDBF**, **MPDBF**, and **2PrDBF** smoothly polymerized to a polymer: maximum molar masses of the soluble fractions were around 5,500 (DP around 20). Solubility was hence improved by introducing alkyl chains at the 2- and 7-positions of the fluorene moiety of **DBF** monomer backbone although the extent of improvement was limited. In a sharp contrast to **PDBF**, **MPDBF**, and **2PrDBF**, **tBuDBF** led to only up to 3-mer by polymerization [117]. The low reactivity is considered to arise from bulkiness of *t*-Bu groups.

Two polymers having polar alkoxy groups were also synthesized from the corresponding monomers, 2,7-methoxydibenzofulvene (**MeODBF**) [116] and 2,7-bis(methoxyethoxy)dibenzofulvene (**MEDBF**) [118]. While solubility of poly(**MeODBF**) was similar to that of poly(**DBF**), poly(**MEDBF**) exhibited good solubility in chloroform. Poly(**MEDBF**) prepared by cationic and radical polymerization was soluble, at least, up to M_n over 100,000. It is noteworthy that the numbers of atoms in a side chain in poly(**PDBF**) and poly(**MEDBF**) are the same (five): polar methoxyethoxy group is far superior than completely aliphatic *n*-pentyl group in preventing chain aggregation in chloroform. The reason for this observation is not immediately clear: however, electrostatic repulsion among the polar methoxyethoxy chains may have a role. The fact that solubility of poly(**MeODBF**) was similar to that of poly(**DBF**) suggests that electrostatic repulsion exerted by the menthoxy group at the side-chain terminal is important.

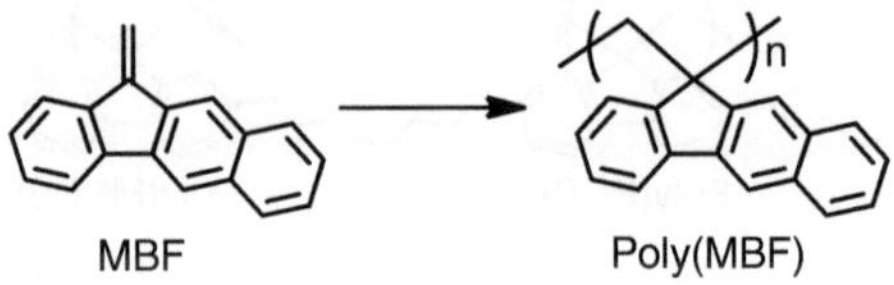

Fig. 31 Polymerization of MBF leading to a π-stacked polymer

A Monomer with Lower Symmetry

Another way to improve solubility of π-stacked polymer is to introduce C_2-*asymmetry* to starting monomer. While fluorene moiety is C_2-symmetric and σ-symmetric, 11-methylenebenzo[b]fluorene (MBF) lack a C_2 axis and a mirror plane. MBF leads to a π-stacked vinyl polymer as well as **DBF** in spite of its higher bulkiness (Fig. 31) [119, 120]. The polymer is soluble up to a DP of ca. 40 (M_n ~9,300 (^{1}H NMR), ~9,500 (light scattering)). For this polymer, stereoisomerism (tacticity) is possible although it has not yet been clarified. Assuming rather random or not-highly-stereotactic main-chain configuration, randomness seems to be effective in preventing chains from aggregating.

Derivatives and Analogues Having Thiophene Moieties

Among various main-chain conjugated polymers, one of the most practically used may be the poly(3,4-ethylenedioxythiophene) (PEDOT)-poly(styrenesulfonate) (PSS) mixture [121–124]. PEDOT is a polythiophene derivative that can be doped using sulfonic acid including PSS and redox active metals. Since the fraction of PSS is higher than that of PEDOT in commercial aqueous solutions, the material is acidic. Also due to long main-chain conjugation PEDOT-PSS shows deep blue color. These two characteristics somewhat limit applications of PEDOT-PSS.

Inspired by the fact that polythiophenes show practical properties, we designed two poly(**DBF**) derivatives with thiophene rings directly attached to the fluorene moiety, poly(2-(2,2′-bithiophen-5-yl)dibenzofulvene) (poly(**BT**-**DBF**)) [125, 126] and poly(2-(5′-(2-ethylhexyl)-2, 2′-bithiophen-5-yl)dibenzofulvene) (poly(**ETBF-DBF**)) [125, 126], and a poly(**DBF**) analogue having bithiophene moiteties in the side chain, poly(4-methylenecyclopenta[2,1-b;3,4-b′]dithiophene) (poly(**MCPDT**)) [127]. These polymers have a π-stacked structure (Fig. 32). Conformation of these polymers were clarified by ^{1}H NMR, UV, and fluorescent spectra as well as the of poly(**DBF**). Although photoelectronic properties of these polymers are not yet investigated in detail, we expect that the polymers are candidates of vinyl polymer that can overcome limitations of PEDOT-PSS.

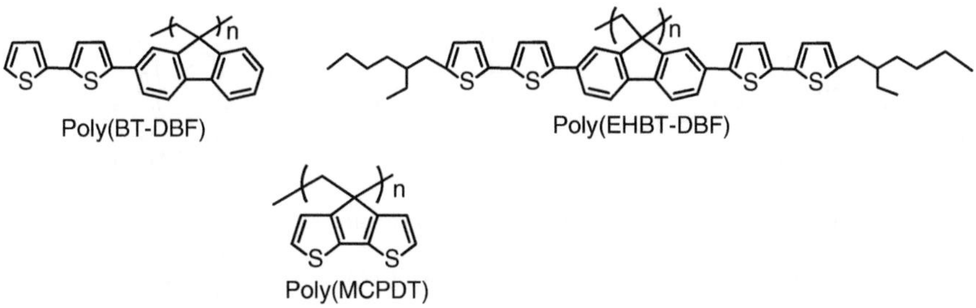

Fig. 32 Thiophene-containing π-stacked polymers

Chirality in π-Stacked Polymers

Asymmetric Polymerization: Optically Active Polymer Without Center of Chirality

An optically active polymer has been obtained from 2,7-di-*n*-pentyldibenzofulvene (**PDBF**) by anionic polymerization using chiral anionic initiators. The optical activity of the polymer arises only from a chiral conformation. In this polymer, the fluorene moieties are in a π-stacked assembly which may be twisted in a single sense.

Synthetic polymers with conformational chirality have become a research field of widespread interest in recent years, and a wide range of polymers with conformational chirality have been synthesized from various types of monomers including vinyl monomers [9, 61–63, 128–136]. The existing examples of optically active vinyl polymers with conformational chirality include isotactic, helical polyolefins bearing asymmetric side chains [133–135] and isotactic, helical polymethacrylates bearing bulky, achiral side chains [61–63, 136]. These polymers have stereocenters in the main and/or side chains. Optically active poly(PDBS) is the first vinyl polymer with conformational chirality bearing no stereocenters in the main and side chains whose chiroptical properties arise only from a chiral conformation.

PDBF was anionically polymerized using the complexes of 9-fluorenyllithium (FlLi) with (−)-sparteine (Sp) and (+)- and (−)-1,4-bis(dimethylamino)-2,3-dimethoxybutane (DDB) in toluene or using *n*-BuLi in tetrahydrofuran (THF) under N_2 atm at −78 °C (Fig. 33). The conditions and results of polymerization are summarized in Table 5. The chiral complexes have been effectively used in helix-inducing polymerization of methacrylates [61–63, 136]. The polymers were partially insoluble in tetrahydrofuran (THF) and chloroform. The THF-insoluble part seems to have the same chemical structure as that of the THF-soluble part based on IR spectra; the insoluble part probably has a higher molecular weight as in the case of poly(**DBF**) with no substituent.

The introduction of two pentyl groups per fluorene unit seems not to affect the formation of π-stacked conformation. Similarly to poly(**DBF**) without substituents, THF-soluble poly(**PDBF**)s showed remarkable hypochromism in absorption spectra and exclusive excimer emission in fluorescent spectra as shown in Fig. 34.

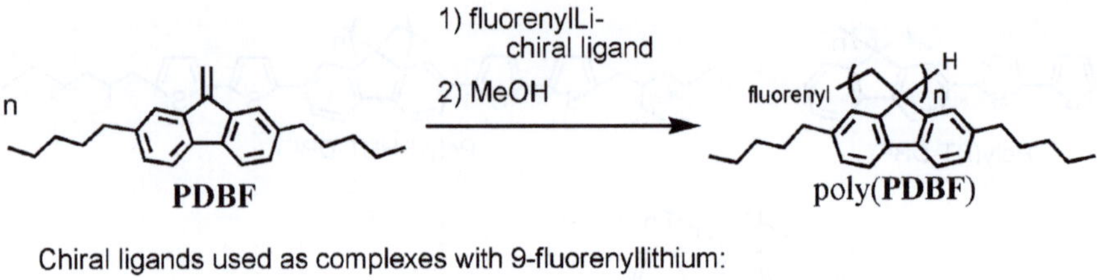

Fig. 33 Asymmetric anionic polymerization of **PDBF**

Table 5 Anionic polymerization of PDBF at –78°C[a]

Run	Initiator	Solvent	Conv.[b] (%)	THF-insol.[c] part Yield (%)	THF-sol. part[c] Yield (%)	M_n[d]	M_w/M_n[d]
1	(+)-DDB-FlLi	Toluene	40	3	24	5,290	1.03
2	(–)-DDB-FlLi	Toluene	50	2	49	5,710	1.16
3	Sp-FlLi	Toluene	73	21	50	5,950	1.05
4	*n*-BuLi	THF	>99	Trace	84	4,750	1.2

[a]Conditions: time = 24 h, monomer amount = 1.00–1.02 mmol, [monomer] = 0.20–0.21 M, [ligand]/[Li] = 1.2, $[\text{monomer}]_o/[\text{RLi}]_o = 20$

[b]Determined by ^{1}H NMR analysis of the reaction mixture

[c]The quenched reaction mixture was first separated into MeOH-soluble and MeOH-insoluble parts, and the insoluble part was then extracted using THF

[d]Determined by SEC using a Viscotek TDA300 detector which collects right-angle laser light scattering, viscosity, and RI information

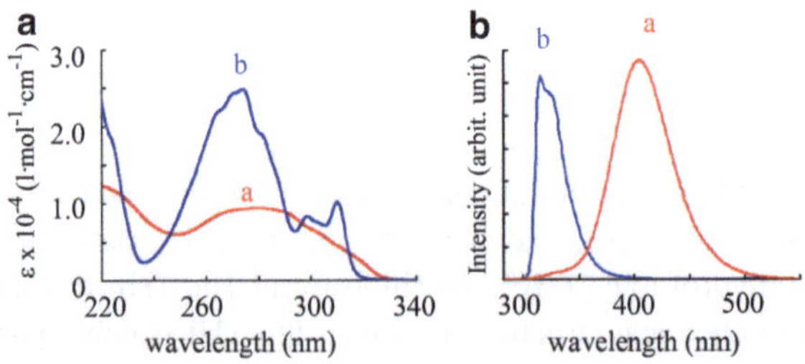

Fig. 34 Absorption (*left*) and emission (*right*) spectra of poly(**PDBF**)s prepared using (–)-Sp-FlLi (run 3 in Table 6) (**a**) and 2,7-di-*n*-pentylfluornene (**b**), a model of monomeric unit in THF at room temperature. Conditions: concentration 1.8×10^{-5} M (**a**), 1.7×10^{-5} M (per chromophore) (**b**); cell length 0.1 mm (absorption), 10 mm (emission); λ_{ex} for emission 282 nm. Reprinted with permission from Nakano et al. [34]. Copyright 2004, The Royal Society of Chemistry

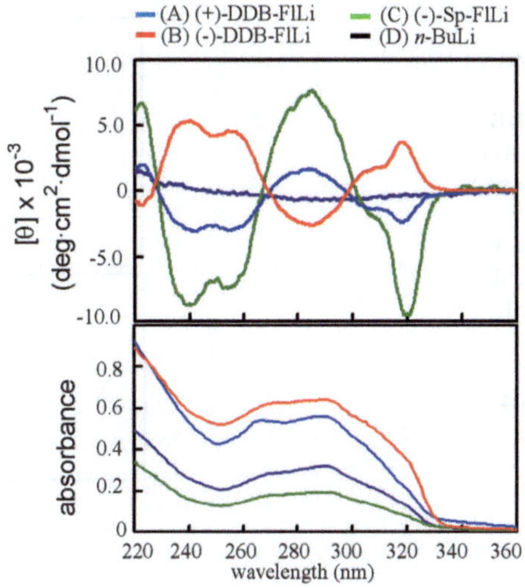

Fig. 35 CD (*top*) and absorption (*bottom*) spectra of THF-soluble poly(**PDBF**)s prepared using (+)-**DDB**-FlLi (M_n 5,290, M_w/M_n 1.03) (run 1 in Table 6) (*A*), (−)-**DDB**-FlLi (M_n 5,710, M_w/M_n 1.16) (run 2 in Table 6) (*B*), (−)-Sp-FlLi (M_n 5,950, M_w/M_n 1.05) (run 3 in Table 6) (*C*), and *n*-BuLi (M_n 4,750, M_w/M_n 1.20) (run 4 in Table 6) (*D*) measured using a JASCO J-820 spectrometer. Concentration of the film samples for molar ellipticity calculation was determined by comparing absorbance at 290 nm of a film with that of a solution in THF (conc. 1.10×10^{-2} M, cell length 0.1 mm, absorbance at 290 nm 0.983). Thickness of film B was 0.10–0.16 μm as measured using a Keyence VK9500 Laser microscope. Reprinted with permission from Nakano et al. [34]. Copyright 2004, The Royal Society of Chemistry

Although the chiral ligands were used for the anionic polymerization, the THF-soluble **PDBF** polymers prepared using the optically active initiators did not show detectable circular dichroism (CD) absorption bands or optical rotation in a THF solution. No CD absorption was observed in solution even at −50 °C. However, thin films cast on a quartz plate from a hexane solution of the polymers (conc. 3 g/L) showed intense CD bands (Fig. 35A–C). The spectra in film of the polymers obtained using (+)-**DDB**-FlLi and (−)-**DDB**-FlLi were almost mirror images in spectral pattern (**A** and **B**). The polymer obtained using (−)-Sp-FlLi showed an intense spectrum with a spectral pattern similar to that of the polymer obtained with (+)-**DDB**-FlLi. The polymer obtained by polymerization with achiral *n*-BuLi did not show CD absorptions in film (**D**). In addition, the film of the polymer synthesized using (−)-**DDB**-FlLi (run 2 in Table 5) showed optical rotatory dispersion (ORD) in the wavelength range shorter than 330 nm: $[\alpha]_{295}$ −25°.

The spectral intensity varied from film to film prepared in repeated experiments. As an example, CD and absorption spectra of the three film samples prepared using the poly(**PDBF**) from run 2 in Table 5 are shown in Fig. 36. The films differed in thickness and roughness of the surface. A smoother film tended to show larger CD band intensities. Film surface morphologies of samples in Fig. 36A, C are shown in Fig. 37a, b, respectively. In most cases, the film surface showed a relatively regular, wrinkled pattern as shown in Fig. 37b. Such macroscopic morphologies might have a close connection to molecular structure (conformation) of the polymer in the film.

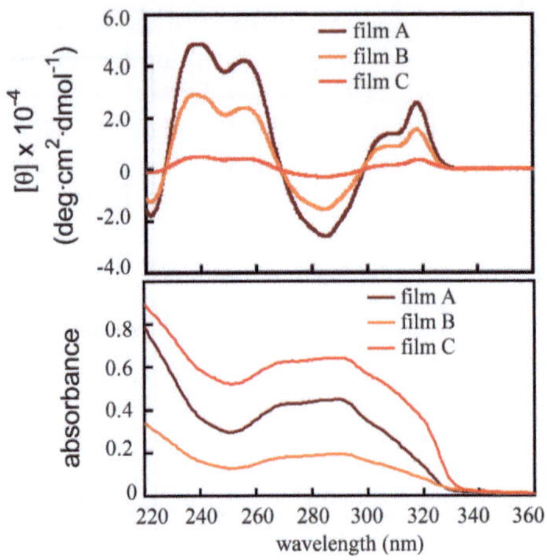

Fig. 36 CD intensity variation of films prepared from poly(**PDBF**) synthesized using (–)-**DDB-FlLi** (run 2 in Table 6): (*A*) a film with a smooth surface, (*B*) a film with a modestly smooth surface (smoothness was between *A* and *C*), and (*C*) a film with a wrinkled surface. Reprinted with permission from Nakano et al. [34]. Copyright 2004, The Royal Society of Chemistry

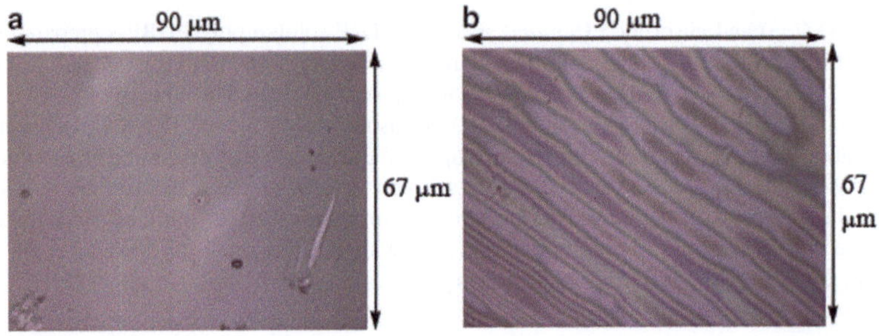

Fig. 37 Surface pictures of films made from poly(**PDBF**) synthesized using (–)-**DDB**-FlLi (run 2 in Table 6): (**a**) a film with a smooth surface and (**b**) a film with a wrinkled surface (wrinkle height 0.3–0.6 **μ**m). Films A and B correspond to spectra A and C in supplementary Fig. 6. This analysis was performed using a Keyence VK-9500 laser microscope (laser wavelength 408 nm). Reprinted with permission from Nakano et al. [34]. Copyright 2004, The Royal Society of Chemistry

However, for a certain polymer, the signs and patterns of the CD spectra were unchanged. In addition, changing the direction and orientation of the film sample did not significantly affect the spectral intensity and pattern. Furthermore, the films did not show birefringence in polarized optical microscopy analysis. These observations rule out the possibility that the CD spectra are due to linear dichroism based on film anisotropy and support that the film spectra reflect molecular chirality. CD absorptions similar to those indicated in film were observed also for a decalin suspension of THF-insoluble polymer (fine power ground with a mortar) synthesized using (–)-Sp-FlLi. This further supports that the CD spectra in film (Fig. 34) do not arise from sample anisotropy.

The chemical structure of the polymers shown in Fig. 33 was confirmed by MALDI mass analysis of THF-soluble polymers (Fig. 38). Therefore, the polymers obtained in this study have no chiral ligand fragment attached to the chain and have no stereocenters in the main or side chain. Hence, the CD absorptions are unambiguously ascribed to a chiral conformation induced during the polymerization process.

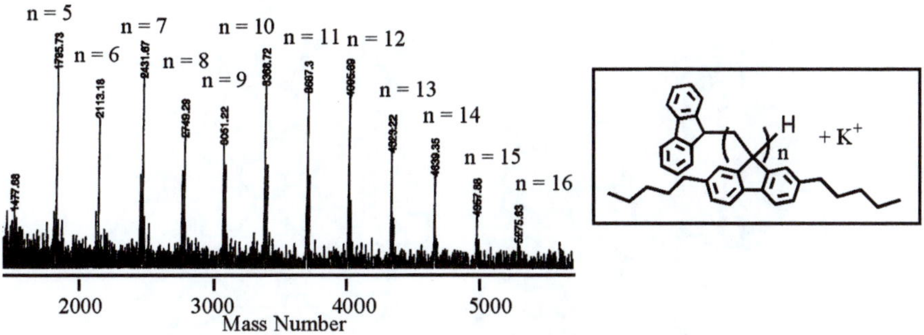

Fig. 38 MALDI mass spectrum of poly(**PDBF**) synthesized using (–)-Sp-FlLi (THF-soluble, MeOH-insoluble part, M_n 5,950 as determined by size-exclusion chromatography) [matrix, 2,5-dihydroxybenzoic acid]. Reprinted with permission from Nakano et al. [34]. Copyright 2004, The Royal Society of Chemistry

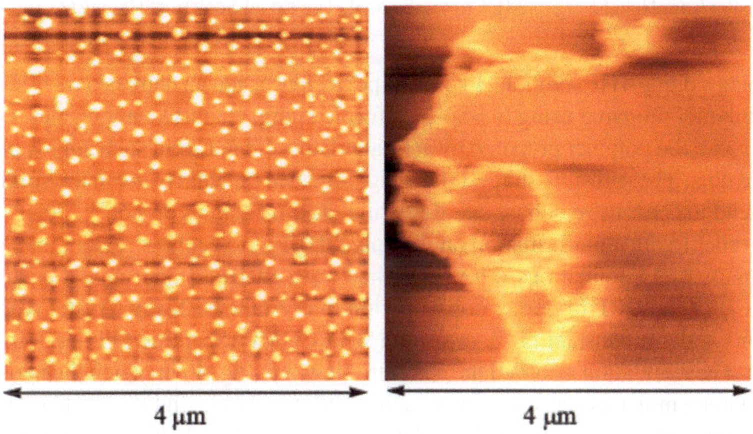

Fig. 39 Atomic force microscopic images (topographic image) of poly(**PDBF**)s synthesized using *n*-BuLi (*left*) and (–)-Sp-FlLi (*right*) (runs 4 and 3, respectively, in Table 6) [Seiko Instruments SPA400-SPI3800N, DFM mode]. Reprinted with permission from Nakano et al. [34]. Copyright 2004, The Royal Society of Chemistry

As described in "X-ray Crystal Analyses" section, **DBF** oligomers have been shown to have a π-stacked, single-handed helical conformation in crystal. A π-stacked structure was indicated also for poly(**PDBF**) by remarkable hypochromicity in absorption and exclusive dimer emission in fluorescence spectra (Fig. 34). Therefore, it will be reasonably assumed that the CD absorptions observed in this work are based on a helical conformation of poly(**PDBF**) with excess single handedness. Because the polymers showing clear CD bands in film did not show chiroptical properties in solution, molecular aggregation may amplify and stabilize the single-handed helical conformation induced by the asymmetric polymerization possibly due to intermolecular cooperative effects in the solid state. Atomic force microscopic (AFM) analyses supported this assumption (Fig. 39). The samples were prepared by

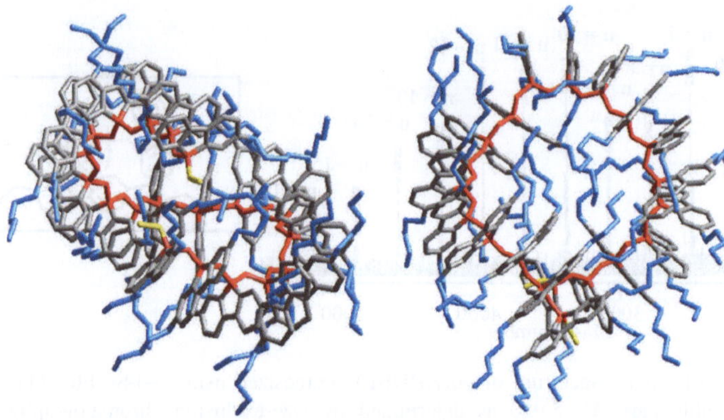

Fig. 40 The conformation of a **PDBF** 20-mer having terminal ethyl groups generated through MD simulation (NVT, 300 K, 8 ns): side and top views. The main chain is shown in *red*, fluorene moieties in *gray*, pentyl groups in *light blue*, and terminal ethyl groups in *yellow*. Reprinted with permission from Nakano et al. [34]. Copyright 2004, The Royal Society of Chemistry

dropping a dilute THF solution (1×10^{-4} M per residue) onto an untreated mica surface. The polymer prepared using (−)-Sp-FlLi formed long, chain-like aggregates. On the other hand, the polymer prepared by achiral *n*-BuLi did not from such aggregates but simple dots. These results suggest that single-handed helical chains in the solid state may tend to form large helical bundles while racemic helices do not.

In order to obtain information on the poly(**PDBF**) molecular structure, molecular dynamics (MD) simulations were performed for a **PDBF** 20-mer model created on the basis of the **DBF** oligomer conformation in crystal (see X-ray Crystal Analyses section) under a constant NVT condition with COMPASS force field at 300 K for 8.6 ns. The simulation afforded the structure shown in Fig. 40 in which the neighboring fluorene moieties stack on top of each other in a chiral, slightly twisted arrangement. The entire chain formed a relatively long-pitched helical structure in the simulation. Hence, the local twist and the helix may contribute to the CD spectra.

The chiral conformation contributing the CD spectra is constructed during the polymerization by the effect of the chiral ligands. The CD spectra and the chiral conformation do not arise from residual chiral ligand which may be contained in the polymer samples in a small quantity. The isolated polymers were washed with methanol to remove the ligand residue, and the ligand residue contents in the MeOH-insoluble polymers were less than 0.01 mol% (per monomeric residue) as determined by ^{1}H NMR. It was confirmed that a film made from a mixture of the achiral polymer synthesized using *n*-BuLi and 0.01 mol% of (−)-Sp did not show clear CD. This rules out the possibility that residual ligand is responsible to the CD spectra in Fig. 35.

The poly(**PDBF**)s prepared using chiral initiators have a chiral conformation in solution although their chiroptical properties are under detectable level, i.e., they are "cryptochiral" [137]. In order to test conformational stability in solution, THF-soluble polymer synthesized using (−)-Sp-FlLi (run 3 in Table 5) was heated in a toluene solution at 110 °C for 16 h. A film made from the polymer after the heat

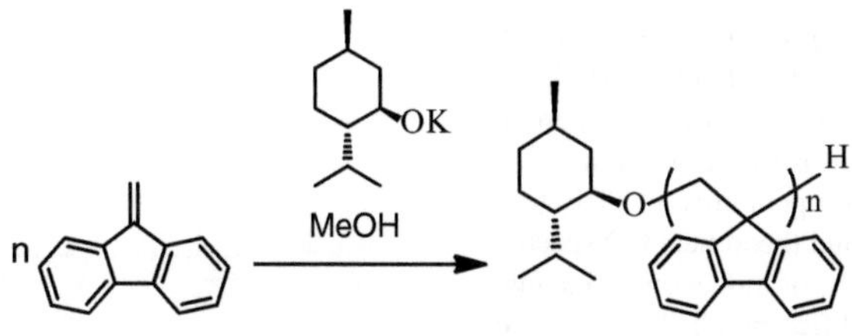

Fig. 41 Polymerization of DBF using potassium menthoxide

treatment still showed intense CD bands, indicating that the conformation induced in polymerization was stable in a solution under these conditions.

To summarize this section, asymmetric anionic polymerization afforded poly(**PDBF**)s showing significant CD absorptions in film although their intensity of chiroptical properties is below detectable level in solution. The polymers belong to a new class of optically active vinyl polymers whose chiroptical properties originate only from a chiral conformation in the absence of any configurational information.

π-Stacked Polymers Bearing Chiral Terminal Group: CPL-Emitting Polymer

Chirality induction in poly(**DBF**) and derivative is also feasible by anionic polymerization using potassium menthoxide prepared by KH and (−)-menthol (Fig. 41) [36]. Optically active poly(**DBF**) obtained by this method indicates intense CD spectra in the wavelength range of the stacked fluorene chromophore (Fig. 42a). As mentioned earlier, a hexamer of **DBF** takes a helical conformation in crystal. Hence, the CD spectra can be ascribed to a preferred-handed helical conformation of the main chain induced by the chiral terminal group. The fact that CD intensity, not the spectral shape, changes depending on the temperature of measurement suggests that the helix is somewhat dynamic and that helical bias depends on temperature. Also, through comparisons of the experimental CD spectral with theoretical spectra obtained by CNDO/S-CI calculations, absolute handedness of the preferred helix was decided to be left (Fig. 42b, c). Theoretical spectral simulations were straightforward in the case of poly(**DBF**) because the polymer has only fluorene chromophore without electronic interactions among different chromophores that may reduce accuracy of simulations.

Since fluorene is generally a good emitter and poly(**DBF**) can facilitate charge transport, application of poly(**DBF**) and derivatives for organic light emitting diodes (OLEDs) would be interesting [37]. Especially, chiral π-stacked polymers may emit circularly polarized light (CPL). From this view, poly(2,7-bis(4-(*t*-butyl)phenyl) dibenzofulvene) (poly(**BBPDBF**)) was synthesized by anionic polymerization using potassium menthoxide. The side-chain 4-(*t*-butyl)phenyl group was introduced to improve fluorescent efficiency by extending π-conjugation of the monomeric unit because the quantum efficiency of poly(**DBF**) is only about 0.06–0.12 [31]. Poly(**BBPDBF**) indicated intense CD spectra in solution and in the solid state (cast film), and it exhibited fluorescence at a quantum yield of 0.40 which is only about half that of 2,7-bis(4-(*t*-butyl)phenyl)fluorene as a monomeric unit model

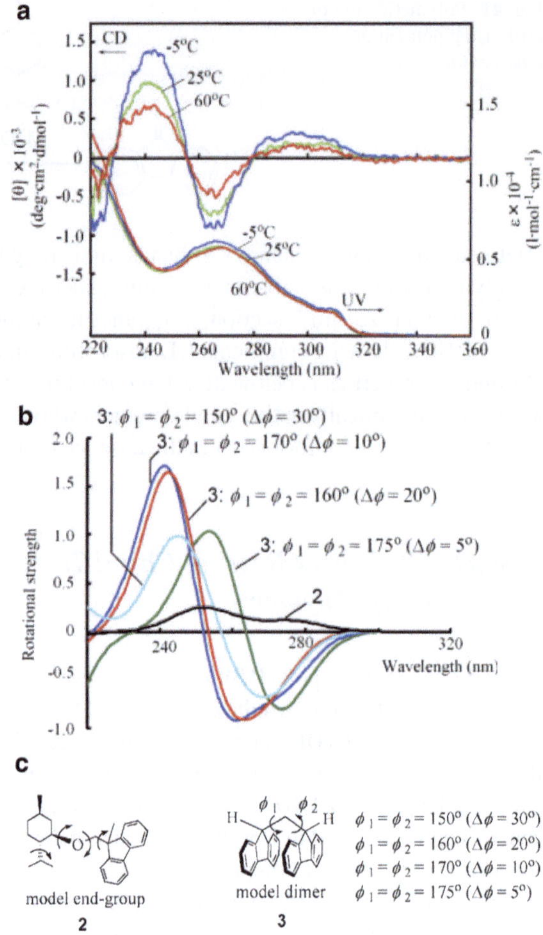

Fig. 42 Experimental CD spectra of poly(**DBF**) having a menthoxy terminal group measured at different temperatures (in THF) (**a**) and theoretical CD spectra (**b**) of model dimer having different dihedral angles (**c**). Reprinted with permission from Nakano et al. [36]. Copyright 2009, Wiley

(0.79). In general, excimer emission quantum yield is much lower than monomer emission quantum yield as can be seen for poly(**DBF**) (0.06–0.12) and fluorene (0.69) [31]. The rather high yield of poly(**BBPDBF**) may mean a lower order in stack: the side-chain phenylene moiety might not be completely stacked with a larger rotational freedom compared with the core fluorene moiety.

Poly(**BBPDBF**) indicated white fluorescent light whose emission range was from 365 nm to beyond 650 nm in the solid state while it showed narrower-range emission in a THF solution. The white fluorescent may arise from interchain excimer formation in addition to intramolecular excimer. The side-chain phenylene moieties of a chain may interact with those with another chain, and the interaction may not be so uniform, resulting in emission sites with various energy levels to make the sum emission appear white. In addition, the emission was circularly polarized in the full range of emission wavelength (Fig. 43). This may be the first, organic white CPL emitter.

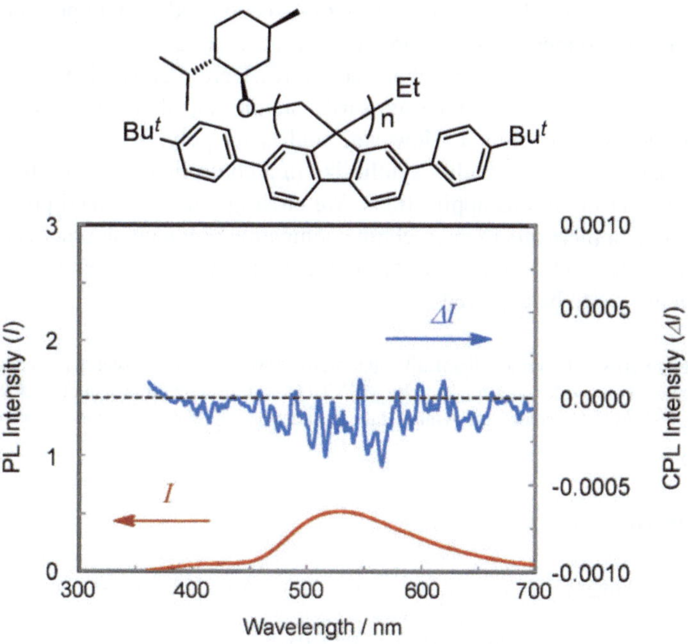

Fig. 43 Structure of and CPL (*blue*) and fluorescent spectra (*red*) of optically active poly(**BBPDBF**). The spectra were taken on a film casted on a quartz plate. Reprinted with permission from Watanabe et al. [37]. Copyright 2009, The Royal Society of Chemistry

Conclusions

DBF exhibits characteristic polymerization behaviors including the high reactivity and the stereochemical features in polymerization which have not been known for other simple vinyl monomers with aromatic side groups such as styrene, vinylfluorenes, and vinylcarbazoles. These characteristics may not be easily imagined from the simple chemical structure of **DBF**.

The π-stacked structure of poly(**DBF**) is not only unique as a basic conformation of a vinyl polymer but also has practical importance in the facts that it can mediate charge transport. While this property has been studied mainly for main-chain conjugated polymers, through studies on poly(**DBF**) and family, we found that vinyl polymers can also be candidates of organic conductive materials. In applications, advantages of vinyl polymers over main-chain conjugation systems are colorlessness or a wider color tuning range, a deeper HOMO level, and higher solubility. Based on these features, π-stacked vinyl polymers may replace inorganic transparent (colorless) electrode materials such as ITO.

In addition, the chiral conformation suggests that poly(**DBF**) and its family may find applications in the fields such chiral recognition, nonlinear optical materials, and chiral catalysis. An application based on the combination of chirality and charge transport is circularly polarized light for OLEDs.

π-Stacked structures have not been generally recognized as an important category of structure in polymer chemistry probably because there have been only a limited number of π-stacked polymers whose structures were clearly identified and also because their syntheses are not necessarily simple and their practical properties have not been well exploited. However, studies on poly(**DBF**) and family have revealed that they can be readily synthesize in a controlled way and that they are highly potential in various applications. We envisage that π-stacked polymers can be excellent complements to main-chain conjugation polymers in practical polymer materials science and, further, that they could exhibit their own characteristic properties that are yet to be explored.

Acknowledgments The authors thank all collaborators involved in our research projects described in this chapter. This work was supported by MEXT (Japan) through Grants-in-Aid 25288051 and 25620091, JST ACT-C Project, the Asahi Glass Foundation, and the Sumitomo Foundation.

References

1. Fegan A, White B, Carlson JCT, Wagner CR (2010) Chem Rev 110:3315
2. Fujita M (1999) Acc Chem Res 32:53
3. Fujita M, Tominaga M, Hori A, Therrien B (2005) Acc Chem Res 38:371
4. Linton B, Hamilton AD (1997) Chem Rev 97:1669
5. Li W-S, Aida T (2009) Chem Rev 109:6047
6. Rosen BM, Wilson CJ, Wilson DA, Peterca M, Imam MR, Percec V (2009) Chem Rev 109:6275
7. Leininger S, Olenyuk B, Stang PJ (2000) Chem Rev 100:853
8. Chakrabarty R, Mukherjee PS, Stang PJ (2011) Chem Rev 111:6810
9. Yashima E, Maeda K, Iida H, Furusho Y, Nagai K (2009) Chem Rev 109:6102
10. Ashkin A (1970) Phys Rev Lett 24:156
11. Ashkin A, Dziedic JM (1987) Science 235(4795):1517
12. Neuman KC, Blocka SM (2004) Rev Sci Instrum 75:2787
13. Ferraris J, Cowan DO, Walatka V, Perlstein JH Jr (1973) J Am Chem Soc 95:948
14. Jerome D, Mazaud M, Ribault M, Bechgaard K (1980) J Phys Lett 41:L955
15. Jerome D (2004) Chem Rev 104:5565
16. Kato R (2004) Chem Rev 104:5319
17. Kobayashi A, Fujiwara E, Kobayashi H (2004) Chem Rev 104:5243
18. Ravy S (2004) Chem Rev 104:5609
19. Talham DR (2004) Chem Rev 104:5479
20. Bendikov M, Wudl F, Perepichka DF (2004) Chem Rev 104:4891
21. Watson JD, Crick FHC (1953) Nature 171:737
22. Nelson JC, Saven JG, Moore JS, Wolynes PG (1997) Science 277:1793
23. Hill DJ, Mio MJ, Prince RB, Hughes TS, Moore JS (2001) Chem Rev 101:3893
24. Lokey RS, Iverson BL (1995) Nature 375:303
25. Nguyen JQ, Iverson BL (1999) J Am Chem Soc 121:2639
26. Wang W, Li L-S, Helms G, Zhou H-H, Li ADQ (2003) J Am Chem Soc 125:1120
27. Li ADQ, Wang W, Wang L-Q (2003) Chem Eur J 9:4594
28. Wang W, Han JJ, Wang L-Q, Li L-S, Shaw WJ, Li ADQ (2003) Nano Lett 3:455
29. Nakano T, Takewaki K, Yade T, Okamoto Y (2001) J Am Chem Soc 123:9182
30. Nakano T, Yade T (2003) J Am Chem Soc 125:15474

31. Nakano T, Yade T, Fukuda Y, Yamaguchi T, Okumura S (2005) Macromolecules 38:8140
32. Nakano T, Nakagawa O, Yade T, Okamoto Y (2003) Macromolecules 36:1433
33. Nakano T, Yade T, Yokoyama M, Nagayama N (2004) Chem Lett 33:296
34. Nakano T, Nakagawa O, Tsuji M, Tanikawa M, Yade T, Okamoto Y (2004) Chem Commun 144
35. Nakano T, Yade T, Ishizawa H, Nakagawa O, Okamoto Y (2002) ACS Polym Prep 43:609
36. Nakano T, Tanikawa M, Nakagawa O, Yade T, Sakamoto T (2009) J Polym Sci Part A Polym Chem 47:239
37. Watanabe K, Sakamoto T, Suzuki M, Fujiki M, Nakano T (2011) Chem Commun 47:10996
38. Nakano T (2010) Polym J 42:103
39. Rathore R, Abdelwahed SH, Guzei IA (2003) J Am Chem Soc 125:8712
40. Stevenson CD, Kiesewetter MK, Reiter RC, Abdelwahed SH, Rathore R (2005) J Am Chem Soc 127:5282
41. Murphy CJ, Arkin MR, Jenkins Y, Ghatlia ND, Bossman SH, Turro NJ, Barton JK (1993) Science 262:1025
42. Elias B, Shao F, Barton JK (2008) J Am Chem Soc 130:1152
43. Genereux JC, Barton JK (2010) Chem Rev 110:1642
44. Greenhow EJ, McNeil D, White EN (1952) J Chem Soc 1952:986
45. More O'Ferrall RA, Slae S (1969) J Chem Soc Chem Commun 486
46. Wieland H, Probst O (1937) Liebigs Ann Chem 530:274
47. Evans A, George D (1961) J Chem Soc 1961:4653
48. Alwyn G, Evans G, George DB (1962) J Chem Soc 1962:141
49. Yuki H, Hotta J, Okamoto Y, Murahashi S (1967) Bull Chem Soc Jpn 40:2659
50. Richards DH, Scilly NF (1969) J Polym Sci Polym Lett 7:99
51. Dewar MSJ, Zoebisch EG, Healy EF, Stewart JJP (1985) J Am Chem Soc 107:3902
52. Ueda M, Mano M, Mori H, Ito H (1991) J Polym Sci A Polym Chem 29:1779
53. Ueda M, Takahashi M, Suzuki T (1983) J Polym Sci Polym Phys Ed 20:1139
54. Ueda M, Mori H (1990) J Polym Sci A Polym Chem 28:1779
55. Suenaga J, Sutherlin DM, Stille JK (1984) Macromolecules 27:2913
56. Md. Sheikh RK, Imae I, Tharanikkarusu K, LeStrat VM-J, Kawakami Y (2000) Polym J 32:527
57. Zundel T, Baran J, Mazureki M, Wang J-S, Jerome R, Teyssie P (1998) Macromolecules 31:2724
58. Okamoto Y, Habaue S, Isobe Y, Suito Y (2003) Macromol Symp 195:75
59. Ray B, Isobe Y, Morioka K, Habaue S, Okamoto Y, Kamigaito M, Sawamoto M (2003) Macromolecules 36:543
60. Matsumoto A, Tanaka T, Tsubouchi T, Tashiro K, Saragai S, Nakamoto S (2002) J Am Chem Soc 124:8891
61. Nakano T, Okamoto Y, Hatada K (1992) J Am Chem Soc 114:1318
62. Nakano, Okamoto Y (2001) Chem Rev 101:4013
63. Okamoto Y, Nakano T (1994) Chem Rev 94:349
64. Matsumoto A, Odani T (2001) Macromol Rapid Commun 22:1195
65. Schmidt GMJ (1971) Pure Appl Chem 27:647
66. Wegner G (1977) Pure Appl Chem 49:443
67. Baessler H (1984) Adv Polym Sci 63:1
68. Enkelmann V (1984) Adv Polym Sci 63:91
69. Tieke B (1985) Adv Polym Sci 71:79
70. Hasegawa M (1995) Adv Phys Org Chem 30:117
71. Farina M (1984) In: Atwood JL, Davis JDE, MacNicol DD (eds) Inclusion compounds, vol 3. Academic, London, p. 297
72. White DM (1960) J Am Chem Soc 82:5678
73. Minagawa M, Yamada H, Yamaguchi K, Yoshii F (1992) Macromolecules 25:503
74. Allcock HR, Ferrar WT (1982) Macromolecules 15:697
75. Allcock HR, Silverberg EN, Dudley GK, Pucher SR (1994) Macromolecules 27:7559

76. Matsumoto A, Ishizu Y, Yokoi K (1998) Macromol Chem Phys 199:2511
77. Zhang X, Hogen-Esch TE (2000) Macromolecules 33:9176
78. Koch W, Holthausen MC (2000) A chemist's guide to density functional theory. Wiley-VCH, New York, pp 197–216
79. Bouman TD, Hansen AE (1988) Chem Phys Lett 149:510
80. Becke AD (1988) Phys Rev A38:3098
81. Lee C, Yang W, Parr RG (1988) Phys Rev B 37:785
82. Sekine Y, Brown M, Boekelheide V (1979) J Am Chem Soc 101:3126
83. Allinger NL, Yuh YH, Lii J-H (1989) J Am Chem Soc 111:8551
84. Halgren TA (1999) J Comput Chem 20:730
85. Sun H (1998) J Phys Chem 102:7338
86. Mohamadi F, Richards NGJ, Guida WC, Liskamp R, Lipton M, Caufield C, Chang G, Hendrickson T, Still WC (1990) J Comput Chem 11:440
87. Berendsen HJC, Postma JPM, van Gunsteren WF, DiNola A, Haak JR (1984) J Chem Phys 81:3684
88. Fletcher R, Reeves CM (1964) Comput J 7:149
89. Tinoco I (1960) J Am Chem Soc 82:4785
90. Rohdes W (1961) J Am Chem Soc 83:3609
91. Ellis JR (1986) In: Skotheim TA (ed) Handbook of conducting polymers, vol 1. Dekker, New York, Chapter 13
92. Houk KN, Lee PS, Nendel M (2001) J Org Chem 66:5517
93. Horrocks DL, Brown WG (1970) Chem Phys Lett 5:117
94. Matsuda M, Watanabe A (1987) In: Hogen-Esch TE, Smid J (eds) Recent advances in anionic polymerization. Elsevier, New York, p. 73
95. Hirayama F (1965) J Chem Phys 42:3163
96. Strohriegl P, Grazulevicius JV (1997) In: Nalwa HS (ed) Handbook of organic conductive molecules and polymers, vol 1. Wiley, New York, Chapter 11
97. Pearson JM, Stolka M (1981) Poly(N-vinylcarbazole). Gordon and Breach, New York, Chapter 4
98. Borsenberger PM, Weiss DS (1993) Organic photoreceptor systems for imaging systems. Dekker, New York
99. Shirota Y (2000) J Mater Chem 10:1
100. Strohriegl P (2002) Adv Mater 14:1439
101. Gill W (1972) J Appl Phys 43:5033
102. Hoofman RJMO, de Haas MP, Siebbeles LDA, Warman JM (1998) Nature 392:54
103. Grozema FC, Siebbeles LDA, Warman JM, Seki S, Tagawa S, Scherf U (2002) Adv Mater 14:228
104. Grozema FC, van Duijnen PT, Berlin YA, Ratner MA, Siebbeles LDA (2003) J Phys Chem A 107:5976
105. Forero S, Nguyen PH, Brütting W, Schwoerer M (1999) Phys Chem Chem Phys 1:1769
106. Kepler RG, Zeigler JM, Kurtz SR (1987) Phys Rev B 35:2818
107. Yokoyama M, Akiyama K, Yamamori N, Mikara H, Kusabayashi S (1985) Polym J 17:545
108. Fujino M, Mikawa H, Yokoyama M (1982) Photogr Sci Eng 26:84
109. Okahata Y, Kobayashi T, Tanaka K, Shimomura M (1998) J Am Chem Soc 120:6165
110. Fink H-W, Schönenberger C (1999) Nature 398:407
111. Porath D, Bezryadin A, de Vries S, Dekker C (2000) Nature 403:635
112. Braun E, Eichen Y, Sivan U, Ben-Yoseph C (1998) Nature 391:775
113. Debije MG, Milano MT, Benhard WA (1999) Angew Chem Int Ed 38:2752
114. Maiya BG, Ramasarma T (2001) Curr Sci 80:1523
115. Coropceanu V, Nakano T, Gruhn NE, Kwon O, Yade T, Katsukawa K-i, Bredas J-L (2006) J Phys Chem B 110:9482
116. Nakano T, Nakagawa O, Tsuji M, Yade T (2003) Polym Prep Jpn 52(2):1272
117. Yade T, Nakano T, Polym J (2006) Sci A Polym Chem 44:561
118. Li G, Nakano T, Tanaka K, Higashimura H (2012) Polym Prep Jpn 61(1):401; 62(2):2752

119. Nakano T, Yade T (2005) Polym Prep Jpn 54(1):224
120. Nakano T, Yade T (2005) Polym Prep Jpn 54(2):2496
121. Winder C, Sariciftci NS (2004) J Mater Chem 14(7):1077
122. Nikolou M, Malliaras GG (2008) Chem Rec 8(1):13
123. Yoshioka Y, Jabbour GE (2007) In: Skotheim TA, Reynolds JR (eds) Handbook of conducting polymers, 3rd edn, vol 2, pp 3/1–3/21
124. Bao Z (2004) Nature Ma 3(3):137
125. Nakano T, Nishii S, Fukuda Y, Yaegashi T, Katsukawa K-i (2006) Polym Prep Jpn 55(1):317
126. Nakano T, Nishii S, Fukuda Y, Fujiki M, Akimoto S, Sato S (2006) Polym Prep Jpn 55(2):2785
127. Sakamoto T, Yade T, Nakano T (2010) Polym Prep Jpn 59(2):3981
128. Crnelissen JJLM, Rowan AE, Nolte RJM, Sommerdijk NAJM (2001) Chem Rev 101:4039
129. Nakano T, Okamoto Y (2000) Macromol Rapid Commun 21:603
130. Green MM, Park J-W, Sato T, Teramoto A, Lifson S, Selinger RLB, Selinger JV (1999) Angew Chem Int Ed 38:3138
131. Rowan AE, Nolte RJM (1998) Angew Chem Int Ed 37:63
132. Gelman SH (1998) Acc Chem Res 31:173
133. Natta G, Pino P, Corradini P, Danusso F, Mantica E, Nazzanti G, Moraglio G (1955) J Am Chem Soc 77:1708
134. Pino P, Lorenzi GPP (1960) J Am Chem Soc 82:4745
135. Pino P, Lorenzi GP, Lardicci L (1960) Chim Ind (Milan) 42:712
136. Okamoto Y, Suzuki K, Ohta K, Hatada K, Yuki H (1979) J Am Chem Soc 101:4763
137. Mislow K, Bickart P (1976/1977) Isr J Chem 15:1

Reversible Polymerization Techniques Leading to π-Stacked Polymers

Andrea Cappelli, Marco Paolino, Giorgio Grisci, Germano Giuliani, Alessandro Donati, Antonella Caterina Boccia, Filippo Samperi, Raniero Mendichi, and Salvatore Vomero

Introduction

Fulvene, benzofulvene (**BF**), and dibenzofulvene (**DBF**) derivatives (Fig. 1) have captured the attention of chemists for more than one century. Fulvene is a cyclic isomer of benzene and one among the classical nonalternant π-electron systems that has catalyzed a great deal of attention for its potential tendency to fulfill the Huckel $4N+2$ rule [1, 2]. A zwitterionic character has been proposed for fulvene in its singlet ground state (S_0) and, in the opposite direction, in the lowest triplet state (T_1). On the basis of this behavior, fulvene was argued to act as an aromatic chameleon capable of adapting to several excited states [3]. The zwitterionic character of fulvene ground state was assumed to result from intramolecular charge transfer from the exocyclic double bond to the five-member ring (which acquires a pseudo-aromatic cyclopentadienide character) and can be affected to a variable extent by the presence and nature of substituents. For instance, electron-donating groups in position 6 or electron-withdrawing groups on the ring are capable of reinforcing the dipolar character, stabilizing the fulvene, and decreasing the tendency to oxygenation, dimerization, and polymerization. In fact, the dipole moment of fulvene is reported to be equal to 0.42 D and becomes 1.44 D in 6, 6-dimethylfulvene and 4.5 D in 6-dimethylaminofulvene [4, 5].

DBF is an isomer of phenanthrene, in which the fulvene moiety is incorporated in a tricyclic system. This provides to the fulvene moiety some peculiar features.

A. Cappelli (✉) • M. Paolino • G. Grisci • G. Giuliani • A. Donati • S. Vomero
Dipartimento Farmaco Chimico Tecnologico and European Research Centre for Drug Discovery and Development, Università degli Studi di Siena, Via A. Moro, 53100 Siena, Italy
e-mail: andrea.cappelli@unisi.it

A.C. Boccia • R. Mendichi
Istituto per lo Studio delle Macromolecole (CNR), Via E. Bassini 15, 20133 Milano, Italy

F. Samperi
Istituto di Chimica e Tecnologia dei Polimeri (CNR), Via Gaifami 18, 95125 Catania, Italy

T. Nakano (ed.), *π-Stacked Polymers and Molecules: Theory, Synthesis, and Properties*,
DOI 10.1007/978-4-431-54129-5_2, © Springer Japan 2014

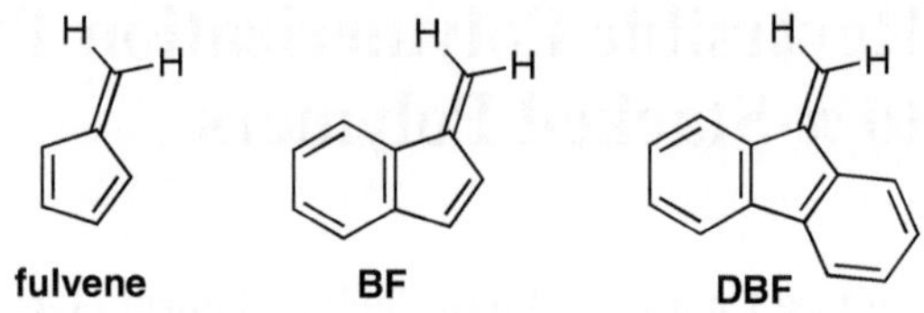

Fig. 1 Structures of fulvene, benzofulvene (**BF**), and dibenzofulvene (**DBF**)

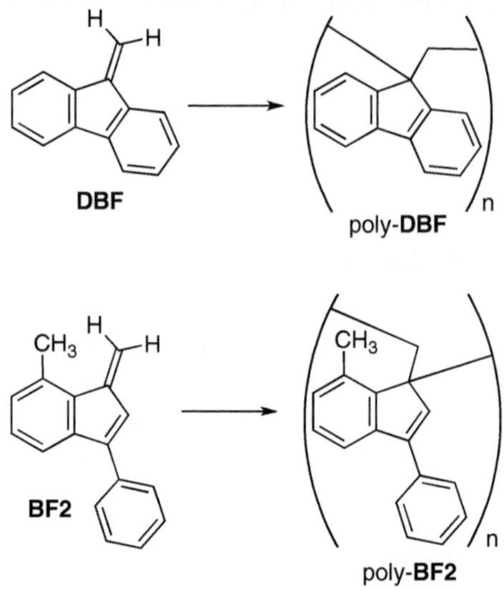

Fig. 2 Polymerization of dibenzofulvene (**DBF**) and benzofulvene derivative **BF2**

For instance, **DBF** was found to polymerize under a variety of conditions including anionic, cationic, and radical catalyses to give poly-**DBF** and the properties of polydibenzofulvene derivatives (e.g., poly-**DBF**) have been largely investigated by Nakano and coworkers [6–14]. Intriguingly, poly-**DBF** was found to behave as a new synthetic "π-way" molecule showing a hole drift mobility only slightly lower than that of Se [13], an inorganic semiconductor [6, 7, 11].

Benzofulvene derivatives are isomers of corresponding naphthalene derivatives that have been relatively less studied than their monocyclic and tricyclic counterparts. A particular mention can be made for 7-methyl-1-methylene-3-phenylindene (**BF2**, Fig. 2), a benzofulvene derivative described by Londergan and coworkers, that was reported to polymerize under free radical catalysis to afford a vinyl polymer containing pendant aromatic chromophores [15].

After a Decade from the Discovery of Poly-BF1, the History of Polybenzofulvene Derivatives

Serendipity has often played a key role in research and, in perfect agreement with this role, it has had a fundamental responsibility in the discovery of the

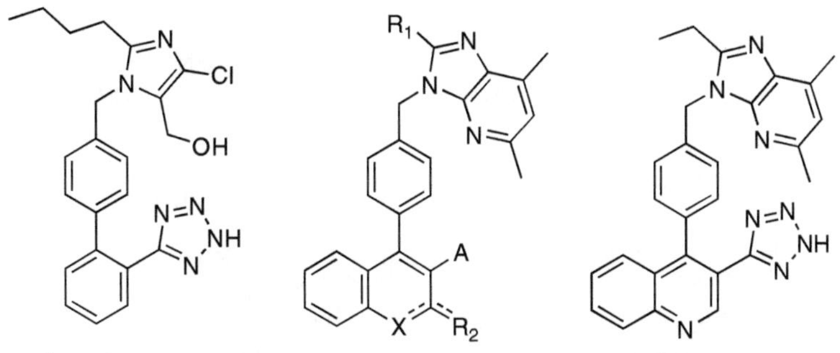

Fig. 3 Structures of the AT_1 receptor antagonists losartan, **4-PQ** derivatives, and **CR3210**

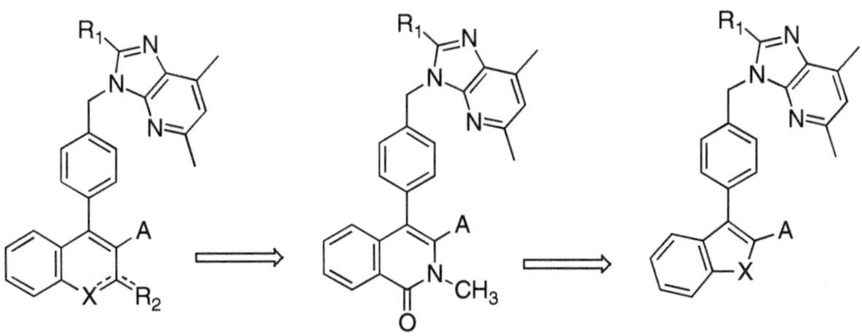

Fig. 4 Design of **4-PIQ** derivatives and **1-PI** derivatives from angiotensin II AT_1 receptor antagonists **4-PQ**

thermoreversible spontaneous polymerization of benzofulvene derivatives. However, "serendipity doesn't come from being the first to see something, but from being the first to see it in a new way." [16]

At the beginning of the twenty-first century, our research group was involved in a large research program devoted to the discovery of new antihypertensive drugs [17–19]. The program of structural modification of losartan (Fig. 3) derivatives led us to the design, synthesis, and biological evaluation of a series of 4-phenylquinoline compounds (**4-PQ** derivatives).

Certain members of this family of **4-PQ** derivatives showed in vitro properties comparable to those shown by losartan, but the candidate selected for further preclinical studies (CR3210) showed a relatively low oral bioavailability and a rapid excretion [18].

Thus, the program of structural modification was pursued with the synthesis of both 4-phenylisoquinolinone (**4-PIQ**) derivatives and 1-phenylindene (**1-PI**) derivatives (Fig. 4).

During the synthetic work devoted to the preparation of target carboxylic acid **1-PI-1** (Fig. 5), the cleavage of corresponding *tert*-butyl ester **1-PI-2** in standard

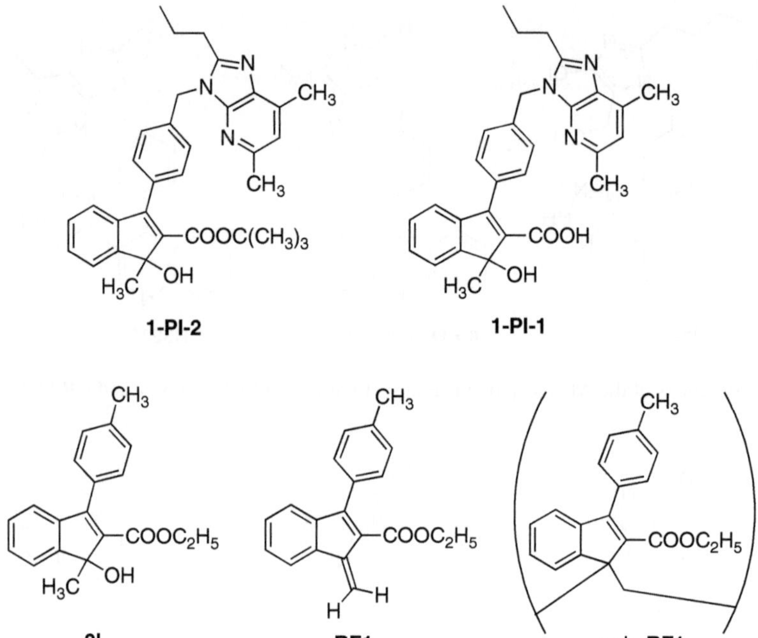

Fig. 5 Structures of indenol derivatives, **1-PI-1, 1-PI-2, 2I**, benzofulvene derivative **BF1,** and poly-**BF1**

acid conditions was found to be rather problematic because of the formation of polymeric materials of which isolation and characterization were difficult. We realized that the contemporary presence of acid (i.e., COOH) and basic (i.e., the imidazopyridine) moieties made it difficult to study the formed macromolecule. Therefore, the structure of **1-PI-2** was simplified by eliminating the imidazopyridine moiety and the ester group was stabilized (to the acid conditions) by transforming the *tert*-butyl into ethyl group. These structural modifications led to the synthesis of indenol **2I** and of benzofulvene derivative **BF1**, as well as to the discovery of its spontaneous polymerization into poly-**BF1** [20]. The properties of poly-**BF1** (e.g., formation, thermoinduced depolymerization, structure, and aggregation) were studied in detail and there was consistent evidence suggesting a vinyl polymer structure stabilized by aromatic interactions [21].

Therefore, the strategy of structure simplification/stabilization showed once more to provide a wonderful tool for rationalizing and solving complex problems. Once it was understood that indenol **2I** was liable to dehydrate in acid conditions to provide **BF1**, which in turn was liable to spontaneous polymerization by solvent removal, the chemistry of indenol **1-PI-2** was reinvestigated (Scheme 1) [18].

Thus we found that the cautious reaction of indenol **1-PI-2** with formic acid gave expected acid **1-PI-1** containing trace amounts of benzofulvene derivative **BF-AT1**. On the other hand, substantial amounts of the latter were obtained by dehydration of indenol **1-PI-2** or acid **1-PI-1** in the presence of *p*-toluenesulfonic acid (PTSA) in $CHCl_3$ (or $CDCl_3$). Surprisingly enough, benzofulvene derivative **BF-AT1** was

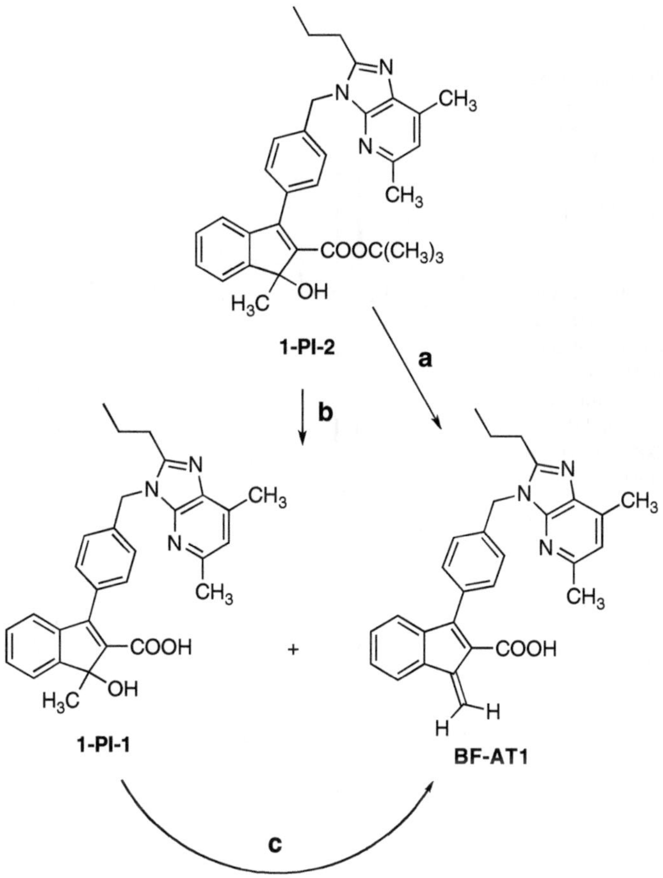

Scheme 1 Chemistry of indenol **1-PI-2**. Reagents: (*a*) PTSA, $CHCl_3$, (*b*) HCOOH, and (*c*) PTSA, $CDCl_3$

stable as a pure crystalline solid, but polymerized spontaneously when the mixture of the dehydration reaction of **1-PI-1** was concentrated without the elimination of PTSA. In order to understand this apparently peculiar behavior, the ethyl ester of acid **1-PI-1** was synthesized and showed the same spontaneous polymerization behavior as **BF1**. The stability of pure **BF-AT1** was assumed to be related to the probable intermolecular interaction of its COOH with the imidazole nitrogen in the solid state otherwise prevented by the presence of PTSA, which was assumed to protonate the heterocyclic nitrogen atoms. Poly-**BF-AT1** showed a thermoinduced depolymerization behavior similar to that shown by poly-**BF1**. The depolymerization process at 55 °C was very slow and became significantly more rapid at 80 °C; at 140 °C it was even faster since it was almost complete after 1 h heating. At 120 °C the unzipping process showed an intermediate velocity and reached an apparent equilibrium after 4 h heating (Fig. 6). Thus, poly-**BF-AT1** displayed a thermoreversible polymerization behavior similar to that shown by poly-**BF1**, but characterized by a faster kinetics.

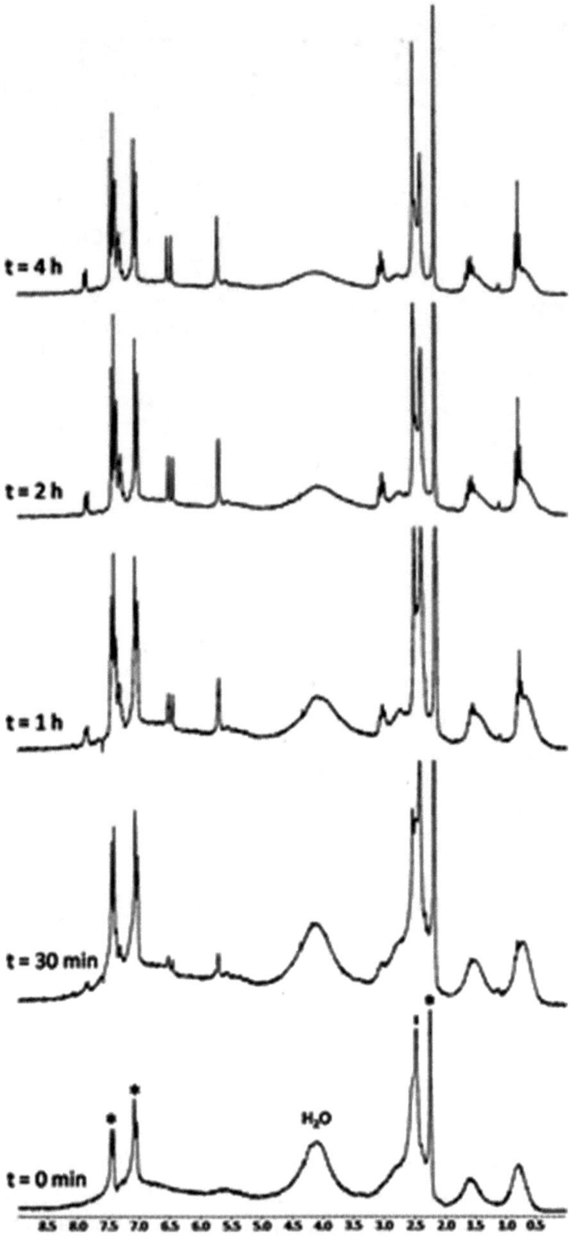

Fig. 6 Thermoinduced depolymerization of poly-**BF-AT1**, followed by ^{1}H-NMR. A solution of 2.5 mg of poly-**BF-AT1** •PTSA in 0.5 mL of DMSO-d_6 was heated at 120 °C and ^{1}H-NMR spectra (200 MHz) were recorded at regular time intervals. The *arrow* indicates the solvent peak, "H_2O" the water peak, and the asterisk the signals attributable to PTSA. Reprinted with permission from Cappelli et al. [18]. Copyright 2006, American Chemical Society

This was considered to be an intriguing result because monomer **BF-AT1** was found to show an angiotensin II AT_1 receptor affinity in the nanomolar range. Moreover, binding studies suggested that depolymerization occurred also in buffer at pH 7.4. In conclusion, **BF-AT1** could be considered as a polymerizing AT_1

receptor ligand capable of forming a thermoreversible polymer (poly-**BF-AT1**) and of being released from the latter by a temperature-dependent kinetics until a temperature-dependent equilibrium was reached as it occurs in equilibrium polymerization processes. This appeared to open a new perspective in the development of novel polymeric prodrugs based on an original release mechanism and increased our research efforts in the field.

In those years, the maximum priority was assigned to the exploration of the chemistry of benzofulvene derivatives. We were aware that the rational design of new bioactive monomers should be based on the precise knowledge of the chemistry of benzofulvene derivatives. Therefore, a comprehensive program of structural modification of **BF1** scaffold was planned in order to investigate on the role of substituents' feature on the polymer properties such as formation, molecular mass distribution (MMD) (often referred as molecular weight distribution, MWD), structure, thermoreversibility, and aggregation.

A careful analysis of **BF1** structure led to the consideration of the ester group as candidate for systematic variations. In fact, its electron-withdrawing effect was considered important in determining the reactivity of this benzofulvene derivative (e.g., the polarization of the exocyclic double bond and the capability of stabilizing a negative charge in transition states and of destabilizing cationic or radical species). Moreover, the presence of two unusually shielded aromatic protons at ca. 6 ppm suggested a key role of specific aromatic interactions in polymer formation and structure. In other words, our working hypothesis was the following: the phenyl group is important for polymer formation and structure and should be left untouched with its ability to establish aromatic interactions; the ester group could modulate polymer formation and structure and could be changed.

With this working hypothesis in mind we designed the series of benzofulvene derivatives **BF3** reported in Table 1 [22].

The preparation of these benzofulvene derivatives allowed the synthetic methodology to be optimized for a good number of members of the family. The most important achievements were the easy access to indenol precursors by *C*-alkylation of the corresponding indenone derivatives with trimethylaluminum and the easy dehydration to the benzofulvene derivatives with some limitations due to the instability of certain functional groups in the methylation/dehydration conditions. The study ultimately demonstrated that most of the properties (i.e., formation, MWD, structure, thermoreversibility, and aggregation) of poly-**BF3** derivatives were modulated by the stereoelectronic features of the substituents present in the indene scaffold. For instance, the steric hindrance at the 2-position of the indene nucleus played a role in inhibiting spontaneous polymerization and in affecting MMD and structure of the corresponding polymers. The carboxylic group was found to be the most favorable (as substituent at the 2-position) with respect to several features of the polybenzofulvene derivatives. In fact, NMR studies suggested that poly-**BF1** and its 4′-desmethylderivative poly-**BF3k** were the most structurally homogeneous polymers of the series, they showed sharp differential scanning calorimetry (DSC) peaks characterized by high ΔH values, and their thermoinduced depolymerization was almost complete after 6 h heating at 150 °C.

Table 1 Benzofulvene derivatives **BF3**

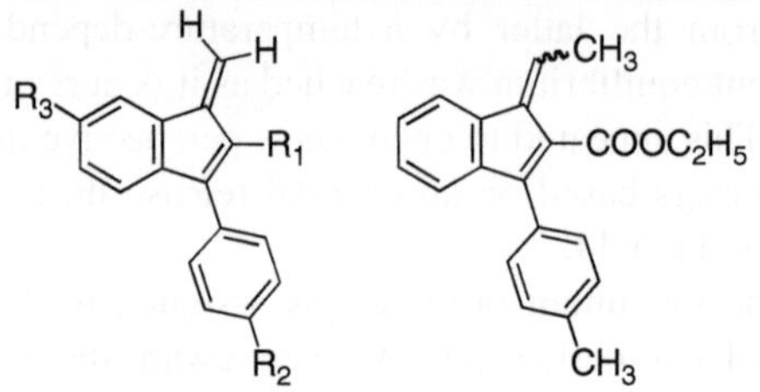

BF3 derivatives **BF3y**

Monomer	R_1	R_2	R_3
BF3a	H	H	H
BF3c	F	H	H
BF3d	Cl	H	H
BF3e	Br	H	H
BF3f	CH_3	H	H
BF3g	CN	H	H
BF3h	CN	CH_3	H
BF3i	CN	H	CH_3
BF3j	C≡C-2-Pyr	H	H
BF3k	$COOC_2H_5$	H	H
BF3ma	$COOC(CH_3)_3$	CH_3	H
BF3n	$CON(CH_3)_2$	CH_3	H
BF3o	$C(CH_3)_3$	H	H
BF3y			

Since poly-**BF1** obtained by spontaneous polymerization regularly showed very high average molecular masses (MM, i.e., number average molar mass, M_n, and weight average molar mass, M_w), the use of an anionic initiator such as phenyl lithium (PhLi) was considered in order to synthesize poly-**BF1** showing low molecular weight and oligomers (i.e., dimers) to be used as model molecules in the study of the poly-**BF1** structure by NMR [23].

As expected, the anionic polymerization of **BF1** produced a mixture of oligomers and a polymeric material in different proportions, which were regulated by the amount of the initiator used. The polymer was separated from lower oligomers on the basis of the solubility in *n*-hexane and the soluble material was purified by chromatographic techniques to isolate activated monomers and dimers. The intensive use of different analytic techniques such as NMR spectroscopy, absorption and emission spectroscopy, and mass spectrometry allowed the structure of dimers and of polymer to be elucidated, and the results were convergent in confirming for poly-BF1 a vinyl (1, 2) polymer chaining stabilized by means of aromatic stacking interactions. Moreover, NMR studies showed that the anionic polymerization of **BF1** afforded a polymer showing NMR features very similar to those shown by the polymer obtained by spontaneous polymerization, with some differences involving ^{13}C-NMR signals assigned to carbon atoms belonging to the pendent phenyl group.

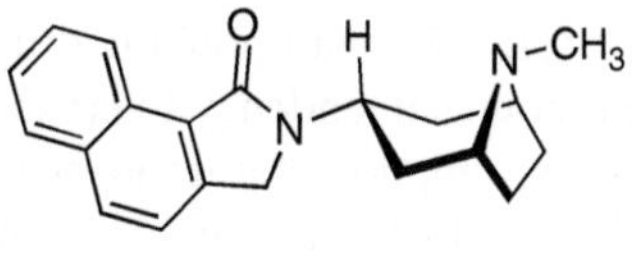

CR3124

Fig. 7 Structure of serotonin 5-HT_3 receptor antagonist **CR3124**

These were assumed to be related to a different tacticity. As expected, the MMD of poly-BF1 prepared by anionic polymerization featured a significantly lower molecular weight with respect to those shown by poly-**BF1** prepared by spontaneous polymerization. However, a discrepancy was observed in the molecular weight between the measured value and the one calculated on the basis of the initiator amount. The discrepancy was assumed to be related to the existence of equilibrium among the various activated oligomers that lowers the number of the growing intermediates.

Interestingly, scanning electron microscopy (SEM) showed different aggregation behaviors for the polymers obtained by the two different polymerization methods that were probably related to the difference in the molecular weight. The polymer prepared by anionic polymerization showed lower molecular weight and the tendency to aggregate into nanospheres and microspheres, whereas the polymer obtained by spontaneous polymerization was characterized by high molecular weight and the tendency to generate larger fractal-like aggregates.

As nanoparticles, microspheres, and nanoporous materials were used in the development of advanced drug delivery systems, we considered the aggregation features of polybenzofulvene derivatives to be promising for their application in controlled drug release. Among the members of the family the most promising polymer was believed to be poly-**BF3a**, which showed a low molecular weight and the clear tendency to aggregate in nanospheres, and was selected for application studies [24].

Thus, different batches of poly-**BF3a** were prepared and their MMD was characterized to show that spontaneous polymerization of **BF3a** could be considered a reproducible process suitable for the preparation of this polymer free of catalyst or initiator impurities and showing a weight average M_w in the range of 4.0–7.9 kg/mol and a polydispersity index ($PDI = M_w/M_n$) in the range of 1.2–2.0. The polymer showed high hydrophobicity, a complex structure, a flat DSC curve, and an appreciable thermal stability and was devoid of the tendency to the thermoinduced depolymerization. After intensive studies, we found that the formation of nanoparticles in the continuous excess of the bad solvent (ethanol) led to favorable shape and dimensions. Furthermore, the nanoprecipitation of poly-**BF3a** in the presence of poly(ethylene glycol) (i.e., PEG1000) gave the best results in terms of nanoparticle morphology.

The inclusion of serotonin 5-HT_3 receptor antagonist **CR3124** (Fig. 7) as a model drug in the formulation was found to produce nanoparticles showing a variable degree of cementation ending in the formation of a nanocomposite material in which **CR3124** played the role of the matrix and poly-**BF3a** nanoparticles were the filler.

CR3124 is a lipophilic drug possessing a quite extended aromatic moiety and a pronounced tendency to crystallize in a stable form. It was selected on the basis of the assumption that its aromatic moiety could interact by means of specific interactions with the aromatic moieties of poly-**BF3a**. DSC studies suggested that the interaction of **CR3124** with the surface of poly-**BF3a** nanoparticles was capable of stabilizing the amorphous state of the drug. However, the low wettability of the binary system determined aggregation of the formulation in water. The introduction of PEG1000 in the formulation led to the constitution of a ternary nanocomposite material formed by a binary matrix showing higher wettability leading to a faster drug release, a greater released amount, and a higher stability of the suspension with respect to those obtained in the absence of PEG1000. Therefore, the interaction of poly-**BF3a** with **CR3124** appeared to stabilize the amorphous form of the drug and to modulate its dissolution kinetics from a very fast release (in the presence of PEG1000) to the slow release observed in the absence of PEG1000. These results suggested that drug molecules possessing physicochemical features similar to those of **CR3124** could benefit from the interaction with poly-**BF3a** nanoparticles.

However, we realized that the high hydrophobicity of polybenzofulvene derivatives could hamper their application in the development of advanced drug delivery systems because the pronounced tendency towards the aggregation in water could prove a very limiting feature. Therefore, also on the basis of the beneficial effect of PEG on the properties of the above nanocomposite drug delivery system, we planned the covalent modification of poly-**BF1** with PEG side chains [25].

PEG is one of the most widely used polymers in the pharmaceutical field because it shows favorable features such as water solubility and biocompatibility, so that it has been used to confer solubility in the physiological environment and prevent the interaction with plasma proteins and cells, and is therefore used as shielding agent for in vivo delivery of several bioactive compounds [26, 27].

The covalent conjugation of polymeric matrices with PEG (permanent PEGylation) has been carried out by polymerization of preformed PEGylated monomers such as oligo(ethylene glycol) methacrylate (OEGMA) [28–34]. In general, homo- or copolymerization of macromonomers ("grafting through" method) has been used in the synthesis of cylindrical polymer brushes, which have received a great deal of attention in very recent times [35].

In the design of covalent conjugates between poly-**BF1** and PEG, we envisioned that the employment of the "grafting through" method with monodisperse methyl-end-capped oligo(ethylene glycol) (MOEG) rather than polydisperse PEG chains should simplify the study of the resulting polymers. In the first approach, two conjugates of **BF1** (MOEG-**BF1**) were prepared with the aim of evaluating the propensity of the resulting macromonomers towards spontaneous polymerization and of characterizing the properties of the resulting polymers. As the macromonomers 2-MOEG-3-**BF1** (Fig. 8) and 2-MOEG-9-**BF1** showed the tendency to polymerize spontaneously with the same modalities shown by **BF1**, the presence of MOEG chain did not appear to alter the molecular recognition events leading to polymerization.

Moreover, poly-2-MOEG-3-**BF1** and poly-2-MOEG-9-**BF1** showed characteristic NMR features, absorption/emission spectra, and thermoinduced depolymerization behavior very similar to those shown by poly-**BF1** obtained by spontaneous polymerization.

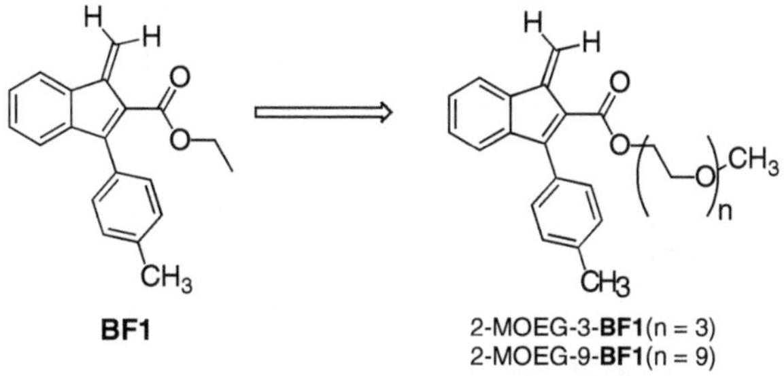

Fig. 8 Design of macromonomers 2-MOEG-3-**BF1** and 2-MOEG-9-**BF1**

The polymer bearing longer MOEG side chains (poly-2-MOEG-9-**BF1**) showed an amphiphilic character and interacts with water to give a strong physical hydrogel. The system was characterized by combining rheology and NMR transverse (spin–spin) relaxation measurements in order to determine the pore size distribution in the hydrogel. This was not perfectly homogeneous and could be considered as a sort of highly concentrated suspension of homogeneous particles generating small (intra-particle) pores and larger (extra-particle) channels. Owing to the small pore size, we assumed that small molecules could migrate at different rates through the intra-particle spaces, while the activity of larger molecular systems (e.g., large proteins, enzymes) could be limited to the extra-particle channels.

The hydrogel was then loaded with human immunoglobulin G (HIgG) and was found to be capable of releasing the loaded protein in physiological mimicking media for time periods not inferior to 48 h [36]. Because of its large dimensions, the loaded protein was assumed to be unable to locate inside the hydrogel particles, but to interact mainly with the surface of the channels pervading gel matrix. Finally, biological studies suggested potential biocompatibility features for this polymer, which could be thus considered a promising material for the development of advanced drug delivery systems such as gastro-resistant oral formulations for sustained release of proteins or in situ gel-forming preparations from which the loaded protein can be released slowly upon injection.

Based on the encouraging results obtained with poly-2-MOEG-9-**BF1**, the design of benzofulvene macromonomers was developed further by shifting the attachment point used for the covalent conjugation between the MOEG chain and benzofulvene moiety. Thus, a small series of benzofulvene macromonomers was designed by placing MOEG chains of different length at the 6-position of **BF3k** benzofulvene scaffold (Fig. 9) [37].

All the designed benzofulvene derivatives were found to polymerize by solvent removal with the same modalities shown by **BF1**, and the analysis of the NMR spectroscopic data of the resulting polymers suggested a different arrangement (flexibility or conformational behavior) of the main chain induced by the steric bulk of the MOEG side chains.

SEC-MALS analysis was consistent in showing perceptible differences in the conformational behavior of the backbone. In particular, the polymer bearing the longest MOEG side chains was significantly more flexible than the other poly-**BF3k** derivatives, probably because of the differences in the main chain arrangement induced by the steric repulsion of the MOEG chains.

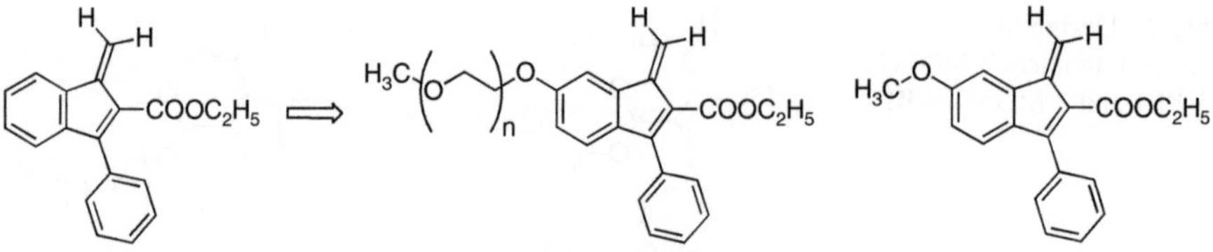

Fig. 9 Design of macromonomers 6-MOEG-2-**BF3k**, 6-MOEG-3-**BF3k**, and 6-MOEG-9-**BF3k** and structure of reference monomer 6-MO-**BF3k**

Because of the presence of relatively long MOEG-9 side chains, poly-6-MOEG-9-**BF3k** displayed an amphiphilic character similar to poly-2-MOEG-9-**BF1**. However, poly-6-MOEG-9-**BF3k** was found to differ significantly from poly-2-MOEG-9-**BF1** in the aggregation propensity. In particular, the former appeared to be less liable to aggregation with respect to the latter, and its hydrogel could be dispersed in water by ultrasound exposure.

These results ultimately suggested that polymer–solvent and polymer–polymer interactions could be modulated by shifting the attachment point used for the covalent conjugation between the MOEG chain and the indene moiety.

In the aim of both expanding the molecular diversity of the side chains attached to the polybenzofulvene backbone and developing general procedures for the preparation of a broad range of polybenzofulvene molecular brushes, the chemical design was developed further by introducing the concept of "click chemistry" in the preparation of polybenzofulvene brushes.

Click chemistry reactions [38] were defined by Sharpless and coworkers as efficient and versatile chemical transformations capable of producing chemical substances rapidly, quantitatively, without by-products and side reactions, under mild reaction conditions, and with broad tolerance to functional groups and applicability. The most popular click chemistry reactions are the Cu(I)-catalyzed 1,3-dipolar cycloaddition between organic azides and alkynes (CuAAC) [39] and the addition of thiols to alkenes (thiol-ene coupling) [40, 41].

The employment of click chemistry reactions in macromolecular science has allowed the rapid synthesis of a large array of polymers [42–51]. Indeed, the presence of clickable groups along the backbone, in the side chains, or at chain termini can be exploited for post-polymerization modifications (i.e., functionalization) leading to polymers bearing functionalities incompatible with the polymerization conditions [42].

Molecular brushes can be synthesized by different approaches, of which the polymerization of macromonomers (grafting through approach) shows appealing features such as well-defined grafting density and side chain length but also drawbacks such as a variable polymerization degree and a limited versatility (i.e., the functional groups in the side chains must be compatible with the preparation and polymerization of the macromonomer). On the other hand in the "grafting onto"

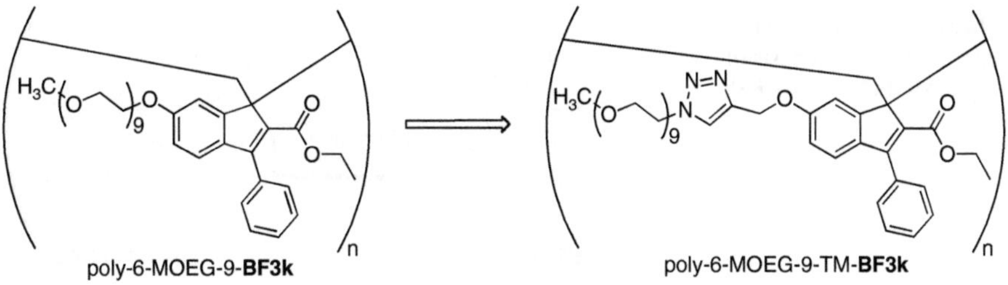

Fig. 10 Design of macromonomers poly-6-MOEG-9-TM-**BF3k** from poly-6-MOEG-9-**BF3k**

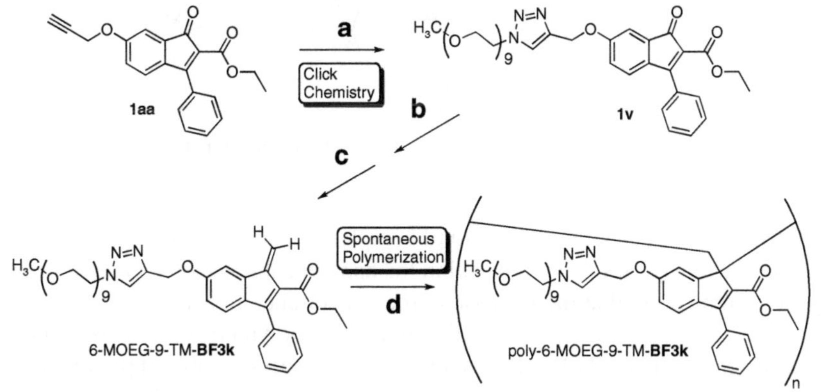

Scheme 2 "Grafting through" approach to the synthesis of poly-6-MOEG-9-TM-**BF3k**. Reagents: (*a*, click chemistry) $CH_3(OCH_2CH_2)_9N_3$, $Cu(PPh_3)_3Br$, DIPEA, THF; (*b*) $Al(CH_3)_3$, CH_2Cl_2; (*c*) PTSA, $CHCl_3$; and (*d*, spontaneous polymerization) solvent elimination

route, previously prepared side chains are linked to an already formed polymeric backbone by means of suitable coupling reactions. Thus, the coupling step plays a key role in determining important characteristics such as grafting density and versatility. The click-type reactions possess all the necessary prerequisites of high yield, selectivity, and tolerance to functional groups to candidate themselves as the choice method. The application of click chemistry reactions to the post-polymerization modification of a macromolecule requires the insertion of clickable groups into the polymeric backbone.

Thus, the implementation of the "click chemistry" concept in the preparation of polybenzofulvene brushes led to the design of poly-6-MOEG-9-TM-**BF3k** (Fig. 10) from closely related poly-6-MOEG-9-**BF3k**. In the newly designed polymer, triazolomethyl (TM) bridges acted as spacers between the MOEG side chains and the polybenzofulvene backbone.

The polymer was firstly synthesized by adopting a "grafting through" approach in which the click chemistry CuAAC reaction represented the key step in the assembling of the precursor of macromonomer 6-MOEG-9-TM-**BF3k** (Scheme 2) [52].

Scheme 3 "Grafting onto" approach to the synthesis of poly-6-MOEG-9-TM-**BF3k**

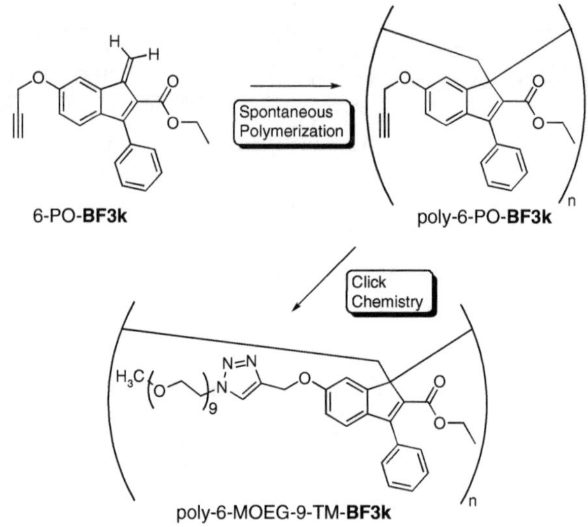

Notwithstanding its relative structural complexity, the macromonomer polymerized spontaneously by solvent removal in the apparent absence of catalysts or initiators. It is noteworthy that the increase in the structural complexity of benzofulvene monomers is, however, compatible with spontaneous polymerization. This suggested the existence of a particularly effective recognition process capable of pre-organizing the monomer molecules for the polymerization. The insertion of a TM spacer in the side chains of poly-6-MOEG-9-**BF3k** leading to poly-6-MOEG-9-TM-BF3k did not affect the structure of polybenzofulvene backbone, but appeared to modify its interaction with water. Poly-6-MOEG-9-TM-**BF3k** produced clear water solutions, but transmission electron microscopy (TEM) analysis and dynamic light scattering (DLS) studies showed the presence (in the solutions) of macromolecular aggregates with average dimensions in the range of 200–300 nm and suggested thermoresponsive colloidal properties for poly-6-MOEG-9-TM-**BF3k** macromolecules [52].

Subsequently, poly-6-MOEG-9-TM-**BF3k** was synthesized by post-polymerization modification of poly-6-PO-**BF3k** (Scheme 3).

This polymer bearing clickable propargyl groups was obtained from the corresponding benzofulvene monomer bearing a propargyloxy group at the 6-position of the indene nucleus (6-PO-**BF3k**) by spontaneous polymerization in the apparent absence of catalysts.

The persistence of acetylenyl moieties in the final polymer spontaneous polymerization was considered the final strong evidence of the existence of a very selective recognition process capable of pre-organizing the monomer molecules for spontaneous polymerization.

CuAAC reaction was used to transform poly-6-PO-**BF3k** into poly-6-MOEG-9-TM-BF3k showing spectroscopic features virtually indistinguishable from those of the homopolymer obtained by spontaneous polymerization of macromonomer 6-MOEG-9-TM-**BF3k**. The very high grafting density obtained by CuAAC modification of poly-6-PO-**BF3k** was also substantiated by the lack of the signals attributable to the propargyl moiety and by the results of the depolymerization studies. However, SEC-MALS measurements suggested that the conversion of poly-6-PO-BF3k into poly-6-MOEG-9-TM-**BF3k** led to a decrease in the molecular weight and to an increase in the polydispersity index. Random ruptures appeared to occur along the backbone to form macromolecules showing a molecular weight quite similar to that shown by poly-6-MOEG-9-TM-**BF3k** obtained by the "grafting through" approach. This behavior could be considered to be the consequence of the balance between the increasing steric hindrance around the polybenzofulvene backbone and its relative stability to covalent bond scission. The high grafting density caused MOEG-9 side chains to repel each other and stretch the polybenzofulvene backbone up to its scission, which can be facilitated by the relative stability of indenyl radical formed by homolytic cleavage.

The polymer prepared by the "grafting onto" approach (poly-6-MOEG-9-TM-BF3k-GO) formed thermoresponsive colloidal dispersions similar to those obtained with the polymer synthesized by the "grafting through" approach (poly-6-MOEG-9-TM-**BF3k**-GT), but TEM analysis and DLS studies suggested that poly-6-MOEG-9-TM-**BF3k**-GO was molecularly dissolved at room temperature.

The absorption/emission spectroscopic features of poly-6-MOEG-9-TM-**BF3k** (similar to those fluorescence microscopy reagent DAPI) allowed the interaction of the polymer with live cells to be studied by fluorescence microscopy. Since biological studies suggested potential biocompatibility features for this polymer, we are confident that our polybenzofulvene derivatives bearing clickable groups can be functionalized by a suitable strategy to obtain nanostructured materials potentially useful in a wide range of biological and biotechnological applications.

Thus after a first phase devoted to the exploration of the chemistry of benzofulvene monomers and polymers, our research interests moved towards the exploration of the potential applications of these polymers. Besides the potential use as drug carriers (which were more congenial to our expertise) we considered the potential application of polybenzofulvene derivatives in the development of optoelectronic devices. We were indeed aware that conjugated polymers have been intensively investigated because of their potential usefulness in the fabrication of optoelectronic devices such as light-emitting diodes (LEDs), field effect transistors (FETs), and photovoltaic cells. Charge and energy are transported through a conjugated polymer backbone or through aromatic rings belonging to the polymer backbone by means of a hopping mechanism. A large variety of π-conjugated polymers have been synthesized, but little consideration has been devoted to polymers showing stacked arrays of π-electron systems along the polymer backbone (through-space conjugated polymers) [53]. One of the most typical π-stacked polymers is poly(*N-vinylcarbazole*) (**PVK**, Fig. 11), which has been demonstrated to transport charge via intra-backbone stacked orbitals that result from face-to-face conformation of the carbazole moieties, and shows photoconductive, photorefractive, hole-transporting, energy-donating, and carrier-switching properties. Therefore, **PVK** has been used in xerography, holographic three-dimensional display, LED, nonvolatile memory, etc. [54]. Polydibenzofulvene (poly-**DBF**) derivatives are π-stacked polymers

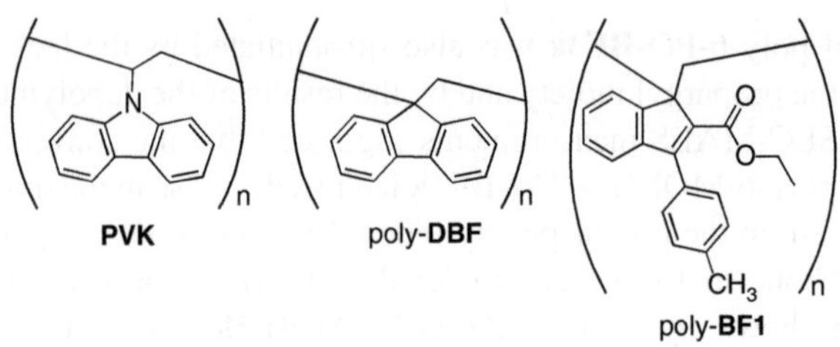

Fig. 11 Structures of **PVK**, poly-**DBF**, and poly-**BF1**. Reprinted with permission from Cappelli et al. [55]. Copyright 2012, The Royal Society of Chemistry

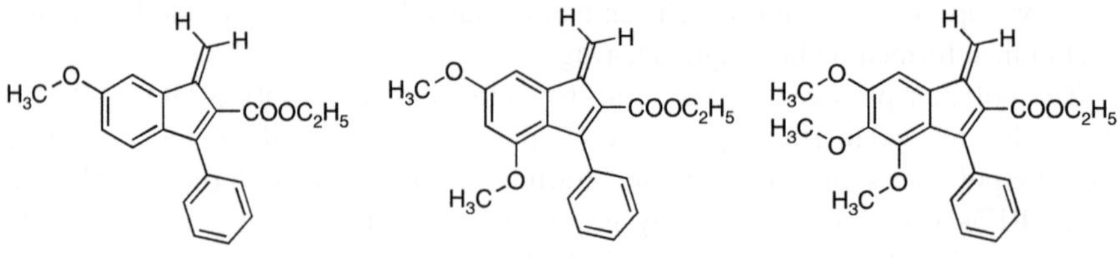

Fig. 12 Structures of benzofulvene monomers 6-MO-**BF3k**, 4,6-MO-**BF3k**, and 4,5,6-MO-**BF3k**

showing a close similarity with **PVK**, but differing from the latter for the absence of configurational isomerism [6]. Poly-**DBF** does not possess stereocenters and its stacked structure did not appear based on the complete face-to-face stacking of the fluorene units (chromophores), but rather on the slightly twisted stacking of the chromophores leading to a helical conformation and a high hole mobility [14].

Our studies have demonstrated that polybenzofulvene derivatives were π-stacked polymers, but nothing was known about their electronic properties such as potential charge-transporting features with respect to the electron distribution in the monomeric unit of these π-stacked polymers.

On the basis of the well-known electron-releasing properties of methoxy group, the structure of benzofulvene derivative 6-MO-**BF3k** was modified by inserting one or two additional methoxy groups to obtain the benzofulvene derivatives bearing two (4,6-MO-**BF3k**, Fig. 12) or three (4,5,6-MO-**BF3k**) methoxy substituents on the benzene ring in order to evaluate the photophysical and electronic properties of the corresponding polymers and ultimately the effect of the presence of the methoxy groups on them [55].

The analysis of the photophysical features shown by the three poly-**BF3k** derivatives suggested the presence of a high degree of π-stacking among the monomeric units in the polymers. This was confirmed by the significantly hypochromism the large Stokes shift. Interestingly, the absorption and emission features of the polymers in the solid state were found to be very similar to those shown in diluted solutions. In particular, the stable photoluminescence quantum yield constituted a promising feature for potential applications in the development of light-emitting devices. Finally, the significant hole mobility shown by poly-4,5,6-MO-**BF3k** along with the apparent role of the methoxy substituents in increasing the charge mobility, the thermoreversible polymerization features, the good solubility in the most common organic solvents, and the consequent processability (e.g., filmability) opened new routes in the development of materials potentially useful in optoelectronics.

Synthetic Approaches to Benzofulvene Compounds

For the preparation of benzofulvene derivatives, a number of synthetic pathways have been described in the literature [2]. For instance, the simplest benzofulvene derivative **BF** was prepared by dehydration of α-benzofulvanol (Scheme 4) [2, 56] as well as by elimination of acetic acid from the corresponding acetic ester (Scheme 4) [57]. Benzofulvene derivative **BF** was obtained as greenish yellow leaflets melting at 37 °C and was reported to be extremely unstable and polymerizing readily when kept in vacuo over H_2SO_4 or in sealed tubes [56].

However, the most general and fairly obvious method for the preparation of benzofulvene derivatives is represented by the reaction of the appropriate indenone derivative with the Grignard reagent followed by the dehydration of the corresponding indenol formed. This synthetic procedure was applied to the preparation of **BF1** as reported in Scheme 5. Thus, the reaction of indenone derivative **1l** with methylmagnesium bromide afforded the expected indenol **2l** with an acceptable yield (31 %).

Dehydration of **2l** in chloroform in the presence of PTSA gave a compound less polar (by thin layer chromatography, TLC) than **2l** that was proved to be of difficult isolation because the solvent removal led to the obtainment of a polymeric material containing only trace amounts of the expected **BF1**, as demonstrated by ^{1}H-NMR spectroscopy. Therefore **BF1** was shown to be a quite elusive chemical entity and its identification by NMR spectroscopy required dehydration in $CDCl_3$, allowing the ^{1}H-NMR analysis to be carried out without the solvent elimination. In this way, a clear ^{1}H-NMR spectrum compatible with the structure of **BF1** was obtained without special purifications.

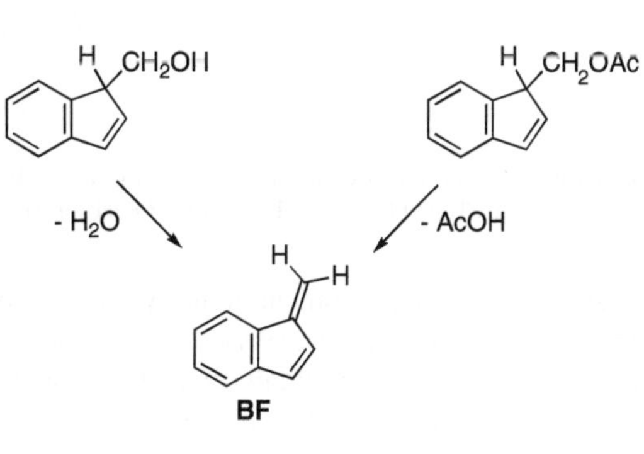

Scheme 4 Synthesis of the simplest benzofulvene derivative **BF**

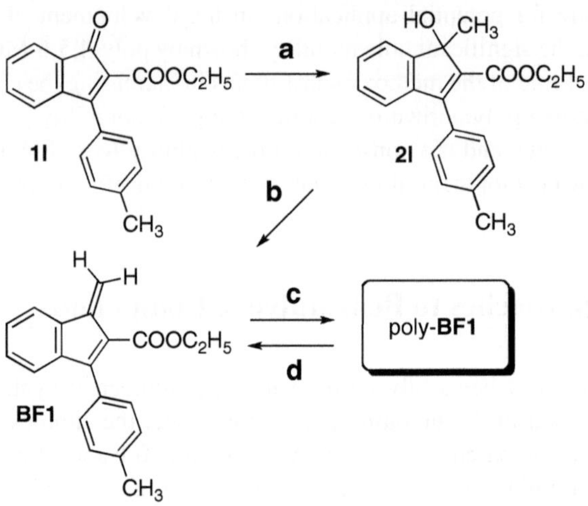

Scheme 5 First approach to the synthesis of our prototypical benzofulvene derivative **BF1**. Reagents: (*a*) CH_3MgBr, $C_2H_5OC_2H_5$; (*b*) PTSA, $CHCl_3$; (*c*) solvent elimination; and (*d*) solvent, heating

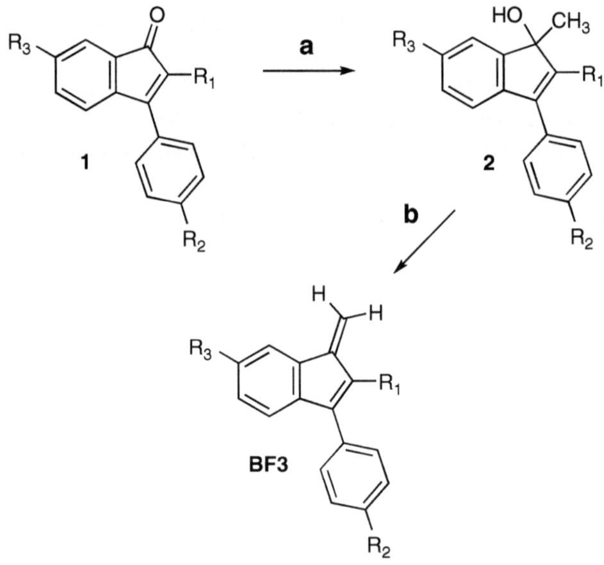

Scheme 6 General procedure for the synthesis of benzofulvene derivatives. Reagents: (*a*) $Al(CH_3)_3$, CH_2Cl_2; (*b*) PTSA, $CHCl_3$ (or HCOOH for **2j**)

A significant improvement in the synthetic approach to benzofulvene derivatives was achieved with the discovery that **1l** could be easily methylated by means of a solution of trimethylaluminum (yield 82 %), and this reactant has been largely used in the course of our studies for alkylating a large number of different indenone derivatives (Scheme 6 and Table 2).

Table 2 Preparation of some relevant benzofulvene derivatives

R_1	R_2	R_3	Indenone	Preparation	Indenol (yield)	BF derivative
H	H	H	**1a**	[58]	**2a** (87 %)	**BF3a**
$Si(CH_3)_3$	H	H	**1b**	[58]	**2b** (83 %)	**BF3a**[a]
F	H	H			**2c** (36 %)[b]	**BF3c**
Cl	H	H	**1d**	Scheme 11	**2d** (84 %)	**BF3d**
Br	H	H	**1e**	Scheme 11	**2e** (92 %)	**BF3e**
CH_3	H	H	**1f**	Scheme 11	**2f** (67 %)	**BF3f**
CN	H	H	**1g**	[59]	**2g** (94 %)	**BF3g**
CN	CH_3	H	**1h**	[22]	**2h** (84 %)	**BF3h**
CN	H	CH_3	**1i**	[22]	**2i** (86 %)	**BF3i**
C≡C-2-Pyr	H	H	**1j**	Scheme 11	**2j** (79 %)	**BF3j**[c]
$COOC_2H_5$	H	H	**1k**	[60]	**2k** (82 %)	**BF3k**
$COOC_2H_5$	CH_3	H	**1l**	[20]	**2l** (82 %)	**BF3l**
$COOC(CH_3)_3$	CH_3	H	**1m**	[18]	**2m** (75 %)	**BF3ma**[d]
$CON(CH_3)_2$	CH_3	H	**1n**	[22]	**2n** (83 %)	**BF3n**
$C(CH_3)_3$	H	H	**1o**	[58]	**2o** (77 %)	**BF3o**
$COO(C_2H_4O)_3CH_3$	CH_3	H	**1p**	[25]	**2p** (82 %)	2-MOEG-3-BF1
$COO(C_2H_4O)_9CH_3$	CH_3	H	**1q**	[25]	**2q** (72 %)	2-MOEG-9-BF1
$COOC_2H_5$	H	OCH_3	**1r**	Scheme 9	**2r** (66 %)	6-MO-**BF3k**
$COOC_2H_5$	H	$O(C_2H_4O)_2CH_3$	**1s**	Scheme 10	**2s** (82 %)	6-MOEG-2-**BF3k**
$COOC_2H_5$	H	$O(C_2H_4O)_3CH_3$	**1t**	Scheme 10	**2t** (73 %)	6-MOEG-3-**BF3k**
$COOC_2H_5$	H	$O(C_2H_4O)_9CH_3$	**1u**	Scheme 10	**2u** (74 %)	6-MOEG-9-**BF3k**
$COOC_2H_5$	H	$OCH_2(C_2HN_3)(C_2H_4O)_9CH_3$	**1v**	Scheme 10	**2v** (44 %)	6-MOEG-9-TM-**BF3k**
$COOC_2H_5$	H	OCH_2CCH	**1aa**	Scheme 10	**2aa** (92 %)	6-MO-**BF3k**
$COOC_2H_5$	H	$OCH_2CH{=}CH_2$	**1ab**		**2ab** (76 %)	6-AO-**BF3k**

[a]**BF3a** was obtained instead of **BF3b** because indenol derivative **2b** desilylated spontaneously in the dehydration conditions

[b]Indenol **2c** was prepared from **2e** by lithium–halogen exchange and subsequent fluorination with *N*-fluorobenzenesulfonimide (NFSI)

[c]Indenol **2j** was dehydrated with formic acid instead of PTSA

[d]**BF3ma** was obtained instead of *t*-butyl ester **BF3m** because indenol derivative **2m** cleaved spontaneously in the dehydration conditions

The data shown in Table 2 suggest that *C*-methylation of indenones **1** with trimethylaluminum was an important improvement in the synthesis of indenol derivatives **2** because the yields were higher with respect to those obtained by means of other standard methods. Moreover in most cases, the indenol derivative was used in the following step of the synthesis without any further purification. The introduction of a fluorine atom at the 2-position of the benzofulvene nucleus of **BF3c** was carried out at the indenol stage. Thus, indenol **2e** underwent lithium–halogen exchange with *t*-BuLi in THF and subsequent fluorination with *N*-fluorobenzenesulfonimide (NFSI).

In general, dehydration of indenols **2** was carried out in chloroform (or $CDCl_3$ so that the reaction could be followed by ^{1}H-NMR analysis) in the presence of PTSA in order to obtain a stock solution (about 0.1 M) of the expected *trans*-diene

Scheme 7 Synthesis of benzofulvene derivatives 4,6-MO-**BF3k** and 4,5,6-MO-**BF3k.** Reagents: (*a*) $Al(CH_3)_3$, CH_2Cl_2; (*b*) PTSA, $CHCl_3$. Substituents: for compounds **1w**, **2w**, **3w**, **4w**, and 4,6-MO-**BF3k**, R=H; for compounds **1x**, **2x**, **3x**, **4x**, and 4,5,6-MO-**BF3k**, R=OCH_3. Reprinted with permission from Cappelli et al. [55]. Copyright 2012, The Royal Society of Chemistry

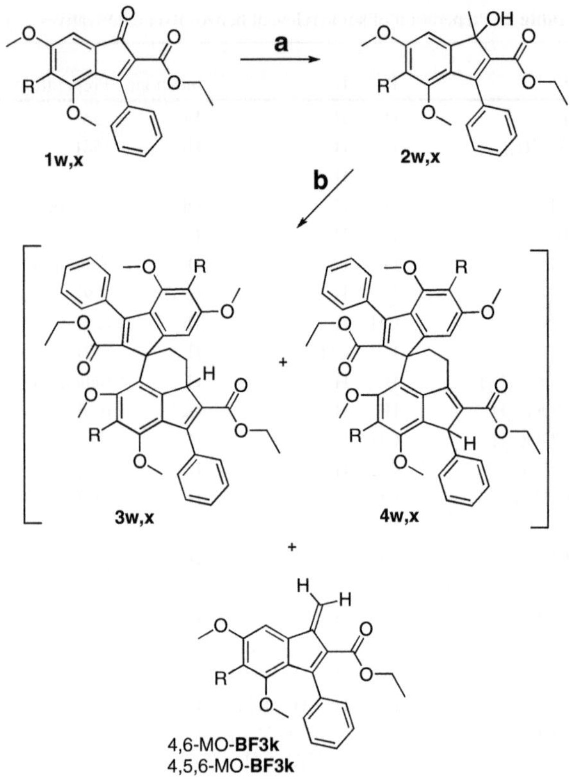

derivative **BF3**. However, the dehydration of **2b** led to **BF3a** because **2b** desilylated spontaneously in the dehydration conditions, whereas indenol **2m** cleaved spontaneously in the dehydration conditions leading to **BF3ma**. Moreover, the dehydration of indenol **2j** required the use of formic acid instead of PTSA because the latter led to the formation of a number of by-products.

The synthetic work performed in the preparation of benzofulvene derivatives shows that the methylation/dehydration procedure is viable in producing benzofulvene monomers bearing a large variety of substituents on the indene nucleus with the limitations due to the instability of functional groups in the methylation/dehydration conditions. Furthermore, our work showed that the insertion of additional methoxy substituents in the benzene ring of monomethoxylated benzofulvene derivative 6-MO-**BF3k** leading to 4,6-MO-**BF3k** and 4,5,6-MO-**BF3k** was able to produce differences in the chemical reactivity indicating a different electron density in the substituted benzene ring and in particular at the 7-position of the indene nucleus. In fact, the dehydration reaction on both substrates **2w** and **2x** (Scheme 7) afforded, beside the expected monomers, a mixture of dimers, some of which (i.e., **3w,x** and **4w,x**) were identified by X-ray crystallography.

However, the formation of dimers was minimized by carrying out the dehydration reaction with low concentrations of indenol and by following the reaction

Scheme 8 Synthesis of **BF3y** isomers. Reagents: (*a*) $Al(C_2H_5)_3$, CH_2Cl_2; (*b*) PTSA, $CDCl_3$ or $CHCl_3$. Reprinted with permission from Cappelli et al. [22]. Copyright 2007, American Chemical Society

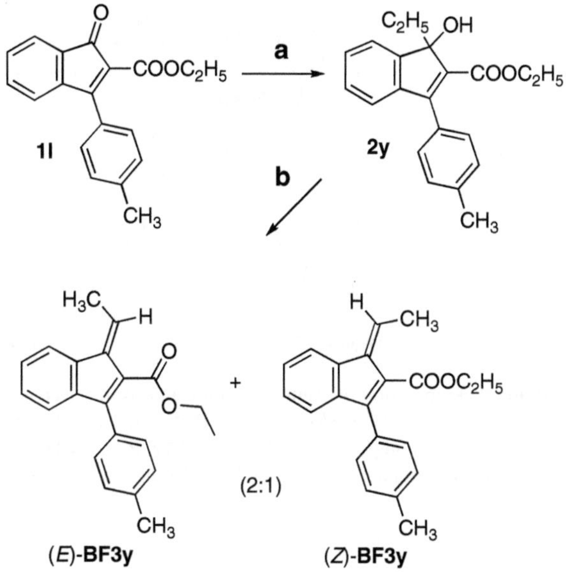

course by thin layer chromatography (TLC). It is noteworthy that the dimer formation appeared to require the presence of PTSA as in the reaction conditions. Furthermore, the solvent removal from the solutions of purified monomers produced only the corresponding polymers without dimers. These results suggest that PTSA plays a key role in the dimerization process.

The alkylation of indenone **1l** with triethylaluminum gave ethylindenol derivative **2y** (Scheme 8), which was dehydrated in the usual conditions to give the mixture of (ca. 2:1) *E* and *Z* isomers of ω-methylbenzofulvene derivative **BF3y**.

The required indenone derivatives were prepared by different methods according to the substituents present in the indene nucleus. Some selected strategies among the most significant ones are reported in the following schemes.

Methoxyindenones **1r,w and 1x** (Scheme 9) were prepared from benzyl chlorides **5r,w and 5x** and ethyl benzoylacetate to provide compounds **6r,w and 6x**, which were cyclized with polyphosphoric acid (PPA) or sulfuric acid to obtain indene derivatives **7r,w and 7x**. Finally, compounds **7r,w and 7x** were oxidized with selenium dioxide to provide indenone derivatives **1r,w and 1x** [55].

The presence of methoxy groups in the indenone nucleus can be profitably used to introduce a large variety of substituents in the final benzofulvene scaffold. As a matter of fact, the demethylation of 6-methoxyindenone derivative **1r** with BBr_3 afforded 6-hydroxyindenone derivative **1z**, which could be alkylated with a wide range of alkyl bromides (Scheme 10). For instance, **1z** was alkylated with activated methyl-end-capped oligo(ethylene glycol) (MOEG) side chains composed of two, three, or nine monomeric units to obtain indenone derivatives **1s,t and 1u**, which were the precursors of macromonomers 6-MOEG-2-**BF3k**, 6-MOEG-3-**BF3k**, and 6-MOEG-9-**BF3k** [37]. On the other hand, the alkylation of **1z** with propargyl

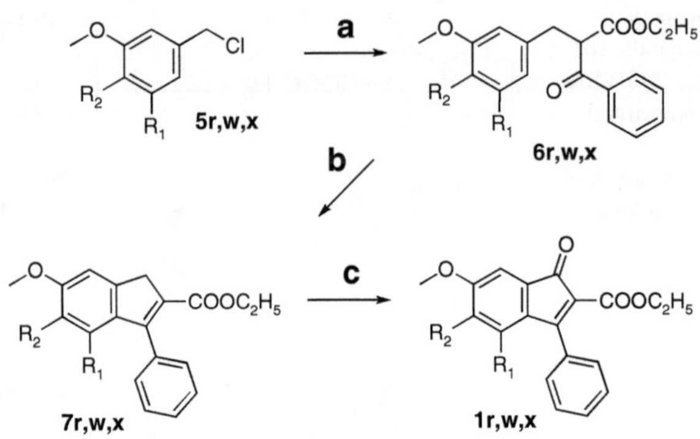

Scheme 9 Preparation of methoxyindenone derivatives **1r,w,x**. Reagents: (*a*) ethyl benzoylacetate, K_2CO_3, NaI, DMF; (*b*) for **7r,w**, PPA; for **7x**, sulfuric acid; and (*c*) SeO_2, 1,4-dioxane. Substituents: for compounds **1r**, **5r**, **6r**, and **7r**, $R_1=R_2=H$; for compounds **1w**, **5w**, **6w**, and **7w**, $R_1=OCH_3$, $R_2=H$; and for compounds **1x**, **5x**, **6x**, and **7x**, $R_1=R_2=OCH_3$

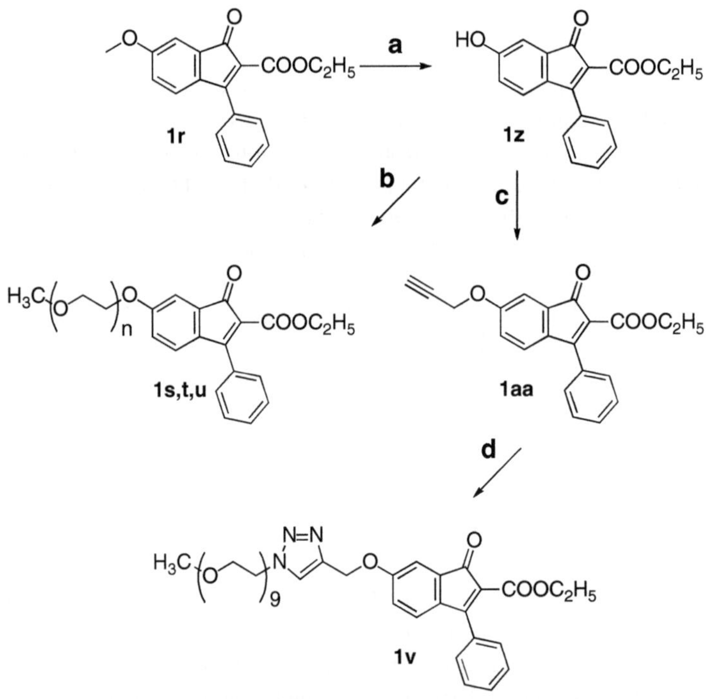

Scheme 10 Preparation of indenones functionalized at the 6-position of the indene nucleus. Reagents: (*a*) BBr_3, CH_2Cl_2; (*b*) $CH_3(OCH_2CH_2)_nBr$, K_2CO_3, NaI, DMF; (*c*) propargyl bromide, K_2CO_3, NaI, DMF; and (*d*) $CH_3(OCH_2CH_2)_9N_3$, $Cu(PPh_3)_3Br$, DIPEA, THF

Scheme 11 Preparation of indenones functionalized at the 2-position of the indene nucleus. Reagents: (*a*) 1-phenyl-2-(trimethylsilyl) acetylene, $Pd(CH_3COO)_2$, Na_2CO_3, n-Bu_4NBr, DMA; (*b*) SO_2Cl_2, CH_3COOH (*c*) NBS, CH_2Cl_2; (*d*) 2-ethynylpyridine, $Pd(PPh_3)_2Cl_2$, CuI, TEA; and (*e*) $Sn(CH_3)_4$, LiCl, $Pd(PPh_3)_2Cl_2$, DMF

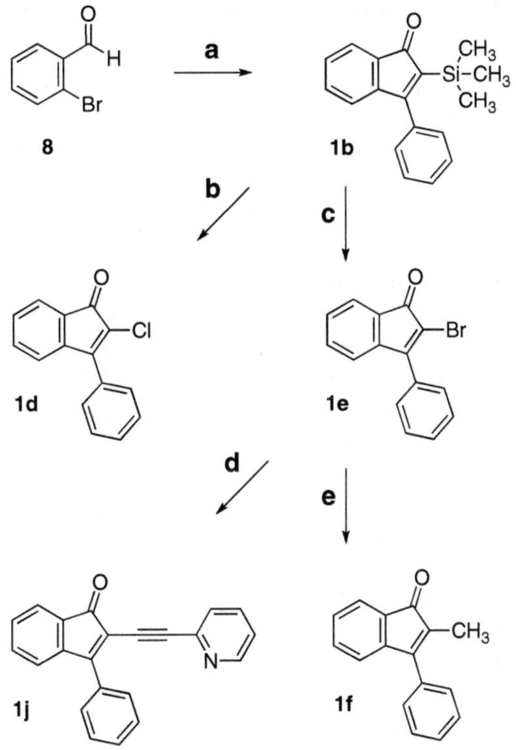

bromide led to 6-propargyloxyindenone derivative **1aa**, which was submitted to a copper(I)-catalyzed alkyne–azide 1,3-dipolar cycloaddition (CuAAC) reaction to obtain indenone **1v** [52]. This compound was the precursor of benzofulvene macromonomer 6-MOEG-9-TM-**BF3k**, which polymerized spontaneously to afford the water-soluble molecular brush poly-6-MOEG-9-TM-**BF3k**.

Finally, a number of indenone derivatives functionalized at the 2-position were prepared by means of the procedure reported in Scheme 11. Trimethylsilylindenone **1b** was obtained by palladium-catalyzed annulation of *o*-bromobenzaldehyde with 1-phenyl-2-(trimethylsilyl)acetylene [58]. The reaction of **1b** with sulfuryl chloride gave chloroindenone **1d**, whereas bromoindenone **1e** was obtained by bromination of **1b** with *N*-bromosuccinimide (NBS). Bromoindenone derivative **1e** was employed in a variety of palladium-catalyzed cross coupling to obtain key indenone intermediates bearing suitable substituents at the 2-position of the indene nucleus. For instance, the coupling of **1e** with tetramethyltin in the presence of $Pd(PPh_3)_2Cl_2$ (Stille coupling) afforded methylindenone **1f**, or with 2-ethynylpyridine in the presence of the same catalyst (Sonogashira coupling) gave indenone **1j**.

Polymerization of Benzofulvene Derivatives

The chemical behavior of benzofulvene derivatives depends on their structure (i.e., the substituents present on the different positions of benzofulvene nucleus) and on the environment surrounding their molecules. In general, benzofulvenes are stable compounds, but their exocyclic double bond appears to express a polar character and a pronounced tendency to interact with nucleophiles [2].

Our prototypical benzofulvene derivative **BF1** was found to be stable in diluted solutions in the most common organic solvents (i.e., 0.1 M, in chloroform). When the solvent was evaporated to increase the concentration up to about 0.4 M, we observed the formation of the corresponding polymer that was almost complete after the total solvent removal. Thus, **BF1** appeared to polymerize spontaneously in the apparent absence of initiators or catalyst by simple solvent removal (concentration of the solution) [20, 21].

On the basis of these results, subsequent studies were finalized to investigate on the effect of the substituents' features on the propensity of benzofulvene monomers to spontaneous polymerization.

Our studies showed that quite a large number of differently substituted benzofulvene derivatives polymerized upon solvent removal to give the corresponding polymers (poly-**BF** derivatives, Table 3), with the exception of 2-dimethylamide derivative **BF3n** and ω-methylbenzofulvene derivative **BF3y**, which did not show tendency to spontaneous polymerization. Furthermore, poly-**BF3o** was obtained only in trace amounts by following the same procedure used for the other polymers [22].

The results shown in Table 3 suggested that spontaneous polymerization of benzofulvene derivatives is compatible with a large variety of substituents showing different stereoelectronic properties, but it appeared to be very sensitive to steric hindrance in the near proximity of the polymerizing functionality. In particular, the dimensions of the substituents at the 2-position of the indene moiety appeared to affect the spontaneous polymerization, but the comparison of the behavior of **BF3g** (R_1 = –C≡N), **BF3j** (R_1 = –C≡C-2-Pyr), and **BF3o** [R_1 = CON(CH_3)$_2$] suggested that the steric hindrance at the 2-position of the indene moiety played a more important role than the total volume of R_1 substituent in inhibiting spontaneous polymerization. Indeed, **BF3j** showed a pronounced tendency to the spontaneous polymerization in spite of the dimensions of its R_1 substituent, probably because the presence of the tube-shaped alkynylene spacer dislocates the hindrance at the 2-position. Moreover, the lack of tendency to polymerize shown by the isomers of ω-methylbenzofulvene derivative **BF3y** showed that the insertion of a small substituent as the methyl group in the polymerizing functionality of our protypical benzofulvene derivative **BF1** is able to depress spontaneous polymerization propensity at all [22].

Very interestingly, the spontaneous polymerization of benzofulvene derivatives appeared to be compatible with the presence in the monomer of relatively long monomethyl oligo(ethylene glycol) (MOEG) side chains. This result has made possible the synthesis of biologically relevant polybenzofulvene molecular brushes (i.e., poly-2-MOEG-9-**BF1**, poly-6-MOEG-9-**BF3k**, and poly-6-MOEG-9-TM-**BF3k**) by means of the so-called grafting through approach [25, 37, 52].

Table 3 Spontaneous polymerization of some relevant benzofulvene derivatives

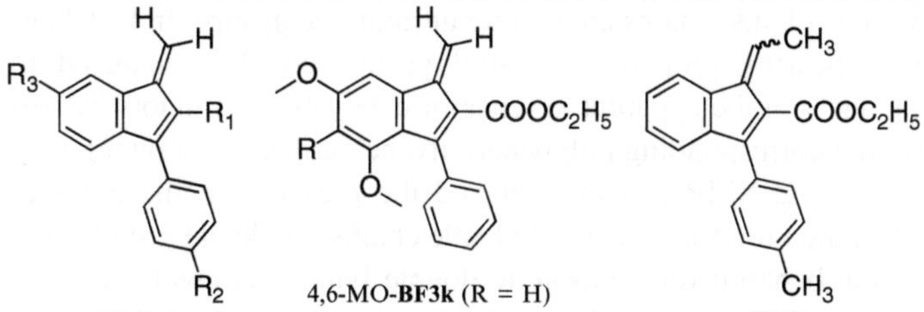

BF derivatives 4,6-MO-**BF3k** (R = H) 4,5,6-MO-**BF3k** (R = OCH_3) **BF3y**

Monomer	R_1	R_2	R_3	Polymer	Yield[a] (%)
BF3a	H	H	H	Poly-**BF3a**	72
BF3c	F	H	H	Poly-**BF3c**	82
BF3d	Cl	H	H	Poly-**BF3d**	77
BF3e	Br	H	H	Poly-**BF3e**	58
BF3f	CH_3	H	H	Poly-**BF3f**	60
BF3g	CN	H	H	Poly-**BF3g**	82
BF3h	CN	CH_3	H	poly-**BF3h**	90
BF3i	CN	H	CH_3	Poly-**BF3i**	87
BF3j	C≡C-2-Pyr	H	H	Poly-**BF3j**	55
BF3k	$COOC_2H_5$	H	H	Poly-**BF3k**	73
BF3l (BF1)	$COOC_2H_5$	CH_3	H	Poly-**BF3l** (poly-**BF1**)	78
BF3ma[b]	$COOC(CH_3)_3$	CH_3	H	Poly-**BF3ma**	74
BF3n	$CON(CH_3)_2$	CH_3	H	–	–
BF3o	$C(CH_3)_3$	H	H	Poly-**BF3o**	2
2-MOEG-3-**BF1**	$COO(C_2H_4O)_3CH_3$	CH_3	H	Poly-2-MOEG-3-**BF1**	75
2-MOEG-9-**BF1**	$COO(C_2H_4O)_9CH_3$	CH_3	H	Poly-2-MOEG-9-**BF1**	92
6-MO-**BF3k**	$COOC_2H_5$	H	OCH_3	Poly-6-MO-BF3k	84
6-MOEG-2-**BF3k**	$COOC_2H_5$	H	$O(C_2H_4O)_2CH_3$	Poly-6-MOEG-2-**BF3k**	53
6-MOEG-3-**BF3k**	$COOC_2H_5$	H	$O(C_2H_4O)_3CH_3$	Poly-6-MOEG-3-**BF3k**	81
6-MOEG-9-**BF3k**	$COOC_2H_5$	H	$O(C_2H_4O)_9CH_3$	Poly-6-MOEG-9-**BF3k**	46
6-MOEG-9-TM-**BF3k**	$COOC_2H_5$	H	$OCH_2(C_2HN_3)(C_2H_4O)_9CH_3$	Poly-6-MOEG-9-TM-**BF3k**	44
4,6-MO-**BF3k**				Poly-4,6-MO-**BF3k**	51[c]
4,5,6-MO-**BF3k**				Poly-4,5,6-MO-**BF3k**	42[c]
6-PO-**BF3k**	$COOC_2H_5$	H	OCH_2CCH	Poly-6-PO-**BF3k**	74
6-AO-**BF3k**	$COOC_2H_5$	H	$OCH_2CH{=}CH_2$	Poly-6-AO-**BF3k**	79
BF3y				–	–

[a]The yields were calculated from the starting indenol derivatives

[b]**BF3ma** was obtained instead of *t*-butyl ester **BF3m** because indenol derivative **2m** cleaved spontaneously in the dehydration conditions

[c]The relatively low yield was due to the formation of dimers during the dehydration reaction (see Scheme 7)

Furthermore, our studies showed the spontaneous polymerization is also compatible with the presence of side chains containing unsaturated groups. In fact, benzofulvene monomers bearing propargyl or allyl groups could be induced to polymerize spontaneously in the apparent absence of catalysts or initiators by solvent removal to give the corresponding polybenzofulvene derivatives showing very high molecular weight. ^{13}C-NMR analysis showed the presence in the polymer spectra of signals attributed to propargyl or allyl side chains (clickable groups) and suggested that only the benzofulvene exocyclic double bond was involved selectively in the vinyl polymerization. Thus, the presence of acetylenyl or vinyl moieties in the side chains of benzofulvene monomers did not affect the monomer interactions leading to spontaneous polymerization. This result was considered as a strong evidence of the existence of a very selective recognition process capable of preorganizing the monomer molecules for spontaneous polymerization.

According to its ability to stabilize a negative charge as indenyl anion, benzofulvene derivative **BF1** was found to interact with phenyl lithium (PhLi, 1 equiv.) to give a complex mixture of adducts (Scheme 12) [23].

Normal-phase chromatography (i.e., flash chromatography) allowed the isolation of two activated monomer isomers (**Mono-1,2** and **Mono-1,4**), of two pure activated dimer isomers (**Dimer-1,2-1,4-Major** and **Dimer-1,2-1,4-minor**) and of the mixture of two other dimers (**Dimer-1,2-1,2-Major** and **Dimer-1,2-1,2-minor**), which were separated by semi-preparative reverse phase HPLC. Furthermore, more polar fractions of the flash chromatography purification gave a complex mixture of low molecular weight poly-**BF1** (e.g., trimers and higher MW oligomers). These results suggested that after the initial nucleophilic attack of PhLi to benzofulvene derivative **BF1**, the activated monomer **Mono-1,2** is able to react with a second **BF1** molecule to give dimers and further to provide oligomers and polymers. In fact, when the same experiments were performed with a low stoichiometric ratio of PhLi (e.g., 0.1 equiv.), substantial amounts of polymeric fractions insoluble in *n*-hexane were obtained. Therefore, the results suggested that benzofulvene derivative **BF1** is able to polymerize in the presence of an anionic initiator such as PhLi to afford a polymer and/or oligomers depending on the amount of the initiator employed. More precisely, stoichiometric amounts of PhLi produced almost exclusively soluble oligomers, while low initiator amounts (i.e., 0.1 equiv.) led to the presence of both oligomers and substantial amounts of poly-**BF1** in the reaction mixture.

The comparative evaluation of the HPLC-MS chromatograms of the oligomer fractions suggested that the dimeric species were among the major oligomers independently of the reaction temperature. We assumed that the conditions of the anionic polymerization lead to an equilibrium (among various oligomers) similar to that shown by **BF1** in the presence of *N*-methylpiperazine (NMP, see below) with the difference that the anionic activated monomer is capable of starting the propagation step leading to poly-**BF1**, which becomes the major species when low stoichiometric ratios of PhLi are used.

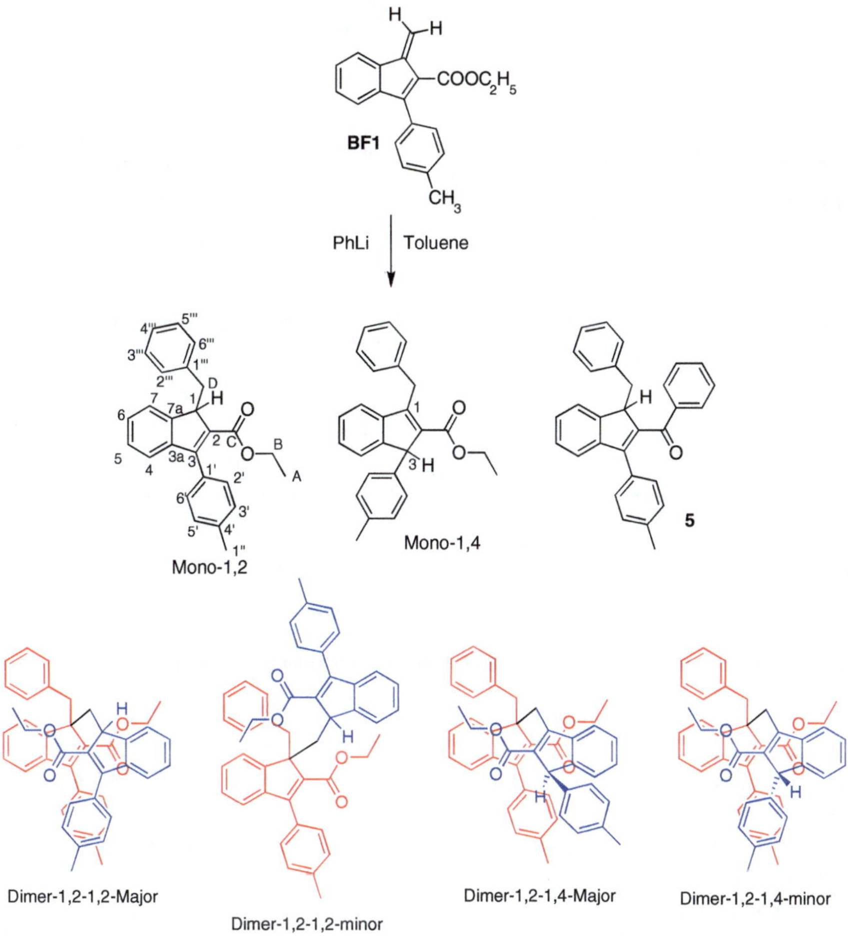

Scheme 12 Reaction of benzofulvene derivative **BF1** with phenyl lithium. The *red structures* in dimers refer to the *bf1* moiety and the blue ones to the *bf2* one. Reprinted with permission from Cappelli et al. [23]. Copyright 2008, American Chemical Society

Polymerization Mechanism

As shown in the previous section, benzofulvene derivatives undergo polymerization in different conditions. The mechanism of anionic polymerization of **BF1** appears to be pretty comprehensible. The exocyclic double bond expresses a polar character and interacts with nucleophiles according to its ability to stabilize a negative charge as an aromatic indenyl anion. Indeed, **BF1** was found to react with NMP to afford the apparently unstable Michael-type adducts **6** and **7** (Scheme 13) analogous to that described for **DBF** [61, 62].

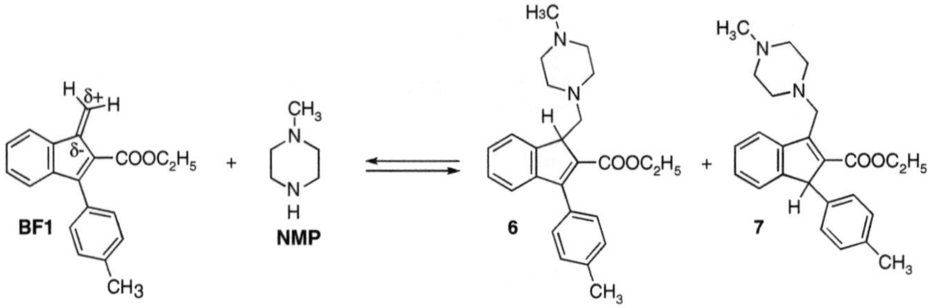

Scheme 13 Reaction of benzofulvene derivative **BF1** with *N*-methylpiperazine (NMP)

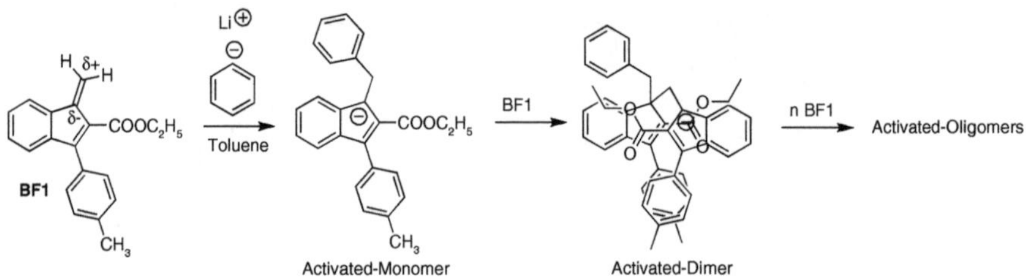

Scheme 14 Anionic polymerization of **BF1** with phenyl lithium as the initiator

In particular, when solutions of **BF1** in $CDCl_3$ were treated with increasing amounts of NMP (1, 2, 3, or 5 equiv.) and quickly submitted to ^{1}H-NMR analysis, the comparison of the resulting spectra showed that with 1 equiv. of NMP, the concentration of **6** and **7** in the mixture was low, but increased with the increasing NMP concentration. The mixture containing 2 equiv. of NMP was submitted to ^{1}H-NMR analysis at different times after NMP addition and the comparison of the spectra suggested that about 24 h was sufficient to reach the equilibrium. In these conditions, **BF1** was unable to polymerize even in concentrated solutions. In fact, when a solution of **BF1** in chloroform containing an equimolar amount of NMP was concentrated (the same experiment without NMP was run in parallel as the control), poly-**BF1** formation was inhibited and we observed the presence of relatively little amounts of piperazinyl derivatives **6** and **7** along with substantial amounts of **BF1**.

On the other hand, **BF1** interacts with a strong nucleophile such as PhLi in the same manner as for NMP, with the difference that the activated monomer **Mono-1,2** (Scheme 14) bearing a negative charge stabilized as an indenyl anion is able to react with a second **BF1** molecule to give dimers and further to provide oligomers and polymer molecules.

Interestingly, the anionic polymerization of **BF1** with PhLi as the initiator occurred in diluted solutions (0.1 M), whereas the spontaneous polymerization required concentrations at least fourfold higher. This difference suggests a key role for the negative charge in promoting aggregation of the monomers around the incipient macromolecules. Another thing worth of nothing is the possible role of the

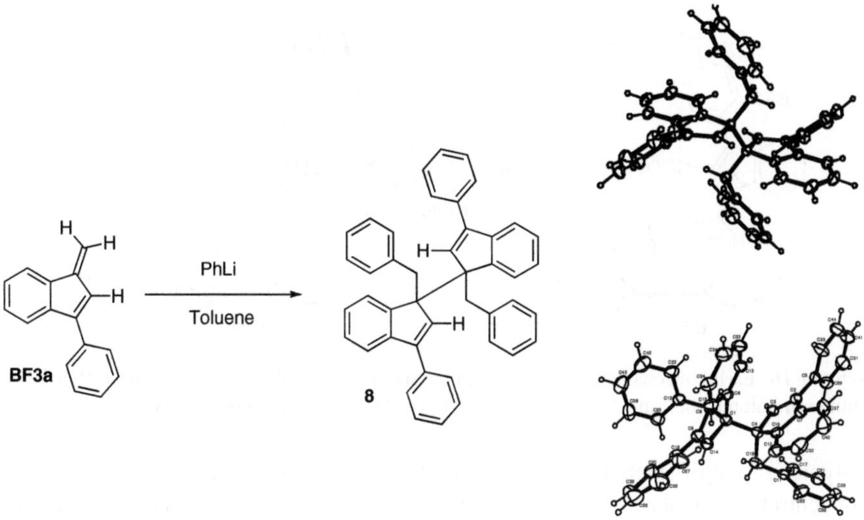

Scheme 15 Reaction of **BF3a** with phenyl lithium

carboxylic group in stabilizing the negative charge. Indeed, when benzofulvene derivative **BF3a** (bearing a hydrogen atom in place of –$COOC_2H_5$ in position 2 of benzofulvene scaffold) was treated with PhLi in the attempt to obtain the corresponding polymer, we obtained a complex reaction mixture containing significant amounts of symmetric dimer **8** (Scheme 15), which was thought to derive from the collapse of the radicals produced from the initial nucleophilic attack of PhLi to **BF3a**.

These results suggested that ionic or radical species can be involved in the polymerization of benzofulvene derivatives according to the nature of substituents present in the benzofulvene scaffold [63].

The mechanism of spontaneous polymerization of benzofulvene derivatives is more difficult to be established with precision.

Spontaneous polymerization has been described to occur in a large number of different systems. For instance, electron-rich olefins react with electron-poor ones in different ways generating different reaction products such as cycloadducts but also alternating copolymers and homopolymers depending on the difference in the electron density between the reactants. The matter has been nicely considered by Hall and Padias, who have elaborated the so-called bond-forming initiation theory as a result of their long-standing interest in the research field [64, 65]. In their reviews, these authors noticed that olefins showing similar polarity usually do not react when mixed, but a moderate difference in electron density is sufficient to produce spontaneous free radical polymerization. The spontaneous formation of a random copolymer between styrene and methyl methacrylate in the apparent absence of initiators was described by Walling in 1949 [66]. However, spontaneous polymerization has been reported to occur with a number of (metha)acrylic acid derivatives in aqueous and organic solutions in mild conditions. The spontaneous

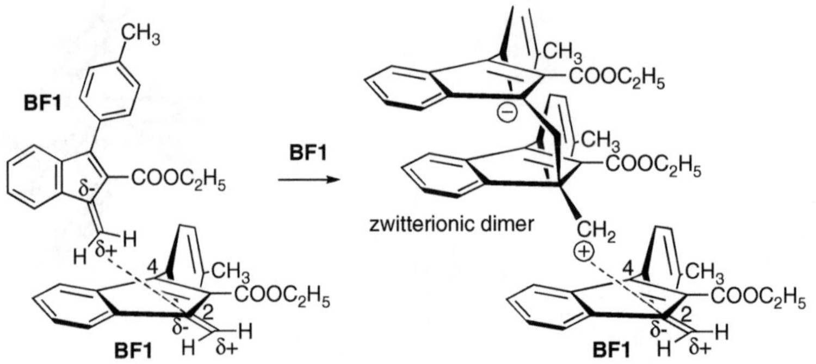

Scheme 16 Earlier mechanistic hypothesis for the spontaneous polymerization of **BF1**. Reprinted with permission from Cappelli et al. [21]. Copyright 2005, Wiley Periodicals, Inc.

polymerization of (metha)acrylamides in water was demonstrated to occur at high concentrations of the monomers through the formation of monomeric associates and the phenomenon of self-organization of the monomers [67]. Finally, alkylcyanoacrylates are among the most reactive monomers known and show the tendency to apparently spontaneous polymerization promoted by adventitious traces of unknown initiators [68].

Our earlier studies led to conceive a mechanistic hypothesis based on the "bond-forming initiation theory" and on the polar character of the exocyclic double bond of **BF1**. Thus, we assumed that the interaction of the strongly polarized exocyclic double bonds of two **BF1** molecules could produce the formation of an activated dimer featuring a zwitterionic tetramethylene moiety (Scheme 16). This zwitterionic dimer intermediate was supposed to show a negative charge stabilized as an aromatic indenyl anion and an unstabilized CH_2^+ which could initiate the cationic polymerization (initiation step) process leading to the corresponding polymer [20, 21].

Because of the steric overcrowding, the zwitterionic intermediate is able to grow (propagation step) only in very concentrated solutions. The exclusive formation of the 1,2-polymer appeared to be justified by the fact that the 2-position (Scheme 16) is evidently less sterically hindered than the 4-position (the freely rotating phenyl ring at the 4-position occupies more space than the CH_2= at the 2-position, which does not occupy space outside the plane of the bicyclic system) and the zwitterionic dimer, or the growing zwitterionic intermediate, sufficiently selective in reacting with the less crowded position of **BF1**.

This mechanistic hypothesis appeared to be supported by the experimental evidence reported below.

The spontaneous polymerization of **BF1** was found to be operative also in the case of the solvent elimination in the presence of a radical inhibitor such as 2,6-di-tert-butyl-*p*-cresol. This result was interpreted in terms of a negligible role of free radical species in the polymerization mechanism, which was assumed to be ionic in nature. Moreover, **BF1** was found to be stable in solution at reflux in the presence of a catalytic amount of dibenzoyl peroxide, so that polymerization of **BF1** could not be induced by free radical species.

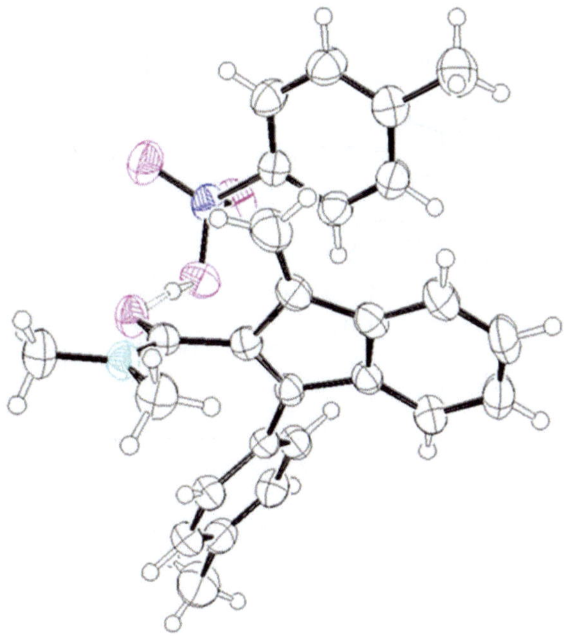

Fig. 13 Crystal structure of the complex between benzofulvene derivative **BF3n** and PTSA. Reprinted with permission from Cappelli et al. [22]. Copyright 2007, American Chemical Society

The spontaneous polymerization of **BF1** was active also in the absence of PTSA, suggesting cationic catalysis to be unnecessary. Furthermore, the formation of the polymer was observed also during the dehydration process, when concentrated solutions were processed.

The formation of poly-**BF1** was depressed by the presence of a nucleophile such as NMP.

The isomers of ω-methylbenzofulvene derivative **BF3y** were lacking of the tendency to polymerize shown by **BF1**. In other words, the presence of a methyl on the exocyclic methylene appeared to prevent the interaction and the consequent formation of the zwitterionic intermediate, which was assumed to initiate the spontaneous polymerization.

Hindered amide derivative **BF3n** did not show the tendency towards spontaneous polymerization and the crystal structure of the complex **BF3n** PTSA showed the interaction of electropositive $=CH_2$ with the π-orbital of PTSA (Fig. 13, the distance between the centroid of PTSA toluene ring and vinylene carbon of **BF3n** was 3.802 Å). This particular cation-π interaction was considered to support the proposed initiation mechanism in view of the fact that a similar interaction between the activated methylene group and the indene π-orbital could be the basis for the formation of the postulated zwitterionic dimer.

However, subsequent experimental evidence has led to reconsider the mechanistic hypothesis described above. For instance, the results obtained in the experiments of spontaneous polymerization of **BF1** in the presence of different vinyl monomers led to the assumption that the molecules of this benzofulvene derivative are capable

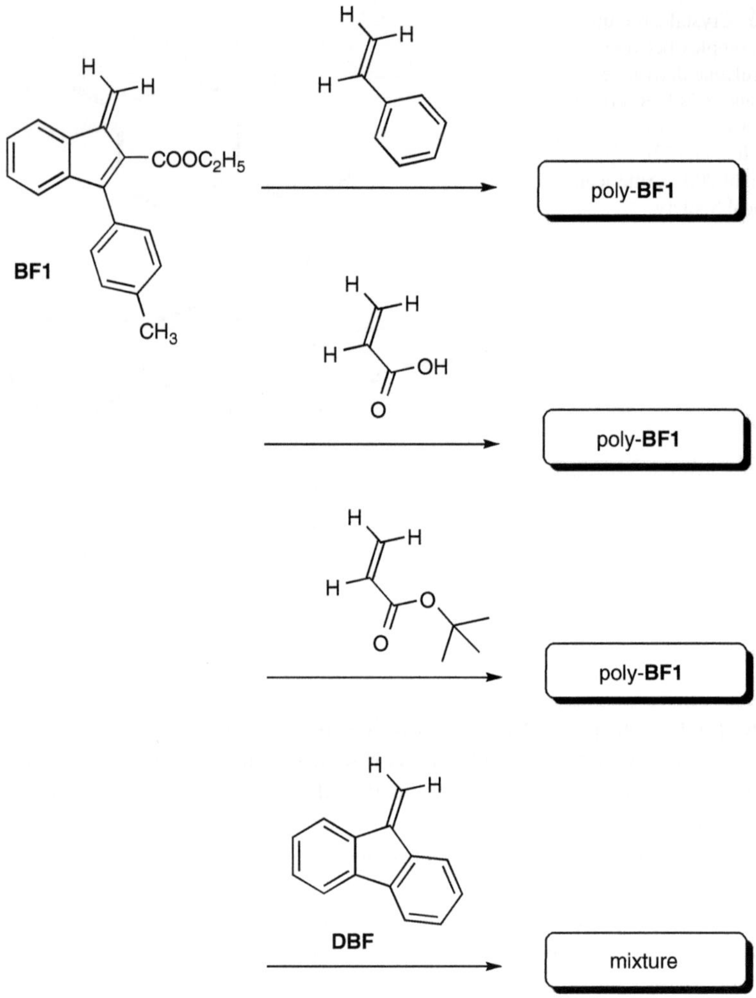

Scheme 17 Spontaneous polymerization of **BF1** in the presence of different vinyl monomers

of recognizing themselves among many other molecules with a surprising precision. In fact, the spontaneous polymerization of **BF1** was found to provide the corresponding homopolymer even if the mixture of the spontaneous polymerization reaction contained different vinyl monomers such as styrene, acrylic acid, *tert*-butyl acrylate, and **DBF** (Scheme 17).

A particular case is represented by **DBF** because MALDI-TOF analysis of the polymerization mixture revealed the presence of the two homopolymers along with a mixture of copolymers (unpublished results).

Therefore, the molecular recognition step seems to play a pivotal role in the monomer aggregation before spontaneous polymerization. We assume that the benzofulvene derivatives form columnar aggregates in concentrated solutions, the

Fig. 14 Comparison of the structure of closely related benzofulvene derivatives

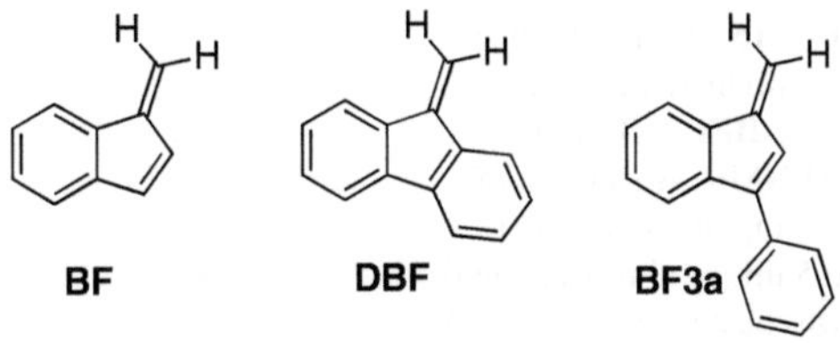

aggregates organize their architecture, and, when the inter-monomer distance becomes compatible with the formation of covalent bonds among the monomers, the spontaneous polymerization starts and goes on rapidly as a cascade.

In this manner, only the monomers capable of interacting with the benzofulvene molecules are included in the polymer. Moreover, selectivity has been shown also between polymerizing moieties present in the same monomer molecule as in the case of spontaneous polymerization of 6-AO-**BF3k** and 6-PO-**BF3k**.

Probably, the freely rotating phenyl ring plays a fundamental role in aggregation, structure organization, polymerization, and perhaps also in determining backbone features such as conformational preferences, stiffness, and stability. Interestingly, the simplest benzofulvene derivatives **BF** and **DBF** (Fig. 14) possess planar structures and are both crystalline solids, while we were unable to crystallize **BF3a** because it polymerized spontaneously. Most probably specific aromatic (T-shaped) interactions are important in governing the behavior of our 3-phenylbenzofulvene derivatives and of the corresponding polymers.

Thermoinduced Depolymerization

One of the most intriguing features shown by polybenzofulvene derivative is the thermoreversibility of the polymerization process. When the solutions of poly-**BF1** in the most common organic solvents showing different boiling points (i.e., ethyl acetate, toluene, and xylene) were heated to reflux, we observed both the depolymerization and the regeneration of the benzofulvene monomer **BF1** [20]. In other words, the solvated conditions led the polymer to release the monomer at relatively low temperatures, as it occurs in an equilibrium polymerization process, in agreement with the fact that the addition of a solvent to a monomer–polymer system increases the configurational entropy of the monomeric state, relative to that of the polymeric state, and lowers therefore the ceiling temperature [69]. The kinetics of the reaction appeared temperature dependent as the depolymerization was faster in xylene than in ethyl acetate. In order to develop a practical methodology to study the depolymerization phenomenon we decided to use ^{1}H-NMR spectroscopy, and owing to the relatively low speed of the process at low temperatures, we chose to perform the experiments after heating for different times in deuterated nitrobenzene at 140–150 °C. The first results obtained by means of this method were encouraging. In fact, the stacking of the ^{1}H-NMR spectra performed at different times in a

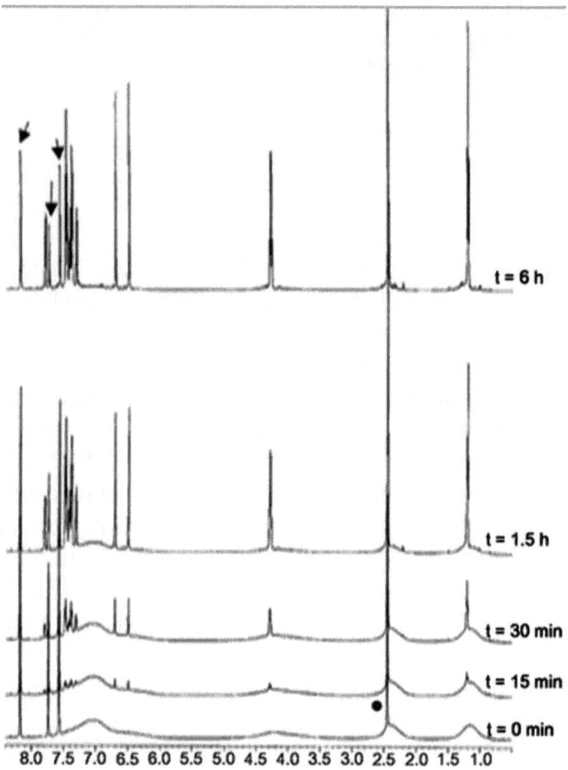

Fig. 15 Thermoinduced depolymerization of poly-**BF1**, followed by ^{1}H-NMR. A solution of 5.0 mg of the polymer in 0.5 mL of nitrobenzene-d_5 was heated at 140 °C and ^{1}H-NMR spectra (600 MHz) were recorded at regular time intervals. *Arrows* indicate the solvent (nitrobenzene) peaks and the *closed circle* indicates the water peak. Reprinted with permission from Cappelli et al. [20]. Copyright 2003, American Chemical Society

single figure (Fig. 15) clearly showed the progressive disappearance of the broad lines typical of the polymer as a consequence of the exposure to high temperature and the progressive increase in intensity of the sharp signals of benzofulvene monomer **BF1** [20].

Thus, the methodology and the way to show the results were considered sufficiently informative about the depolymerization process. Moreover, the integral values of the signals attributable to the vinylene group were compared with that of the lowest field signal of nondeuterated nitrobenzene (considered as the internal standard) in order to appreciate the variation in monomer concentration and to evaluate the trend of the depolymerization reaction. Very interestingly, the comparative study of the depolymerization process of poly-**BF3a,c–k** allowed the effects of the substituents in position 2 of indene nucleus in the trend of the depolymerization reaction to be evaluated [22].

Figure 16 shows that the depolymerization of poly-**BF3k** was nearly complete after 6 h heating at 150 °C (in agreement with the results obtained with closely related poly-**BF1**) and the heating for prolonged times led to the appearance of peaks attributed to products deriving from monomer decomposition (see $t = 24$ h trace).

By the contrast, the polymer bearing a hydrogen atom at the 2-position of the indene moiety (poly-**BF3a**) failed to show any significant depolymerization after

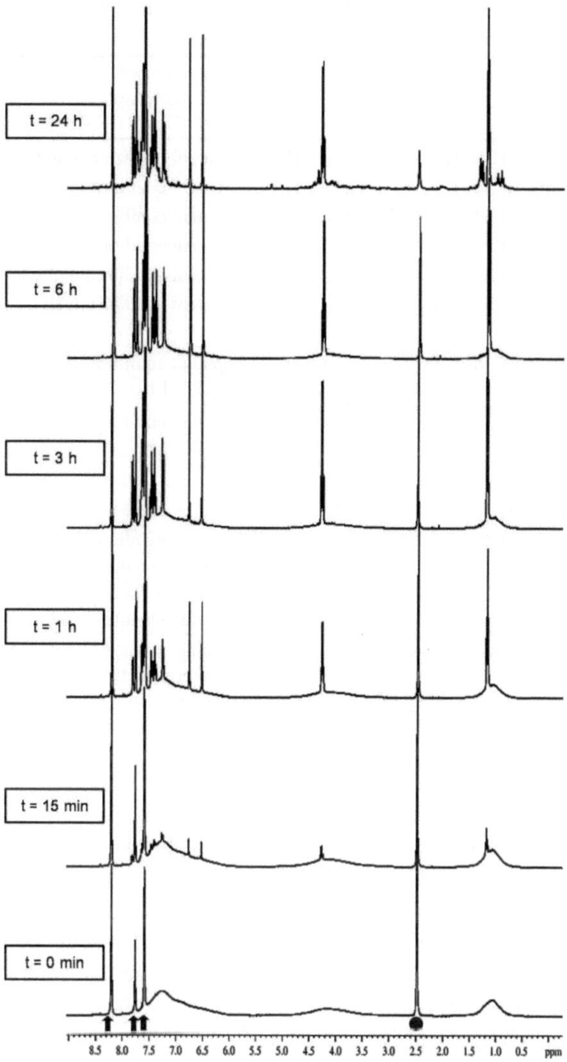

Fig. 16 Thermoinduced depolymerization of poly-**BF3k**, followed by ^{1}H-NMR. A solution of 5.0 mg of the polymer in 0.5 mL of nitrobenzene-d_5 was heated at 150 °C and ^{1}H-NMR spectra (400 MHz) were recorded at regular time intervals. The *arrows* indicate the solvent (nitrobenzene) peaks and the *closed circle* indicates the water peak. Reprinted with permission from Cappelli et al. [22]. Copyright 2007, American Chemical Society

24 h heating at 150 °C. Moreover, the polymers bearing a cyano group at the 2-position of the indene moiety (i.e., poly-**BF3g**,**h**,**i**) showed a particular behavior because their depolymerization was detectable already at room temperature, rapid at 150 °C, but it was not completed after 24 h heating at the same temperature. For the polymers bearing a halogen atom (e.g., poly-**BF3c**,**d**,**e**) or a methyl group in position 2 of the indene moiety (poly-**BF3f**) the thermoinduced depolymerization appeared to be slower and an apparent equilibrium was reached in times variable from 3 to 24 h (Table 4).

Poly-**BF1** was found to depolymerize also in the absence of solvents. Thermogravimetric analysis (TGA) showed that poly-**BF1** in solid state was not

Table 4 Comparison of the thermoinduced depolymerization of poly-**BF3a,c–k** with some significant DSC features

Polymer	R_1	R_2	R_3	Thermoinduced depolymerization at 150 °C by ^{1}H-NMR[a]	Peak n. (appearance)	Peak T (°C)	Peak ΔH (J/g)[b]
Poly-**BF3a**	H	H	H	No depolymerization after 24 h			
Poly-**BF3c**	F	H	H	Apparent equilibrium after 3 h	1 (broad)	202	24.4
Poly-**BF3d**	Cl	H	H	Apparent equilibrium after 6–24 h	1 (broad)	276	51.5
Poly-**BF3e**	Br	H	H	Apparent equilibrium after 3–6 h	1 (broad)	214	70.5
Poly-**BF3f**	CH_3	H	H	Apparent equilibrium after 6–24 h	1 (broad)	242	88.5
Poly-**BF3g**	CN	H	H	Apparent equilibrium after 0.25 h	2 (broad)	199	4.9
						244	35.1
Poly-**BF3h**	CN	CH_3	H	Apparent equilibrium after 0.25–1 h	2 (broad)	188	16.8
						242	37.1
Poly-**BF3i**	CN	H	CH_3	Apparent equilibrium after 0.25–1 h	3 (broad)	212	19.9
						251	–7.5
						327	16.3
Poly-**BF3j**	C≡C-2-Pyr	H	H	Apparent equilibrium after 1–3 h			
Poly-**BF3k**	$COOC_2H_5$	H	H	Depolymerization almost complete after 3–6 h	1 (sharp)	232	98.5
Poly-**BF1**	$COOC_2H_5$	CH_3	H	Depolymerization almost complete after 3–6 h	1 (sharp)	235	94.3

[a]In the thermoreversibility experiments samples of the polymers (5.0 mg) were dissolved into nitrobenzene-d_5 (0.5 mL) and heated in an oil bath thermostated at 150 °C (140 °C for poly-**BF1**); the ^{1}H-NMR experiments were performed at regular time intervals

[b]The positive values refer to the endothermic peaks, and the negative value is exothermic

subject to thermal degradation up to approximately 210 °C (weight loss lower than 2 %), but a fast thermal degradation started around 240 °C with a maximum of the derivative at about 250 °C (Fig. 17). Intriguingly, the comparison of TGA and DTGA (differential thermogravimetric analysis) curves [21] with the DSC (differential scanning calorimetry) curve (Fig. 18) of the same polymer suggested the existence of a relationship between the DSC endothermic peak at 235 °C and the DTGA peak at 240–250 °C. Therefore, the sharp endothermic peak in the DSC curve was attributed to the depolymerization process rather than to the polymer crystallinity [22].

In order to evaluate the effects of the stereoelectronic features of the substituents at the 2-position of indene nucleus in the depolymerization in the absence of solvents, a comprehensive DSC study was performed with the poly-**BF3a,c–k** series [22]. The DSC analysis showed that the thermal behavior of these polymers was affected by the substituent present at the 2-position of the indene nucleus. More precisely, poly-**BF3a** (R_1=H) showed a flat DSC curve, and the halogen-containing polymers (e.g., poly-**BF3c**, R_1=F; poly-**BF3d**, R_1=Cl; poly-**BF3e**, R_1=Br) showed broad endothermic peaks, while in the DSC curve of the polymer bearing a methyl group at the 2-position of the indene nucleus (poly-**BF3f**) the endothermic peak was more evident. The polymers bearing a cyano group (poly-**BF3g,h,i**) show rather complex DSC curves, which are characterized by the presence of two endothermic peaks. Finally, the DSC curve of poly-**BF3j** was even more complex and difficult to explain.

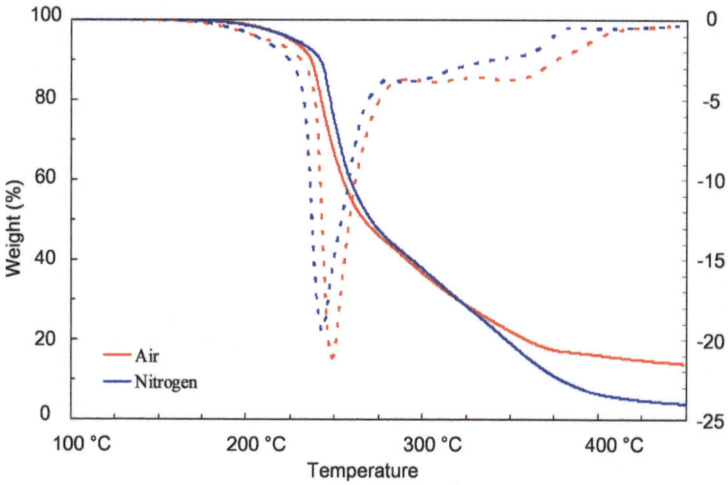

Fig. 17 TGA (*solid line*) and DTGA (*dashed line*) curves obtained with poly-**BF1** in air and nitrogen

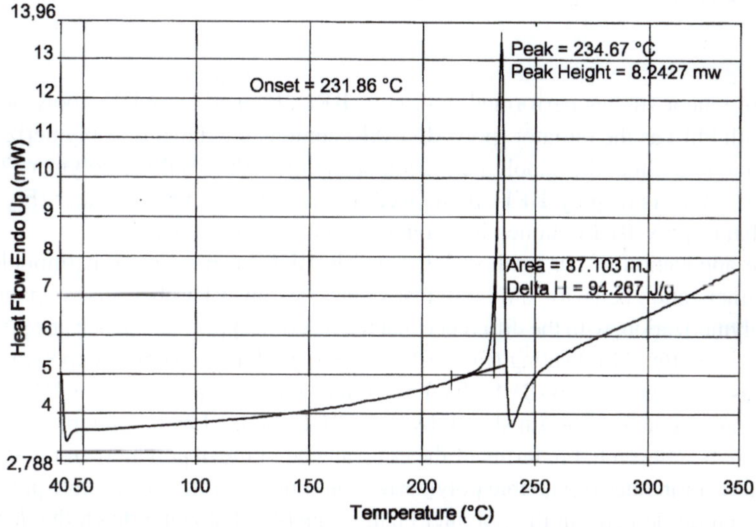

Fig. 18 DSC curve of poly-**BF1**. Reprinted with permission from Cappelli et al. [22]. Copyright 2007, American Chemical Society

The data shown in Table 4 suggested the existence of a relationship between the DSC curve and the thermoinduced depolymerization in nitrobenzene at 150 °C. In fact, in the polymers showing a sharp peak (characterized by a high ΔH value) in their DSC curve, the depolymerization is almost complete after 3–6 h heating, whereas the polymers showing an apparent equilibrium of the thermoinduced depolymerization in nitrobenzene at 150 °C are characterized by broad DSC peaks.

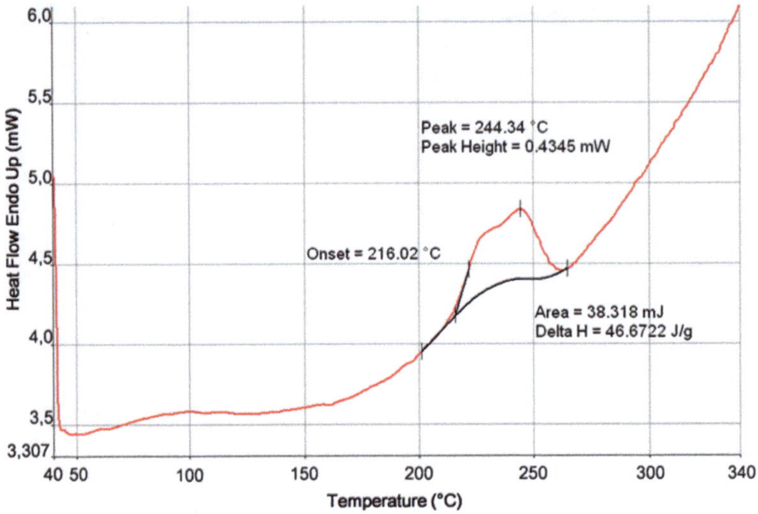

Fig. 19 DSC curve obtained with a sample of poly-**BF1**-AP. Reprinted with permission from Cappelli et al. [23]. Copyright 2008, American Chemical Society

Subsequent studies performed with poly-**BF1** obtained by anionic polymerization with PhLi as the initiator (poly-**BF1**-AP) demonstrated that the polymerization method could affect the depolymerization behavior of the resulting polymer [23].

DSC analysis of poly-**BF1**-AP showed an endothermic peak broader (Fig. 19) than that of poly-**BF1** obtained by spontaneous polymerization.

The application of our methodology to poly-**BF1**-AP revealed a thermoinduced depolymerization behavior similar to that shown by poly-**BF1** obtained by spontaneous polymerization, with the difference that the reaction appeared to reach an apparent equilibrium after 24 h heating (Fig. 20), whereas an almost complete depolymerization was observed for poly-**BF1**-SP after 3–6 h heating in nitrobenzene at 150 °C.

The apparent equilibrium shown by poly-**BF1** obtained by anionic polymerization may be interpreted in terms of the equivalence in the depolymerization and re-polymerization rates (reversible polymerization) but also with the possible presence of "structural defects" in the polymer chains capable of slowing down the depolymerization process. The existence of these "imperfections" was suggested by the comparison of the ^{13}C-NMR spectra of poly-**BF1** obtained in the different conditions (anionic vs. spontaneous polymerization).

We assumed that poly-**BF1**-AP is terminated at both chain ends. The phenyl group of the anionic initiator was believed to be fixed, whereas a relatively acidic proton linked to the indenyl moiety is present at the other end. If the propagation/depropagation reaction requires the presence of a negative charge in the terminal indenyl moiety, the reaction kinetics will depend on the concentration of the deprotonated chains and ultimately on the acid–base equilibrium. In deuterated nitrobenzene, the depolymerization rate is negligible at room temperature probably because

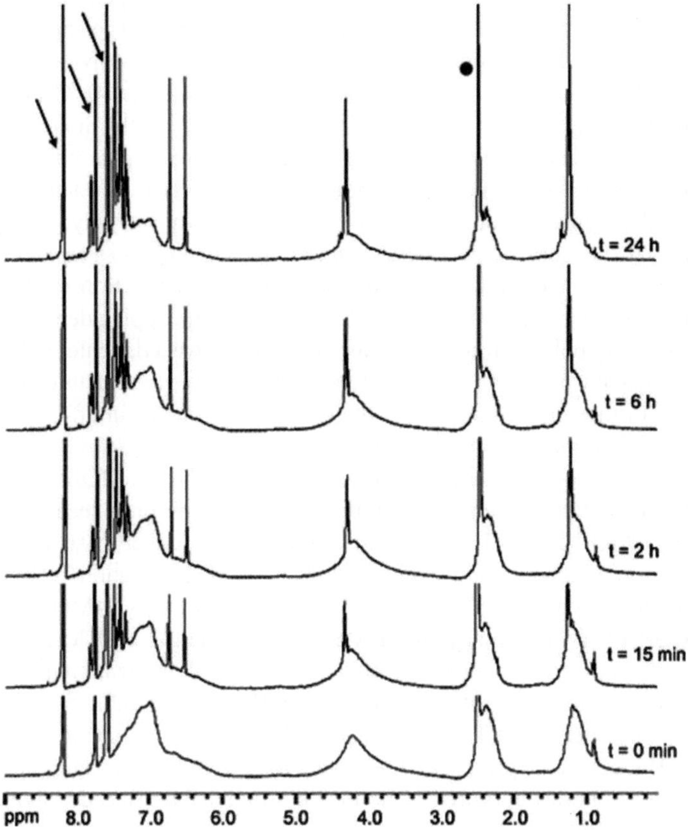

Fig. 20 Thermoinduced depolymerization of poly-**BF1** obtained by anionic polymerization. A solution of 5.0 mg of the polymer in 0.5 mL of nitrobenzene-d_5 was heated at 150 °C and ^{1}H-NMR spectra (400 MHz) were recorded at regular time intervals. In order to appreciate the variation in monomer concentration, the signals attributable to vinylene group were integrated and the integral values were compared with that of the lowest field signal of nondeuterated nitrobenzene, which was considered as an internal standard. Integrals are omitted in the figure for the sake of clarity. The *arrows* indicate the solvent peaks and the *closed circle* the water peak. Reprinted with permission from Cappelli et al. [23]. Copyright 2008, American Chemical Society

of the low amount of the deprotonated chain, while it increases with temperature probably because of the increase in proton mobility. In these conditions, repolymerization of **BF1** on cooling is expected to be a very slow process and spontaneous polymerization usually occurs at concentration values (e.g., >0.4 M in chloroform at room temperature) higher than that used in the thermoinduced depolymerization experiments.

Depolymerization experiments with a solution of poly-**BF1**-AP in deuterated benzene in the presence and in the absence of a base such as DBU showed that depolymerization at room temperature was negligible in the absence of DBU, while it was appreciable in the presence of the base. At room temperature an apparent equilibrium is reached after 5–6 days in the presence of DBU. The apparent equilibrium is characterized by the presence of a prominent amount of polymer with respect to the formed monomer. The depolymerization rate increased at 60 °C and an apparent equilibrium was reached after 6–9 h. A rough estimate of the monomer concentration at the apparent equilibrium at different temperatures supported the reversible nature of poly-**BF1**-AP depolymerization. Thus, poly-**BF1** appeared to show an unzipping depolymerization similar to the one shown by polycyanoacrylates. Indeed, alkyl cyanoacrylate polymers were described to undergo a rapid depolymerization at room temperature as a consequence of the abstraction of the terminal chain proton by base and the monomer produced by the unzipping depolymerization is simultaneously polymerized by the base initiator to give a daughter polymer [70].

The subsequent structure manipulation of benzofulvene structure in **BF1** and **BF3k** gave further information about the role of substituents in affecting the thermoinduced depolymerization behavior.

For instance, the replacement of ethyl group of **BF1** with MOEG chains composed of three (MOEG-3) or nine (MOEG-9) monomeric units leading to macromonomers 2-MOEG-3-**BF1** and 2-MOEG-9-**BF1**, respectively, was demonstrated to have negligible effects on the thermoinduced depolymerization behavior of the corresponding polymers [25]. In fact, when solutions of poly-2-MOEG-3-**BF1** or poly-2-MOEG-9-**BF1** in deuterated DMSO were heated at 150 °C and ^{1}H-NMR spectra were recorded at regular time intervals, the depolymerization trends (almost complete depolymerization after 3–6 h heating, Fig. 21 and Table 5) of both polymers were very similar to that shown by poly-**BF1** obtained by spontaneous polymerization. Therefore, the presence of relatively long MOEG side chains appended to the monomeric unities appeared to be unable to alter the usual depolymerization behavior shown by poly-**BF1**.

DSC analysis of poly-2-MOEG-9-**BF1**showed an endothermic peak (215 °C, 85 J/g) significantly broader than that shown by poly-**BF1** obtained by spontaneous polymerization, whereas TGA showed an initial small weight loss at about 215 °C and a quite sharp maximum of the derivative at 375 °C. The high value of DTGA peak was interpreted in terms of MOEG chain entanglement, which could hamper the monomer release leading to weight loss.

On the other hand, the insertion of substituents at the 6-position of the indene nucleus of **BF3k** produced significant variations in the thermoinduced depolymerization of the corresponding polymers (Table 6) that appeared related to the nature of the substituent [37, 52, 55].

The polymer bearing a methoxy group in the monomeric unit (poly-6-MO-**BF3k**) showed a relatively rapid depolymerization up to an apparent equilibrium (after about 8–24 h heating 150 °C) characterized by the presence of a substantial amount of the monomer. The modification of the methoxy group poly-6-MO-**BF3k** into MOEG chains of different length appeared to slow the depolymerization process progressively with the increase of the chain length.

These results appeared to find a good correlation with the results obtained by means of DSC analysis. Interestingly, the DSC curves of poly-6-MO-**BF3k**,

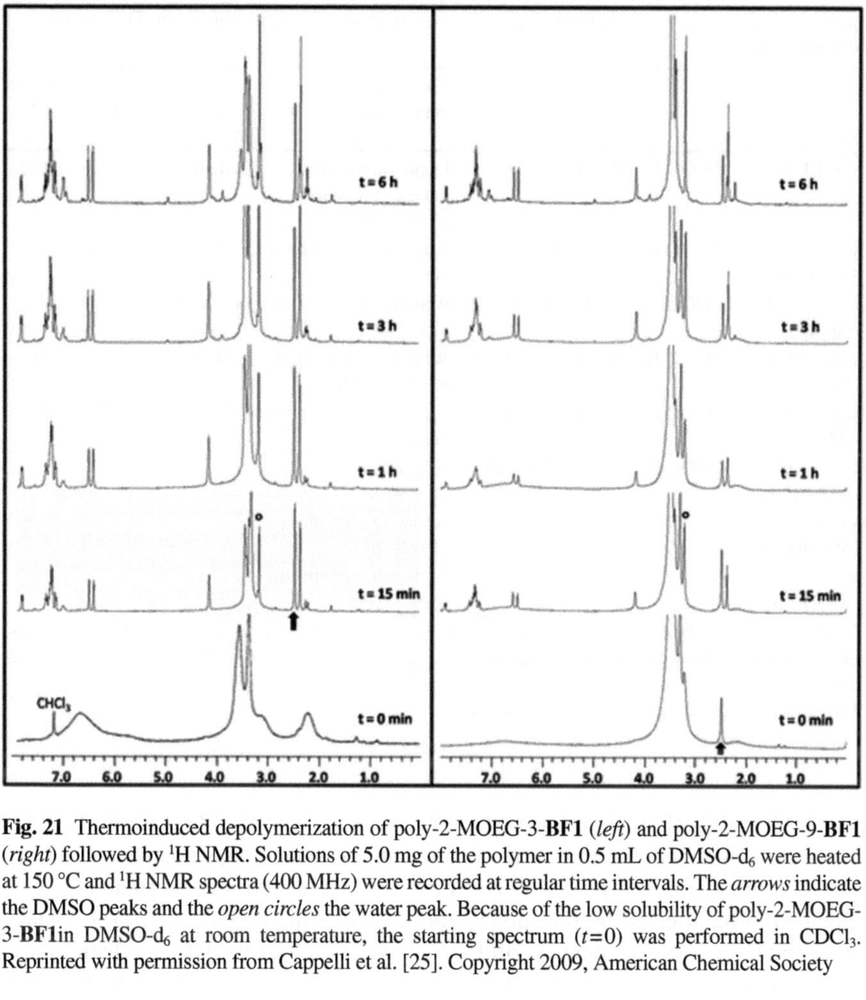

Fig. 21 Thermoinduced depolymerization of poly-2-MOEG-3-**BF1** (*left*) and poly-2-MOEG-9-**BF1** (*right*) followed by ^{1}H NMR. Solutions of 5.0 mg of the polymer in 0.5 mL of DMSO-d_6 were heated at 150 °C and ^{1}H NMR spectra (400 MHz) were recorded at regular time intervals. The *arrows* indicate the DMSO peaks and the *open circles* the water peak. Because of the low solubility of poly-2-MOEG-3-**BF1**in DMSO-d_6 at room temperature, the starting spectrum (t=0) was performed in $CDCl_3$. Reprinted with permission from Cappelli et al. [25]. Copyright 2009, American Chemical Society

Table 5 Comparison of the thermoinduced depolymerization with some significant DSC features of PEGylated poly-**BF1** derivatives

Polymer	R_1	R_2	R_3	Thermoinduced depolymerization at 150 °C by ^{1}H-NMR[a]	Peak n. (appearance)	Peak T (°C)	Peak ΔH (J/g)[b]
Poly-**BF1**	$COOC_2H_5$	CH_3	H	Depolymerization almost complete after 3–6 h	1 (sharp)	235	94
Poly-2-MOEG-3-**BF1**	COO $(C_2H_4O)_3CH_3$	CH_3	H	Depolymerization almost complete after 3–6 h			
Poly-2-MOEG-9-**BF1**	COO $(C_2H_4O)_9CH_3$	CH_3	H	Depolymerization almost complete after 3–6 h	1 (relat. sharp)	215	85

[a]In the thermoreversibility experiments samples of the polymers (5.0 mg) were dissolved into DMSO-d_6 (0.5 mL, nitrobenzene-d_5 for poly-**BF1**) and heated in an oil bath thermostated at 150 °C (140 °C for poly-**BF1**); the ^{1}H-NMR experiments were performed at regular time intervals

[b]The positive values refer to the endothermic peaks

Table 6 Comparison of the thermoinduced depolymerization with some significant DSC features of poly-**BF3k** derivatives

Polymer	R_1	R_2	R_3	Thermoinduced depolymerization at 150 °C by ^{1}H-NMR[a]	Peaks n. (appearance)	Peak T (°C)	Peaks ΔH (J/g)[b]
Poly-**BF3k**	$COOC_2H_5$	H	H	Depolymerization almost complete after 3–6 h	1 (sharp)	232	98.5
Poly-6-MO-**BF3k**	$COOC_2H_5$	H	OCH_3	Apparent equilibrium after 8–24 h	1 (relat. sharp)	231	52
Poly-6-MOEG-2-**BF3k**	$COOC_2H_5$	H	$O(C_2H_4O)_2CH_3$	Apparent equilibrium after 24 h	1 (broad)	224	40
Poly-6-MOEG-3-**BF3k**	$COOC_2H_5$	H	$O(C_2H_4O)_3CH_3$	Apparent equilibrium after 24–48 h	1 (broad)	226	34
Poly-6-MOEG-9-**BF3k**	$COOC_2H_5$	H	$O(C_2H_4O)_9CH_3$	Apparent equilibrium after 48 h	1 (broad)	224	20
Poly-6-MOEG-9-TM-**BF3k**	$COOC_2H_5$	H	$OCH_2(C_2HN_3)(C_2H_4O)_9CH_3$	Apparent equilibrium after 72 h			
Poly-4,6-MO-**BF3k**				Apparent equilibrium after 8–24 h			
Poly-4,5,6-MO-**BF3k**				Apparent equilibrium after 8–24 h			

[a]In the thermoreversibility experiments samples of the polymers (5.0 mg) were dissolved into nitrobenzene-d_5 (0.5 mL) and heated in an oil bath thermostated at 150 °C; the ^{1}H-NMR experiments were performed at regular time intervals

[b]The positive values refer to the endothermic peaks

poly-6-MOEG-2-**BF3k**, poly-6-MOEG-3-**BF3k**, and poly-6-MOEG-9-**BF3k** (Fig. 22) showed a single endothermic peak centered in the same temperature range (224–231 °C) measured for parent poly-**BF3k** (232 °C), and the thermal behavior appeared to be affected by the substituent present at the 6-position [37]. In detail, the thermograms of the polybenzofulvene derivatives bearing MOEG chains show a broad endothermic peaks, which was sharper in the DSC trace of the polymer bearing a methoxy group (poly-6-MO-**BF3k**) and was even sharper in the case of unsubstituted poly-**BF3k**. The heat flows involved in the unzipping reaction were found to fall in the range 20–52 J/g and were significantly lower with respect to unsubstituted poly-**BF3k** (98.5 J/g). This result was rationalized in terms of polymer stiffness. Poly-**BF3k**, showing the highest stiffness, was believed to require the greatest amount of energy to depolymerize. The insertion of the methoxy substituent of poly-6-MO-**BF3k** and the MOEG side chains produces both a change in the backbone arrangement and a decrease of the depolymerization heat as well as a progressive broadening of the DSC peak.

TGA analysis suggested a complex decomposition pathway for poly-**BF3k** derivatives reported in Table 7. The DTGA curves showed that the decomposition process occurred in different steps.

Poly-6-MO-**BF3k**, bearing a methoxy substituent, shows the lower T_i (the onset of thermal decomposition or the temperature corresponding to initial 1 % weight loss)

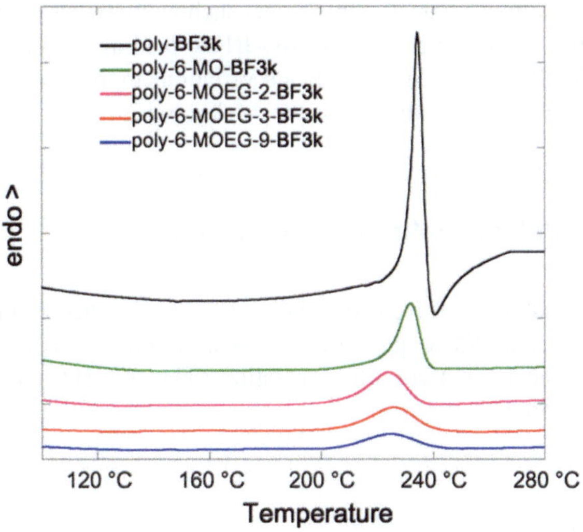

Fig. 22 DSC curves of poly-**BF3k** (*black*), poly-6-MO-**BF3k** (*green*), poly-6-MOEG-2-**BF3k** (*magenta*), poly-6-MOEG-3-**BF3k** (*red*), and poly-6-MOEG-9-**BF3k** (*blue*). Reprinted with permission from Cappelli et al. [37]. Copyright 2010, Wiley Periodicals, Inc.

Table 7 Comparison between DSC and TGA features of poly-**BF3k** derivatives

Polymer	R_1	R_2	R_3	DSC T_p (°C)	TGA T_i (°C)	TGA T_{max1} (°C)	TGA T_{max2} (°C)
Poly-**BF3k**	$COOC_2H_5$	H	H	232	191	243	310
Poly-6-MO-**BF3k**	$COOC_2H_5$	H	OCH_3	231	193	246	380
Poly-6-MOEG-2-**BF3k**	$COOC_2H_5$	H	$O(C_2H_4O)_2CH_3$	224	218	293	377
Poly-6-MOEG-3-**BF3k**	$COOC_2H_5$	H	$O(C_2H_4O)_3CH_3$	226	252	322	428
Poly-6-MOEG-9-**BF3k**	$COOC_2H_5$	H	$O\ (C_2H_4O)_9CH_3$	224	260	402[a]	

[a]Single broad decomposition step

value, which was slightly higher than that obtained with parent poly-**BF3k**. For the polymers containing MOEG side chains, the T_i was found to increase with the increase of the side chain length. Interestingly, a similar trend was observed for the T_{max} (the temperature at which the weight loss rate is maximum) value. The presence of MOEG in the side chain produced a noticeable shift of the DTG peaks towards higher temperatures compared to poly-6-MO-**BF3k**. In particular, the T_{max} value increases as a function of the side chain length. Similarly to PEGylated poly-BF1 derivatives, the high value of DTGA peaks could be explained in terms of MOEG chain entanglement, which may hamper the monomer release.

Moreover, the results of the thermal analysis suggest that during the TGA experiments a variable amount of the formed monomers decompose before their release in the gas stream and that the amount depends on the length of MOEG side chain, which in turn affects the monomer release temperature.

Finally, the insertion of small substituents such as methoxy groups in the benzene ring of the indene nucleus of 6-MO-**BF3k** leading to 4,6-MO-**BF3k** and 4,5,6-MO-**BF3k** did not produce significant variations in the thermoinduced depolymerization of the corresponding polymers (Table 6) [55].

Structure of Polybenzofulvene Derivatives

The structure of polybenzofulvene derivatives has been studied by means of different techniques such as NMR spectroscopy, absorption and emission spectroscopy, and matrix-assisted laser desorption ionization time of flight mass spectrometry (MALDI-TOF MS).

NMR Spectroscopy

Intensive research efforts have been devoted to the NMR characterization of the new benzofulvene polymers in order to determine the nature of some general and peculiar aspects. In particular, poly-**BF1** was studied in details by means of NMR spectroscopy in order to confirm its polymeric nature and to determine its structure. Moreover, an extensive series of polybenzofulvene derivatives have been studied to determine, for each of them: (1) the vinyl nature of the polymers and the presence of aromatic stacking interactions; (2) the role of the electron-withdrawing groups (or donors) present on the benzofulvene fragment in affecting the macrostructure; (3) the influence of a sterically demanding substituents on the conformational behavior.

Firstly, the polymeric nature of poly-**BF1** was assessed based on the following experimental evidence [20].

The ^{1}H-NMR spectrum of poly-**BF1** showed broad signals in the place of the sharp ones typical of **BF1**.

The signals of vinyl $=CH_2$ group of **BF1** (two singlets at 6.35 and 6.55 ppm) became a broad signal (at about 3.2 ppm) attributed to a $-CH_2$ bridge of poly-**BF1** (Fig. 23). Moreover, two ^{13}C-NMR signals (a singlet at 57.5 ppm assigned to a quaternary carbon and a broad signal centered at about 48 ppm assigned to the methylene bridge Fig. 24) appeared in the aliphatic region of poly-**BF1** spectrum to replace the vinylene carbon signals of **BF1** (116.9 and 144.2 ppm). Thus, the conversion of **BF1** vinylene group into a methylene bridge strongly supported the polymeric nature of poly-**BF1** [20].

In principle, two structures are possible for the polybenzofulvene derivatives, depending on the polymer enchainment: 1,2- (vinyl polymer) or 1,4- (diene polymer). Figure 25 shows that the vinyl or diene polymerizations of **BF1** produce different polymer backbones.

In order to discriminate between the vinyl and the diene polymer nature, a combined approach was developed as crucial feature of the study involving poly-**BF1**.

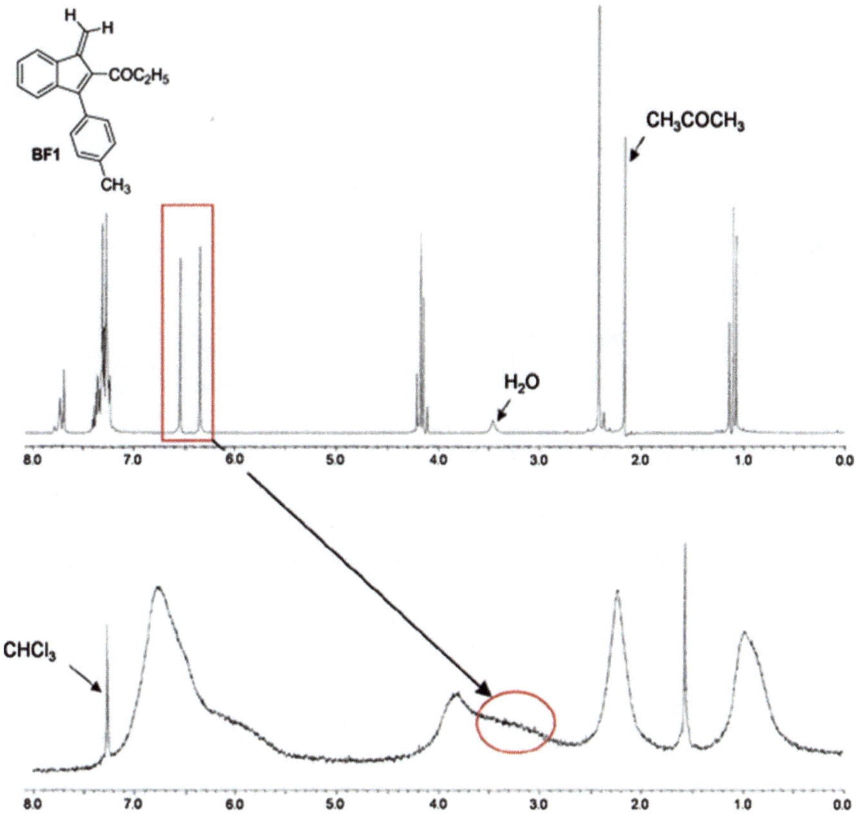

Fig. 23 Comparison of ^{1}H-NMR spectrum (200 MHz) of **BF1** (*top*) with that of poly-**BF1** (*bottom*). Adapted with permission from Cappelli et al. [20]. Copyright 2003, American Chemical Society

The strategy included (1) the synthesis of model molecules representative of the monomeric units in the two different enchainments (1,2- and 1,4); (2) an in-depth characterization by mono- and two-dimensional NMR spectroscopy of model compounds and polymers; (3) analysis of the chemical shifts by means of different approaches [21]:

1. The hydrogenation of **BF1** afforded two model molecules (Scheme 18). Ethyl 1-methyl-3-(4-methylphenyl)-1*H*-indene-2-carboxylate (**1,2-model**) was derived from the hydrogenation at the 1- and 2-positions of the diene fragment and retained the endocyclic double bond in the original position. Ethyl 3-methyl-1-(4-methylphenyl)-1*H*-indene-2-carboxylate (**1,4-model**) was derived from the hydrogenation at the 1- and 4-positions of the diene fragment and showed a double bond shift [21].
2. The structures of 1,2- and 1,4-model compounds were determined by crossing 1D and 2D NMR results. The analysis of ^{13}C-NMR spectra of the model compounds

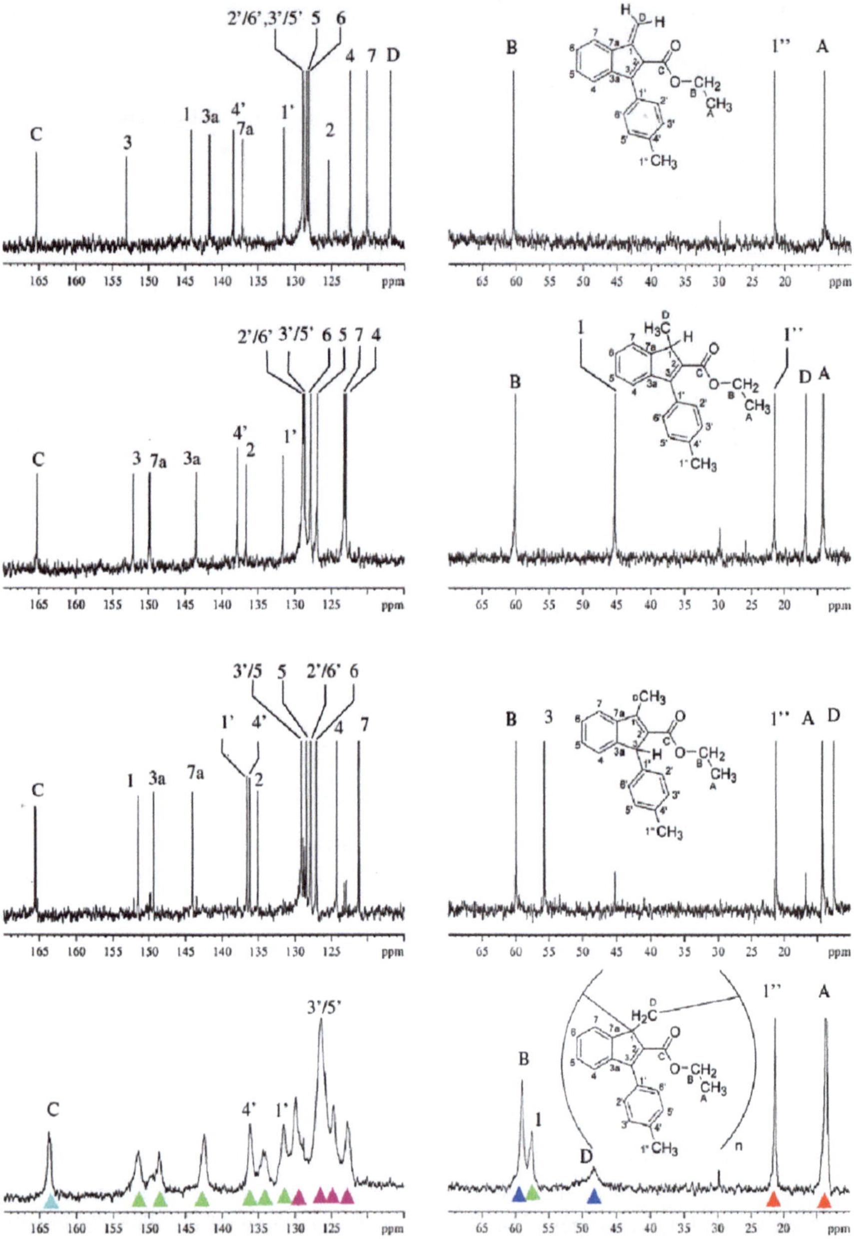

Fig. 24 Comparison of the ^{13}C-NMR (150 MHz, $CDCl_3$) spectra of **BF1**, **1,2-model**, **1,4-model**, and poly-**BF1**. In the poly-**BF1** spectrum the *cyan arrow* indicates the signal attributable to the carbonyl carbon, the *green arrows* those attributable to the quaternary carbons, the *violet arrows* those attributable to aromatic CH, the *blue arrows* the CH_2 signals, and the *red arrows* the CH_3 signals. Reprinted with permission from Cappelli et al. [21]. Copyright 2005, Wiley Periodicals, Inc.

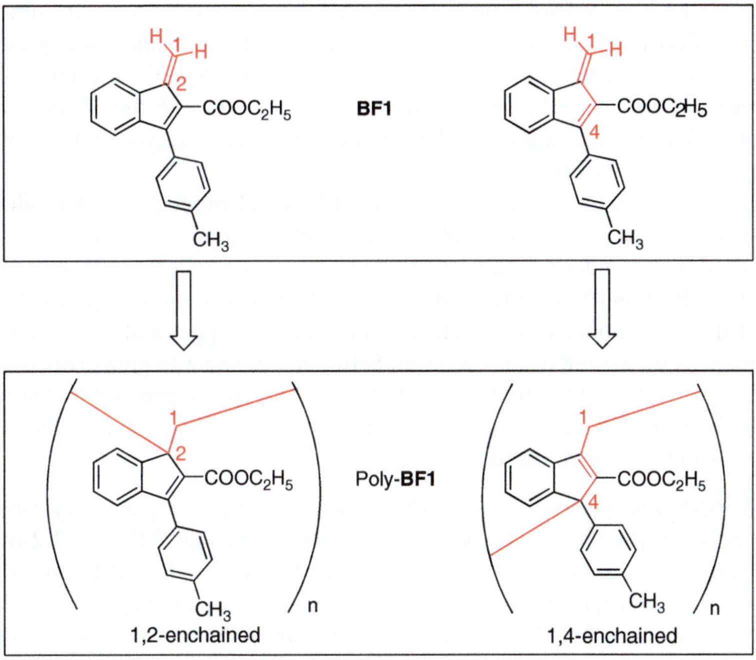

Fig. 25 Polymerization of **BF1** in the two possible enchainments

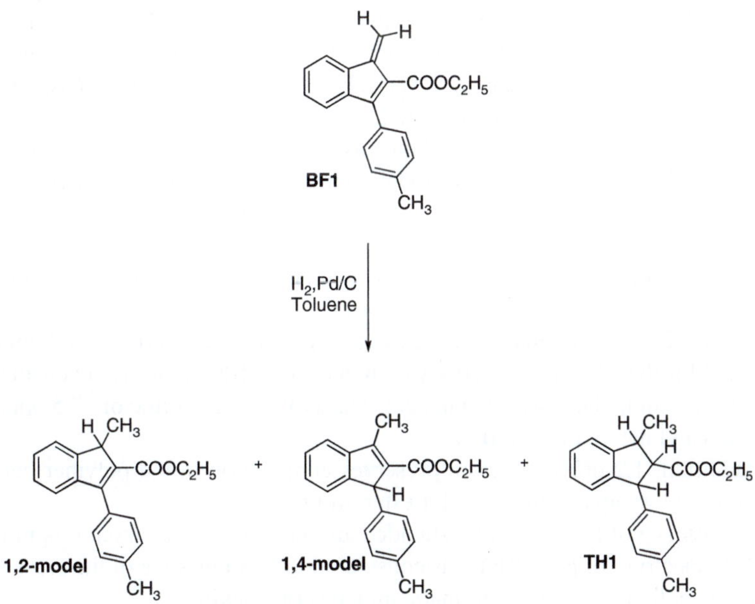

Scheme 18 Catalytic hydrogenation of **BF1** to obtain the two model molecules

showed 18 well-defined signals as expected, having resonances quite similar in terms of chemical shifts. A key role for the structure assignment was played by the long-term correlations in the HMBC experiments. Specifically, the aromatic *ortho* protons (H2′/6′) correlate, through three bonds, with the olefinic carbon atom C3 in the 1,2-structure, while in the 1,4-structure they correlate with the saturated carbon atom C3.

3. The comparison of the ^{13}C-NMR spectra of **BF1**, **1,2-model,** and **1,4-model** suggested that the signal attributable to the carbon atom of the pendent phenyl bonded to the indene moiety (C-1′, Fig. 24) showed a chemical shift almost equal in **BF1** and **1,2-model** (131.5 and 131.6 ppm, respectively), while the C-1′ signal of **1,4-model** was downfield shifted at 136.5 ppm and no signals were present in the region of the spectrum between 130 and 135 ppm. This observation was explained by the fact that C-1′ is bonded to unsaturated C-3 atoms in both **BF1** and **1,2-model**, while in the **1,4-model** the double bond is shifted and C-1′ is linked to a saturated C-3 atom.

The ^{13}C-NMR spectrum of poly-**BF1** showed a signal of a quaternary carbon at 131.5 ppm, in analogy to what was observed in compounds **BF1** and **1,2-model**. This signal was assigned to C-1′ on the basis of HSQC and HMBC experiments since a connection path among C-1″, H-3′/5′, and C-1′ was observed. Therefore, the double bond was assumed to retain its position during the polymerization supporting the vinyl structure (1,2-enchainment) for poly-**BF1**.

Our strategy was comprehensive of chemical shift calculations based on the additive rules by Grant and Paul. Even though the additive method was expected to give large deviations between the experimental and predicted chemical shift values when applied to highly substituted and sterically hindered molecules, the calculated values for the 1,2-enchainment were very close to the experimental ones, confirming once again a vinyl polymer structure for poly-**BF1**. More specifically, the chemical shift of the methylene bridge carbon was calculated starting from the one of the methyl groups of the 1,2- and 1,4-model compounds (16.7 and 12.5 ppm, respectively). The calculated value for the 1,2-enchainment was 45.6 ppm, very similar to the experimental value of 48 ppm, while the one calculated for the 1,4-enchainment was 30.2 ppm, rather different from the experimental value. Similarly, starting from the chemical shift of the quaternary carbons C1 (45.3 ppm) and C3 (55.7 ppm) of the 1,2- and 1,4-model compounds, respectively, a chemical shift of 54.2 ppm was calculated for the aliphatic quaternary carbon of poly-**BF1** in the 1,2-enchainement and of 63.1 ppm in the 1,4-enchainement. The experimental value of 57.5 ppm supports the vinyl nature of poly-**BF1**.

In addition, 1,2-polymerization preserves conjugation in the polymer between the donor aryl group and the acceptor ester group.

The presence of two unusually shielded aromatic protons at about 6 ppm in the ^{1}H-NMR spectrum of poly-**BF1** was considered to be suggestive of the presence of strong interactions among the aromatic moieties (π-stacking).

In order to define if the stereoelectronic nature and/or the steric hindrance of the substituents present on the indene moieties affects the polymer microstructure, different R_1, R_2, and R_3 substituted benzofulvene derivatives were prepared and submitted to a comprehensive NMR study (^{1}H, ^{13}C, and HSQC experiments on the whole set of the polymers) [22].

These studies evidenced that the macromolecular structure of some newly synthesized polymer was rather complex, with the possible presence of enchainment, configurational, and/or conformational isomers. Steric hindrance in proximity of the reactive vinylene group of monomer appeared to play a discriminating role in determining the macromolecular structure of the resulting polymers.

Poly-**BF3k** was found to show the simplest NMR features (indicating good structural homogeneity), which were very similar to the ones shown by poly-**BF1**. This result was consistent with the fact that **BF3k** is the 4′-desmethyl derivative of **BF1** and suggested that the 4′-methyl did not affect the polymer structure. The ^{13}C-NMR spectra of both polymers showed a single peak attributable to the aliphatic quaternary carbon with a chemical shift value (about 58 ppm) very similar to the one calculated for poly-**BF1** in the 1,2-enchainment (54.2 ppm, Table 8). Interestingly, the broad peak assigned to the backbone methylene bridge was resolved in the spectra of both polymers into two broad peaks (ca. 48 and 51 ppm). The different intensity of these broad peaks suggested the contemporary presence of a minor species (m) with a major one (M) and the possible existence of configurational or conformational isomers.

The replacement of the bulky electron-withdrawing group $-COOC_2H_5$ of poly-BF1 and poly-**BF3k** with a small donor methyl as for poly-**BF3f** produced a significant change in the polymer structure. In fact, poly-**BF3f** backbone was assumed to present two different enchainments (1,2- and 1,4-) as suggested by the splitting of the quaternary C1 signal into two peaks having a chemical shift (57.4 and 61.0 ppm) very close to that calculated for poly-**BF1** in the two possible enchainments (54.2 and 63.1 ppm). In other words, some monomeric units of poly-**BF3f** retained the indene double bond in the original position, while in some others it could be shifted as a consequence of a diene (1,4) polymerization.

The introduction of a nitrile group or of a halogen atom at the 2-position of indene nucleus was again more critical in altering the polymer structure. The number of the peaks attributable to aliphatic quaternary carbon and their chemical shift suggested the presence in poly-**BF3g,h** (showing a nitrile group) and poly-**BF3c,d,e** (showing a halogen atom) of enchainment, configurational, and/or conformational isomers.

The maximum of the complexity was observed in poly-**BF3a** (R_1=H), which showed two broad peaks attributable to aliphatic quaternary carbon, while the signals attributable to backbone methylene were dispersed in a wide range of chemical shifts.

A special case was represented by poly-**BF3i** where the small electron-withdrawing CN group at the 2-position of the indene nucleus was associated to the presence of a methyl group at the 6-position. The evident analogy of its NMR features with those of the polymers bearing a bulkier electron-withdrawing ester group (poly-**BF1** and poly-**BF3k**) suggested for the methyl group in the proximity of the reactive exocyclic methylene group in **BF3i** monomer a role in inhibiting diene (1,4) polymerization.

Table 8 Significant NMR features of the poly-**BF3** derivatives

Polymer	R_1	R_2	R_3	Aliphatic quaternary ^{13}C-NMR n.[a]	Aliphatic quaternary ^{13}C-NMR ppm[b]	Backbone CH_2 HSQC n.[c]	Backbone CH_2 ^{13}C-NMR ppm[d]
Poly-**BF3a**	H	H	H	2 broad	55.4, 57.4	>6	33.0–61.0
Poly-**BF3c**	F	H	H	3	49.7(M) 50.9(m) 54.0(m)	5	32.8 40 42.7 44.5 46.1
Poly-**BF3d**	Cl	H	H	4	56.6(M) 56.8(M) 57.7(m) 59.9(m)	5	37.2 38.8 42 44.5 47.7
Poly-**BF3e**	Br	H	H	2	58.1(M) 61.0(m)	4	39.3 42.8 45.8 48.1
Poly-**BF3f**	CH_3	H	H	2	57.4 61	2	30 37.2
Poly-**BF3g**	CN	H	H	4	56.5(M) 57.7(m) 59.6(m) 61.0(m)	5	37.6 39.6 42.4 45.5 48.6
Poly-**BF3h**	CN	CH_3	H	3	56.3(M) 57.4(m) 59.5(m)	4	37.4 42.2 45.9 48.6
Poly-**BF3i**	CN	H	CH_3	1	56.1	2	46.1 49.1
Poly-**BF3j**	C≡C-2-Pyr	H	H	2	57.1(M) 60.9(m)	5	39.2 40 42.2 47.3 49.8
Poly-**BF3k**	$COOC_2H_5$	H	H	1	57.8	NO[e]	48.4(M) 50.8(m)
Poly-**BF1**	$COOC_2H_5$	CH_3	H	1	57.5	NO[e]	47.9(M) 50.8(m)
Poly-**BF3ma**	COOH	CH_3	H	1	57.3	NO[e]	48.2
Poly-**BF2**				[15]	57.8		43.8
Poly-**BF1**				Calcd for 1,2 polymerization [21]	54.2		45.6
Poly-**BF1**				Calcd for 1,4 polymerization [21]	63.1		30.2

[a]Number of ^{13}C-NMR signals attributable to aliphatic quaternary carbon at the 1-position of the indene moiety
[b]The ppm value of the ^{13}C-NMR signal(s) attributable to the aliphatic quaternary carbon
[c]Number of cross-peaks attributable to backbone methylene in the HSQC spectrum
[d]The ppm value of the signal(s) attributable the backbone methylene ^{13}C
[e]NO: not observed

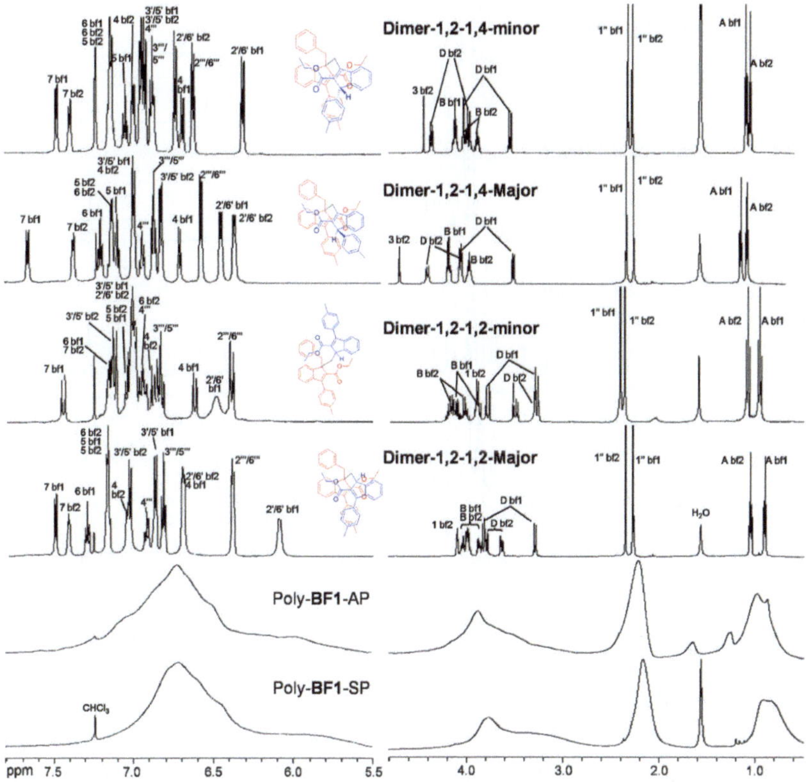

Fig. 26 Comparison of the ^{1}H-NMR spectra of dimers and poly-**BF1** obtained in different conditions. Reprinted with permission from Cappelli et al. [23]. Copyright 2008, American Chemical Society

Subsequently, the anionic polymerization of **BF1** with PhLi as the initiator allowed us to isolate and characterize monomer and dimer species and the corresponding polymer (poly-**BF1**-AP). A careful study performed by NMR spectroscopy on the monomers and dimers, considered as model compounds, was of a great help in the understanding of the poly-**BF1** structure, also providing evidence on π-stacked nature for the polymer [23].

Dimers showed rather complex NMR spectra, but the combined use of COSY, NOESY, HSQC, and HMBC experiments allowed the assignment of the ^{1}H and ^{13}C-NMR spectra of the four dimers isolated from the polymerization mixture to be carried out (Figs. 26 and 27). In the strategy of spectral assignment, the benzyl moiety deriving from the PhLi initiator was used as the starting point.

The analysis of the chemical shift values and NOESY spectra of dimers provided important information about the tendency to accommodate their bicyclic moiety in an antiparallel stacked orientation (Figs. 26 and 27). This observation was considered to be very important for **Dimer-1,2-1,2-Major** in which the tetrahedral geometry of both the indene C-1 atoms appeared to force the indene moieties to stack

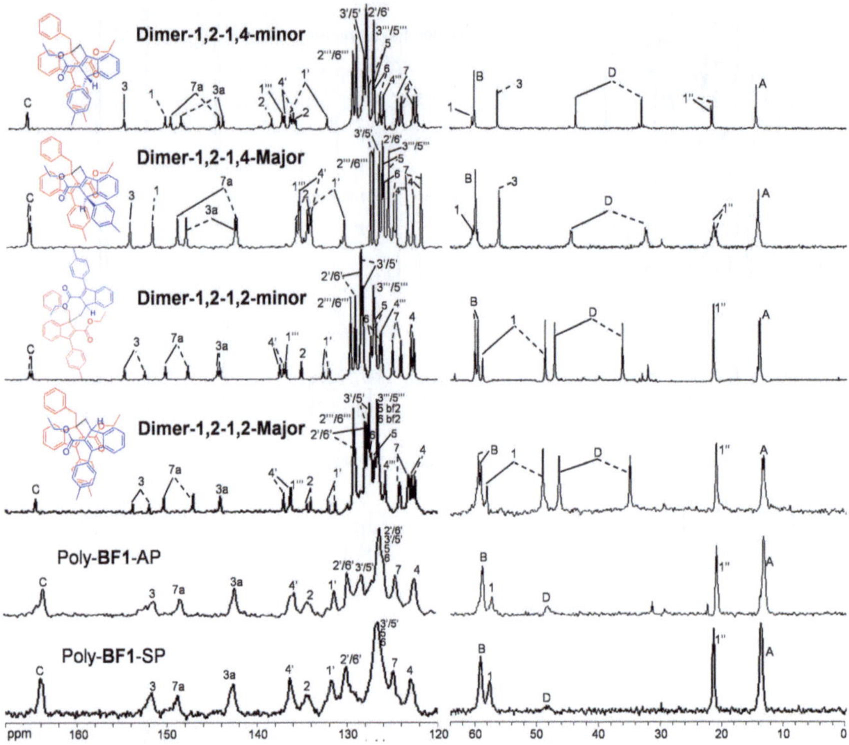

Fig. 27 Comparison of the ^{13}C-NMR spectra of dimers and poly-**BF1** obtained in different conditions. In the spectra of dimmers, *solid lines* indicate the signals attributable to the *bf1* moiety of dimers, while *dashed lines* refer to the *bf2* moiety; when this distinction is not indicated, signals are overlapped. Reprinted with permission from Cappelli et al. [23]. Copyright 2008, American Chemical Society

closely each other and the pendent phenyl groups to interact with the six-member ring of the indene nucleus (or with each other) by means of T-shaped aromatic–aromatic interactions. This assumption was supported by the comparison of the chemical shifts and NOESY contacts for major and minor isomers of the 1,2-1,2 dimer.

More specifically, the chemical shift analysis shows the following evidence:

1. 2′/6′-*bf*1 protons of both species of **Dimer-1,2-1,2-Major** were upfield shifted with respect to the corresponding 2′/6′-*bf*2 protons, indicating that the benzyl group linked to *bf*1 induced a shielding effect on these protons owing to magnetic anisotropy.
2. H-2′/6′-*bf*1 and H-2′/6′-*bf*2 of **Dimer-1,2-1,2-Major** were strongly upfield shifted (0.45 and 0.32 ppm, respectively) with respect to the corresponding H-2′/6′-*bf*1 and H-2′/6′-*bf*2 of **Dimer-1,2-1,2-minor**, indicating these protons of **Dimer-1,2-1,2-Major** were affected by an additional shielding field, owing to magnetic anisotropy of the indene ring.

3. The chemical shift values of 3′/5′-*bf*1 and 3′/5′-*bf*2 protons of **Dimer-1,2-1,2-Major** were less affected (0.15 and 0.10 ppm, respectively) by this shielding effect (with respect to the corresponding **Dimer-1,2-1,2-minor** protons), because of their distal position from the indene ring and benzyl anisotropic cones.

The cross-peaks observed in the NOESY spectrum for the methyl protons of the ethyl group of **Dimer-1,2-1,2-Major** were considered to be diagnostic. In particular, A-*bf*1 and A-*bf*2 protons showed symmetric intra- and inter-residue cross-peaks. Namely, H-A-*bf*1 interacted with both H-2′/6′-*bf*1 and H-3′/5′-*bf*1 (intra-residue) and with H-6-*bf*2 and H-7-*bf*2 (inter-residue) and H-A-*bf*2 interacted with H-2′/6′-*bf*2 and H-3′/5′-*bf*2 (intra-residue) and with H-6-*bf*1 and H-7-*bf*1 (inter-residue). Overall, all these observations supported the hypothesis of an antiparallel stacked orientation of indene moieties of **Dimer-1,2-1,2-Major** with a further stabilization due to "symmetric" (pendant)phenylindene interactions.

The NOESY spectrum of dimer **Dimer-1,2-1,2-minor** was found to feature the interaction between H-1-*bf2* and H-7-*bf1* and the absence of dipolar interactions between ester CH_2–CH_3 and the indene hydrogen atoms. The finding suggested for **Dimer-1,2-1,2-minor** a peculiar arrangement of the two benzofulvene units due to the "isotactic configuration."

Finally, the comparison of the ^{13}C-NMR spectra reported in Fig. 27 demonstrated that the anionic polymerization of **BF1** in the presence of PhLi as the initiator afforded a polymer showing NMR features very similar to those shown by poly-BF1 obtained by spontaneous polymerization. Subtle differences involving the signals attributed to carbon atoms belonging to the pendent phenyl moiety were interpreted in terms of possible existence of isomers of poly-**BF1**-AP showing differences in the environment of the pendent phenyl group. Moreover, the similarities existing between the spectra of poly-**BF1** and those of **Dimer-1,2-1,2-minor** confirmed again the vinyl nature of this polymer and suggested that this dimer could be considered a good candidate as the dimeric unit of this polymer.

The series of benzofulvene derivatives polymerized through spontaneous reaction was then enriched by introducing MOEG side chains having different length in place of the ethyl group of **BF1** to produce 2-MOEG-3-**BF1** and 2-MOEG-9-**BF1** [25].

In order to evaluate if the oligomeric PEG side chains play a role on the polymer structure, a comparative NMR study was performed. The analysis of the ^{13}C-NMR spectra of poly-2-MOEG-3-**BF1** and poly-2-MOEG-9-**BF1** (with that poly-**BF1** obtained by spontaneous polymerization as the reference, Fig. 28) evidenced a stringent similarity in terms of chemical shift values. The aromatic regions of poly-2-MOEG-3-**BF1** and poly-**BF1** spectra were practically indistinguishable and the peaks attributed to carbon atoms C1 and D, the carbon atoms forming the polymer backbone, showed similar chemical shift values. Thus, the presence of the oligomeric PEG side chain appeared to be compatible vinyl (1, 2) polymerization of macromonomers 2-MOEG-3-**BF1** and 2-MOEG-9-**BF1**.

Some differences were observed in the region containing signals of polybenzofulvene backbone and were relative to an increased line broadening. This effect on the ^{13}C spectra was attributed to the different mobility of the polymer backbone that

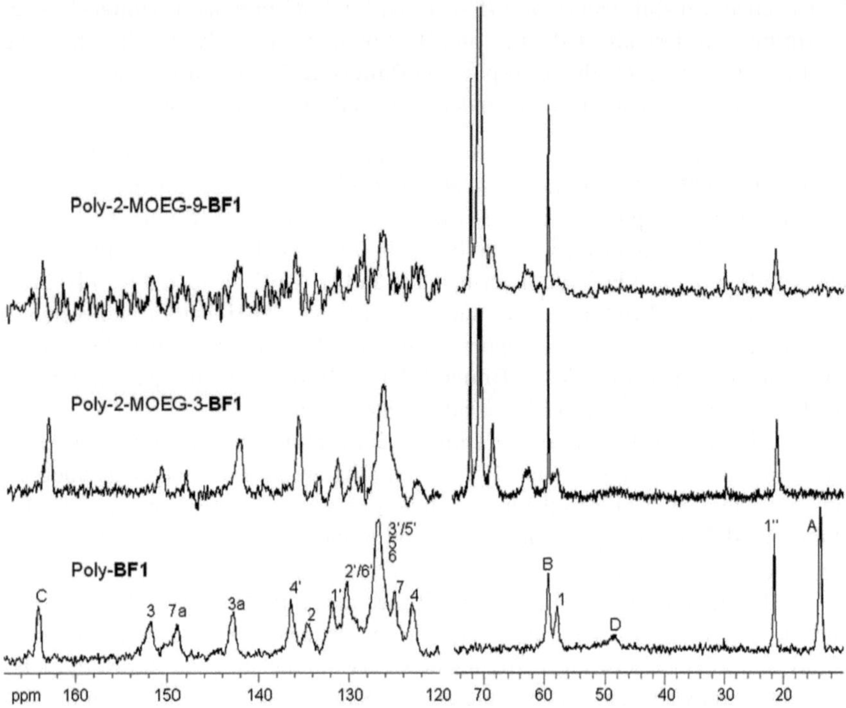

Fig. 28 Comparison of the ^{13}C-NMR ($CDCl_3$) spectra of poly-2-MOEG-3-**BF1,** poly-2-MOEG-9-**BF1**, and poly-**BF1** obtained by spontaneous polymerization. Reprinted with permission from Cappelli et al. [25]. Copyright 2009, American Chemical Society

was closely related to the different MOEG length. We assumed that poly-2-MOEG-9-**BF1** (major line broadening) has a lower mobility with respect to poly-2-MOEG-3-**BF1** (minor line broadening) depending on the different length of MOEG chains.

Then, the synthesis of 6-MO-**BF3k** (the 6-methoxyderivative of benzofulvene monomer **BF3k**) and of three methyl-end-capped oligo(ethylene glycol) ethers (6-MOEG-2-**BF3k**, 6-MOEG-3-**BF3k**, and 6-MOEG-9-**BF3k**) gave the opportunity of proving that the vinyl (1, 2) polymerization mechanism remained operative also in the presence of oligoether side chains at the 6-position of the indene nucleus [37]. The NMR study involving these polymers was based on the extensive characterization of indene derivative 6-MO-**BF3k-MUM** as the model compound for the monomeric unit concerning the polymer obtained by vinyl (1, 2) polymerization of 6-MO-**BF3k**. The strategy used to assign the spectra of the model compound was used in the assignment of the spectra of the polymers. However, the severe line broadening observed in the HMBC experiments hampered the obtainment of all the long-term ^{1}H-^{13}C-connectivities. Thus, the assignment work required the comparison of the ^{13}C-NMR spectra of the polymers with those of the model compound and poly-**BF3k**, also in the light of the 2D collected data. The assignments reported in Fig. 29 demonstrated that the diagnostic signals attributed to poly-6-MO-**BF3k** backbone (C-1: 57.1 ppm and D: 48.5 ppm) showed chemical shift values very

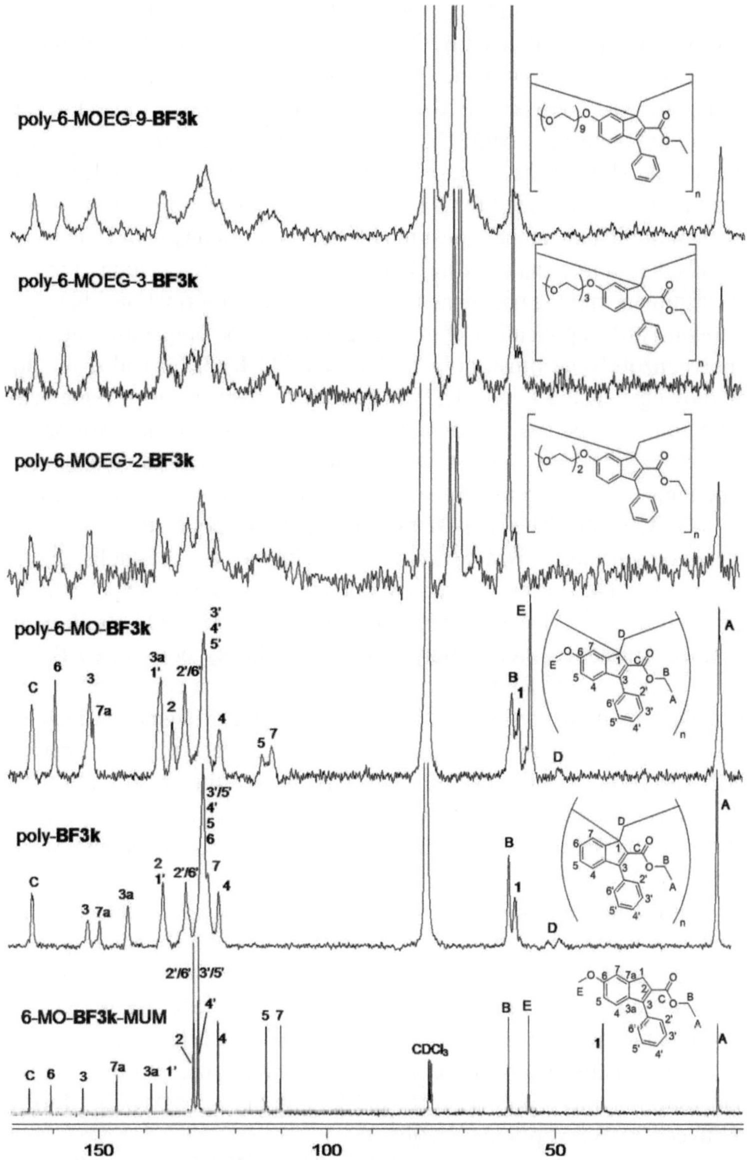

Fig. 29 Comparison of the ^{13}C-NMR spectra ($CDCl_3$) of 6-MO-**BF3k**-MUM, poly-6-MO-**BF3k**, poly-6-MO-**BF3k**, poly-6-MOEG-2-**BF3k**, poly-6-MOEG-3-**BF3k**, and poly-6-MOEG-9-**BF3k**. Adapted with permission from Cappelli et al. [37]. Copyright 2010, Wiley Periodicals, Inc.

similar to those observed for the corresponding signals in poly-**BF3k** spectrum [C-1: 57.8 ppm; D: 48.4 (major), 50.8 (minor)].

Furthermore, the comparison of ^{13}C-NMR spectra (Fig. 29) of poly-6-MO-**BF3k,** poly-6-MOEG-2-**BF3k**, poly-6-MOEG-3-**BF3k**, and poly-6-MOEG-9-**BF3k** showed the existence of important correspondences in both the upfield region containing backbone C-1 and C-D signals and the low-field portion containing C-1′ and C-3 (which may be affected by the potentially competing 1,4-polymerization) and supported the retention of the spontaneous 1,2-polymerization mechanism in these benzofulvene carboxylate derivatives. Finally, this comparative analysis showed that the increase in MOEG chain length produced a broadening and a progressive upfield shift of the signal attributed to C-2′/6′ and a similar broadening involved the signals attributed to C-2 and C-3, thus suggesting that the presence of an increasing steric bulk at the 2-position of the indene moiety could induce a different main chain arrangement affecting both the nonbonded interaction of the pendent phenyl substituent and the backbone flexibility or the conformational behavior.

Analogously, the application of click chemistry to the preparation of the biorelevant polybenzofulvene brush poly-6-MOEG-9-TM-**BF3k** provided the opportunity of proving that the vinyl (1, 2) polymerization mechanism remained again functioning also in the case of elongation of the side chains in position 6 of the indene nucleus [52]. Very interestingly, the increased structural complexity of the macromonomer containing the triazole ring did not affect the monomer interaction leading to poly-6-MOEG-9-TM-**BF3k**. The resulting polymer was exhaustively characterized by means of NMR spectroscopy with the aim of establishing how the introduction of a TM spacer affected the polymer structure. As expected, the comparative analysis of ^{13}C-NMR spectrum of poly-6-MOEG-9-TM-**BF3k** with those of reference substances (e.g., poly-6-MOEG-9-**BF3k**, poly-6-MO-**BF3k**, and model compound 6-MO-**BF3k-MUM**, Fig. 30) showed a remarkable correspondence in the low-field portion containing C-1′ and C-3 signals (which may be affected by the competing 1,4-polymerization). The direct comparison between the ^{13}C-NMR spectrum of poly-6-MOEG-9-TM-**BF3k** and the one of poly-6-MOEG-9-**BF3k** allowed the detection of both the additional signals of TM spacer and the upfield shift of the signal attributed to the methylene carbon of MOEG SC directly attached to the triazole ring.

Based on the interesting results obtained, we extended the strategy established on the extensive characterization of the appropriate model compound for the monomeric unit concerning the polymer obtained by vinyl (1,2) polymerization also to the NMR characterization of the methoxy derivatives of 6-MO-**BF3k** [55]. The two model compounds bearing two (4,6-MO-**BF3k-MUM**) or three (4,5,6-MO-**BF3k-MUM**) methoxy substituents on the benzene ring were analyzed together with the corresponding polymers, poly-4,6-MO-**BF3k** and poly-4,5,6-MO-**BF3k** (Fig. 31), in order to confirm vinyl polymerization.

The assignment work begun from 4,6-MO-**BF3k-MUM**. Specifically, in the aliphatic region, the singlet at 3.77 ppm was assigned to the H-1 protons, because it was the only expected singlet having an integral value corresponding to two protons. Moreover, these signals showed a NOESY contact with a proton in the aromatic region, at 6.71 ppm, assigned to H-7. As a consequence the signal at 6.32 ppm was attributed to the H-5 proton. Once H-7 and H-5 were established, NOESY analysis led to the identification of H-E and H-F. All the other resonances were assigned by means of the ^{1}H-^{13}C heteronuclear two-dimensional techniques.

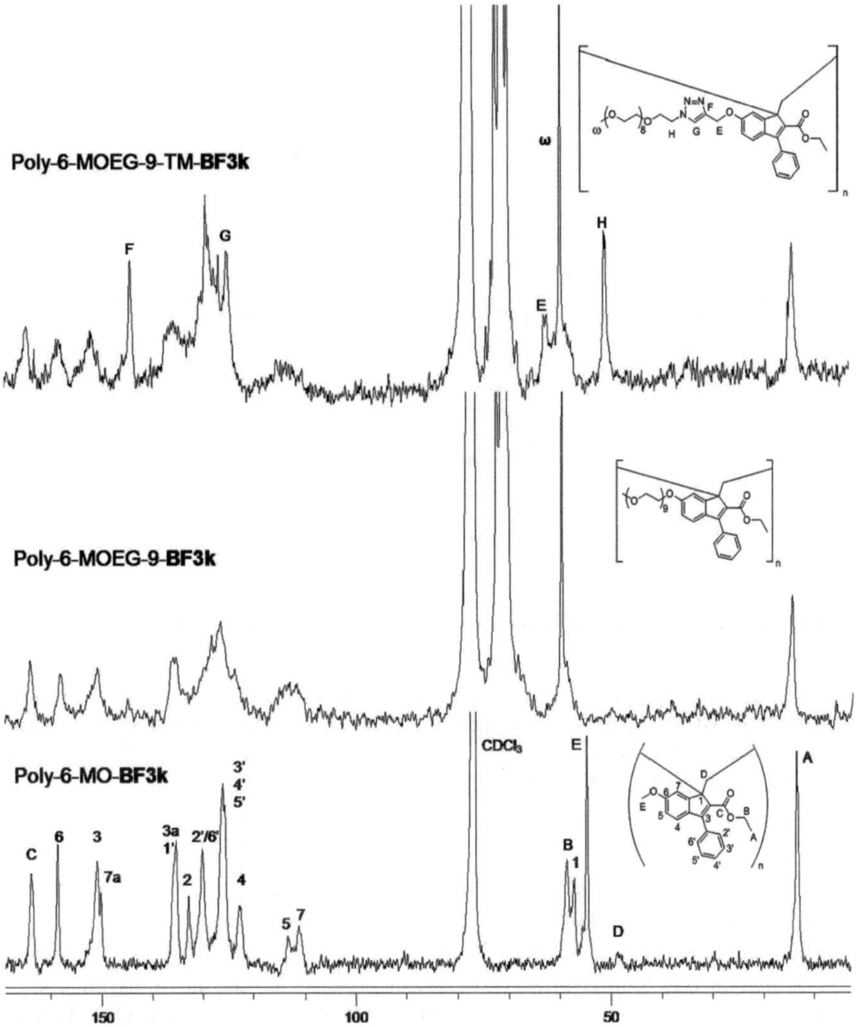

Fig. 30 ^{13}C-NMR spectrum ($CDCl_3$) of poly-6-MOEG-9-TM-**BF3k** compared with those of poly-6-MOEG-9-**BF3k** and poly-6-MO-**BF3k**. Adapted with permission from Cappelli et al. [52]. Copyright 2011, The Royal Society of Chemistry

Owing to the close structural similarity, 4,5,6-MO-**BF3k-MUM** spectra were assigned by following the same strategy used for 4,6-MO-**BF3k-MUM**. The assignment was then checked in a comparative study involving the tree model compounds (6-MO-**BF3k-MUM**, 4,6-MO-**BF3k-MUM**, and 4,5,6-MO-**BF3k-MUM**). The comparison suggested that C-1 and C-3 were the carbon atoms less affected by the introduction of the methoxy groups, while C-1′ belonging to the pendent phenyl group was slightly sensitive (the insertion of a methoxy substituent at the 4-position

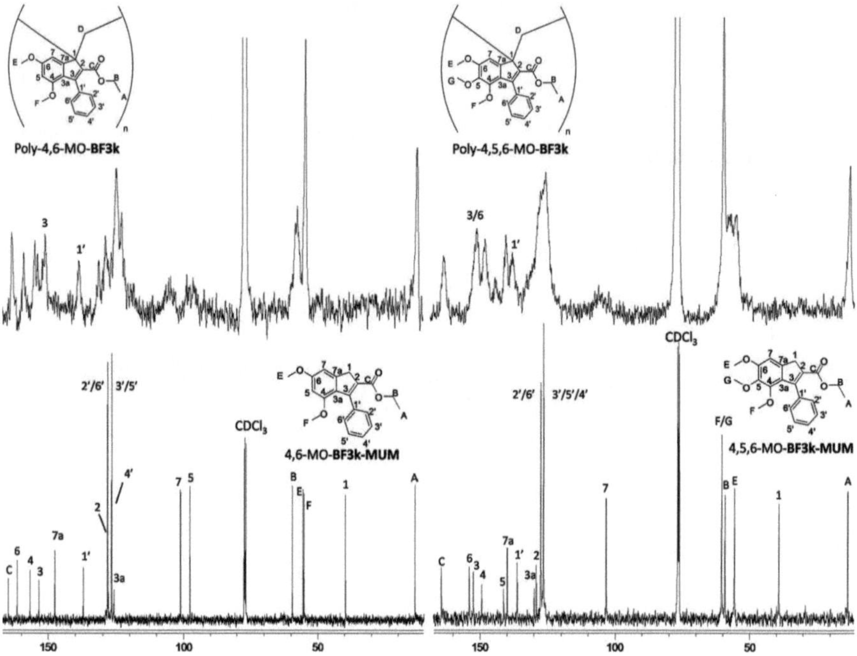

Fig. 31 ^{13}C-NMR spectra ($CDCl_3$) of poly-4,6-MO-**BF3k** and related model compound 4,6-MO-**BF3k-MUM** (*left*); ^{13}C-NMR spectra ($CDCl_3$) of poly-4,5,6-MO-**BF3k** and related model compound 4,5,6-MO-**BF3k-MUM** (*right*). Adapted with permission from Cappelli et al. [55]. Copyright 2012, The Royal Society of Chemistry

of the indene nucleus produced a downfield shift of 2.5 ppm and a successive methoxy insertion in position 5 led to an upfield shift of 0.3 ppm).

The ^{13}C spectra of poly-4,6-MO-**BF3k** and poly-4,5,6-MO-**BF3k** showed a pattern of signals very similar in terms of chemical shifts to that shown by the corresponding model compound (i.e., 4,6-MO-**BF3k-MUM** and 4,5,6-MO-**BF3k-MUM**). In particular, the signals assigned to C-1′ and C-3 (diagnostic of a competing 1,4-polymerization) did not evidence relevant differences in terms of chemical shifts, thus supporting the retention of a 1,2-polymerization mechanism.

Furthermore, the analysis of the evolution of the ^{13}C-NMR features induced by the subsequent methoxy insertion into parent macromolecule poly-**BF3k** (Fig. 31) demonstrated that the changes in the chemical shift values of C-1′ and C-3 signals agreed with those observed in the model compounds and that the insertion of the second methoxy group led to a dramatic change in the NMR features of these polymers. The change was assumed to be related to the alteration in conformational preferences as suggested by the conformation plot (see Sec. 5.1).

Recently, two novel benzofulvene monomers bearing propargyl or allyl groups were synthesized and induced to polymerize spontaneously by solvent removal to give the corresponding polybenzofulvene derivatives bearing clickable propargyl or

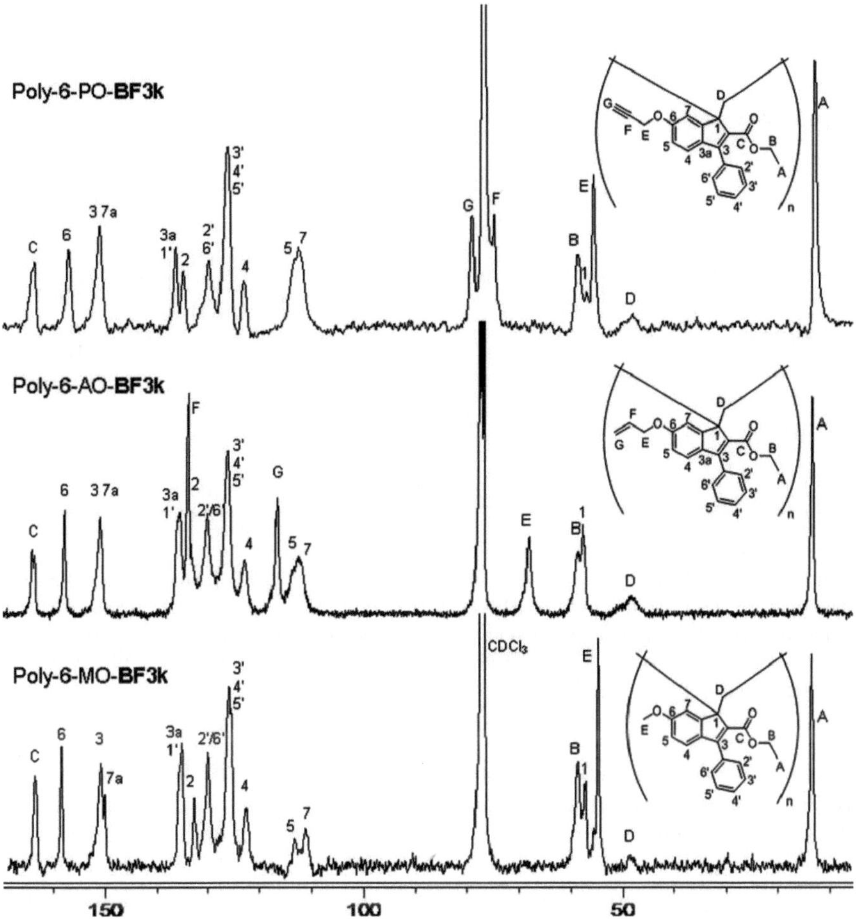

Fig. 32 ^{13}C-NMR spectra ($CDCl_3$) of newly synthesized poly-6-PO-**BF3k** and poly-6-AO-**BF3k**, compared with those of poly-6-MO-**BF3k**

allyl moieties. In the absence of the suitable model compounds, ^{13}C-NMR spectra of poly-6-PO-**BF3k** and poly-6-AO-**BF3k** were evaluated in comparison with that of parent macromolecule poly-6-MO-**BF3k** (Fig. 32) in order to ascertain the correspondence between the structure of these polybenzofulvene derivatives and the presence of intact clickable groups in poly-6-PO-**BF3k** and poly-6-AO-**BF3k**. The comparative analysis of both the low-field portion containing C-1′ and C-3 signals and the upfield region containing C-1 and C-D (which may be affected by the competing 1,4-polymerization) confirmed the retention of the spontaneous 1,2-polymerization mechanism. Moreover, the presence of the additional signals attributed to propargyl (poly-6-PO-**BF3k**) or allyl (poly-6-AO-**BF3k**) side chains demonstrates that 1,2-polymerization involves selectively the benzofulvene exocyclic double bond.

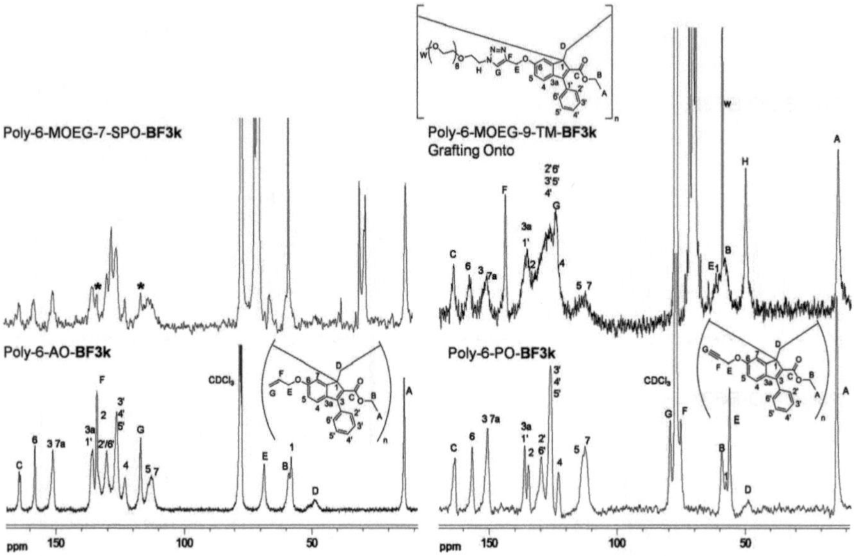

Fig. 33 *Left*: ^{13}C-NMR spectra ($CDCl_3$) of poly-6-MOEG-7-SPO-**BF3k** and poly-6-AO-**BF3k**; the small signals marked with *asterisks* were assigned to the residual vinyl groups remained unmodified after the exposition to the thiol-ene photoreaction conditions. *Right*: ^{13}C-NMR spectra ($CDCl_3$) of poly-6-MOEG-9-TM-**BF3k** obtained by exhaustive functionalization of poly-6-PO-BF3k (grafting onto)

The two polymers bearing clickable moieties (poly-6-AO-**BF3k** and poly-6-PO-**BF3k**) were used as macromolecular precursors in the "grafting onto" approach to the synthesis of polybenzofulvene brushes. The transformation of the carbon–carbon double or triple bonds by coupling with MOEG chains confirmed the availability of these groups to react through click chemistry reactions ("CuAAC" for poly-6-PO-**BF3k** and "thiol-ene" for poly-6-PO-**BF3k**) in a selective manner with the appropriate groups. ^{13}C-NMR spectroscopy was demonstrated to be a powerful tool in the evaluation of the completeness of the coupling reaction (grafting density). The disappearance of the signals of the propargyl and allyl groups of the two macromolecular precursors was complete in the case of poly-6-PO-**BF3k** and partial in poly-6-AO-**BF3k** (Fig. 33).

Absorption and Emission Spectroscopy

DNA is a biopolymer characterized by a backbone formed by alternating phosphate and sugar residues and bearing chromophores (the nucleobases) capable of interacting each other to form a structurally complex supramolecular entity. The double helical structure of DNA is stabilized by two interaction types, namely, hydrogen bond and stacking interactions among the nucleobases, which form a column inside

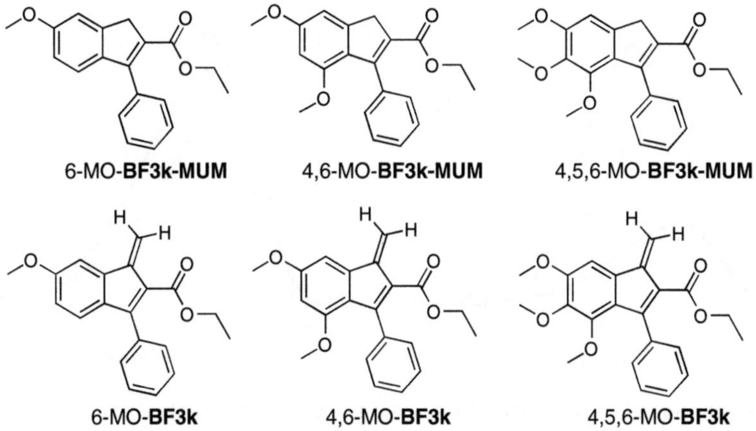

Fig. 34 Structures of model compounds 6-MO-**BF3k-MUM**, 4,6-MO-**BF3k-MUM**, and 4,5,6-MO-**BF3k-MUM** compared with those of benzofulvene derivatives 6-MO-**BF3k**, 4,6-MO-**BF3k**, and 4,5,6-MO-**BF3k**. Reprinted with permission from Cappelli et al. [55]. Copyright 2012, The Royal Society of Chemistry

the double helix. The stacking interactions (also called pi stacking or π–π stacking) are attractive non-covalent interactions among aromatic rings that have stimulated a great research interest, but are not completely understood. The presence of stacking interactions among the nucleobases in the ordered structure of DNA has been connected to the phenomenon called hypochromism, a decrease in the absorption of light by a chromophore in the ordered state (stacked) with respect to the disordered (non-stacked) one. In particular, the hypochromism of a solution of helical calf thymus DNA was found to be 0.37 relative to the constituent mononucleotides [71].

The presence of π-stacking interactions is generally recognized to produce the formation of excimers, which are characterized by redshifted emission spectra with respect to those of the noninteracting species. Thus, hypochromism in the absorption and redshifted emission have been evaluated in polymers bearing different chromophores in order to obtain information on the presence of stacking interactions [7, 15].

Accordingly, the absorption/emission features of our polybenzofulvene derivatives were characterized in order to obtain information on the degree of π-stacking between the chromophore groups.

The most comprehensive study was performed on the methoxy derivatives of poly-**BF3k** and was based on the extensive characterization of indene derivatives 6-MO-**BF3k-MUM**, 4,6-MO-**BF3k-MUM**, and 4,5,6-MO-**BF3k-MUM** (Fig. 34), as the model compounds for the monomeric unit concerning the polymers obtained by vinyl (1,2) polymerization of the corresponding monomers (6-MO-**BF3K**, 4,6-MO-**BF3k**, and 4,5,6-MO-**BF3k**) [55].

The choice of the model compounds for the monomeric unit concerning the polymers obtained by vinyl (1,2) polymerization of the corresponding monomers appeared to be justified by two important considerations. First, NMR studies suggested that vinyl (1,2) polymerization appeared to dominate in a great number of

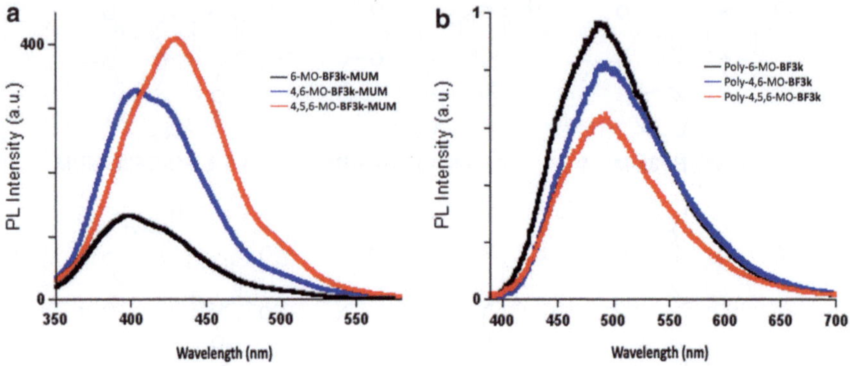

Fig. 35 (**a**) Absorption spectra of model compounds 6-MO-**BF3k-MUM** (*black*), 4,6-MO-**BF3k-MUM** (*blue*), and 4,5,6-MO-**BF3k-MUM** (*red*). (**b**) Absorption spectra of poly-6-MO-**BF3k** (*black*), poly-4,6-MO-**BF3k** (*blue*), poly-4,5,6-MO-**BF3k** (*red*). The spectra were obtained in dichloromethane at room temperature. The concentration in monomer unit for poly-6-MO-**BF3k**, poly-4,6-MO-**BF3k**, and poly-4,5,6-MO-**BF3k** was about 1.78×10^{-5} M. The same concentration was used for model compounds. Reprinted with permission from Cappelli et al. [55]. Copyright 2012, The Royal Society of Chemistry

polybenzofulvene derivatives. Moreover, the study performed on the model compounds for poly-**BF1** monomeric unit (**1,2-model** and **1,4-model**, see "NMR Spectroscopy" section) demonstrated that their absorption and emission features are relatively insensitive to the indene double bond shift [21].

The first step of the study on the methoxy derivatives of poly-**BF3k** concerned the evaluation of the UV–Vis absorption spectra of the polymers in comparison with the corresponding model compounds (Fig. 35). The spectrum of 6-MO-**BF3k-MUM** showed two absorption components, one centered around 240 nm and a second centered at about 310 nm, while in compound 4,6-MO-**BF3k-MUM**, the presence of an additional methoxy group at the 4-position of the indene nucleus caused a redshift of the two absorption peaks to 250 and 330 nm and the appearance of a third component around 300 nm. The insertion of the third methoxy group at the 5-position (as in 4,5,6-MO-**BF3k-MUM**) did not produce the expected further redshifts, but the blueshift of the second component led the component at 300 nm to disappear and the absorption spectrum became analogous to that shown by the model compound bearing a single methoxy group (6-MO-**BF3k-MUM**).

The comparative analysis of the polymer absorption spectra with respect to model compounds (Fig. 35) revealed a significant hypochromism (particularly evident in the component centered around 300 nm) similar to that observed in other π-stacked macromolecules.

The comparison of emission spectra of the three polymers (poly-6-MO-**BF3k**, poly-4,6-MO-**BF3k** and poly-4,5,6-MO-**BF3k**) and of the related monomeric unit models provided interesting information (Fig. 36). The insertion of a second and third methoxy group in the indene nucleus of the monomeric model compounds

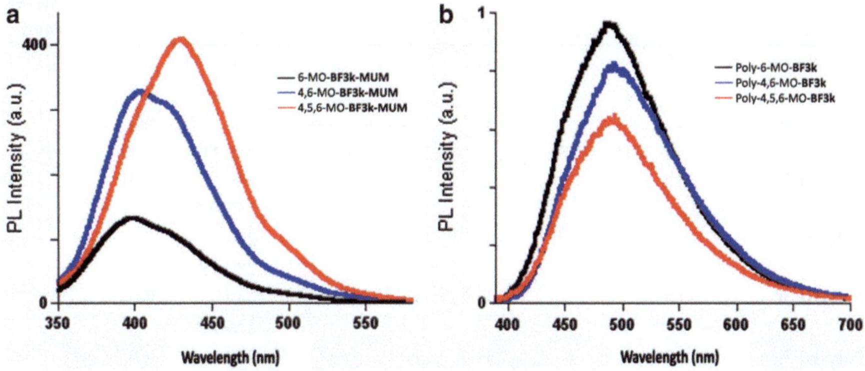

Fig. 36 (**a**) Emission spectra of model compounds 6-MO-**BF3k-MUM** (*black*), 4,6-MO-**BF3k-MUM** (*blue*), and 4,5,6-MO-**BF3k-MUM** (*red*). (**b**) Emission spectra of poly-6-MO-**BF3k** (*black*), poly-4,6-MO-**BF3k** (*blue*), and poly-4,5,6-MO-**BF3k** (*red*). The spectra were obtained in dichloromethane at room temperature. The concentration of model compounds was about 1.78×10^{-5} M and the excitation wavelength was 310 nm. The concentration in monomer unit for poly-6-MO-**BF3k**, poly-4,6-MO-**BF3k**, and poly-4,5,6-MO-**BF3k** was about 1.78×10^{-6} M and the excitation wavelength was 360 nm. Reprinted with permission from Cappelli et al. [55]. Copyright 2012, The Royal Society of Chemistry

caused the emission maximum to shift from 400 nm (6-MO-**BF3k-MUM**) to 405 nm (4,6-MO-**BF3k-MUM**) and 430 nm (4,5,6-MO-**BF3k-MUM**) in conformity with the assumed increase in electron density of the resulting molecular orbitals. On the contrary, the polymers did not show the same trend. The polymer emissions were indeed found to be very similar in shape and were characterized by a peak centered at about 490 nm that resulted redshifted with respect to monomeric unit models. The result was interpreted in terms of a high degree of chromophore stacking, which appeared to overshadow the electronic effect of methoxy groups.

Hypochromism in the absorption and redshifted emission were properly evaluated when the appropriate model (MUM) compounds were promptly available (see Table 9) and were found to be common features shared by the polybenzofulvene derivatives with well-known π-stacked macromolecules. When the study of the appropriate model compound was unfeasible, Stokes shifts (energy difference between the absorption and emission maxima) were used to evaluate the presence of stacking interactions. Indeed, it is generally accepted that excimer interactions and efficient energy transfer contribute to larger Stokes shifts.

The analysis of the data reported in Table 9 demonstrated that most of the polybenzofulvene derivatives showed large Stokes shifts. In particular, the methoxy derivatives of poly-**BF3k** (e.g., poly-6-MO-**BF3K**, poly-4,6-MO-**BF3k**, and poly-4,5,6-MO-**BF3k**) showed shift values around 150 nm that are convergent with the information obtained from the comparative study involving the model compounds in indicating the presence of stacking interactions. Since similar values were found in the remaining polybenzofulvene derivatives, we assumed that the presence of stacking interactions is a common feature of the whole family.

Table 9 Absorption and emission features of some relevant benzofulvene derivatives

Polymer	R_1	R_2	R_3	Abs max (nm)	Em max (nm)	Em MUM[a] (nm)
Poly-**BF3a**	H	H	H	240 275 sh[b]	350 450	
Poly-**BF3c**	F	H	H	240 260 sh[b]	390	
Poly-**BF3d**	Cl	H	H	240 275 sh[b]	395	
Poly-**BF3e**	Br	H	H	240 280 sh[b]	395	
Poly-**BF3f**	CH_3	H	H	240 275 sh[b]	390	
Poly-**BF3g**	CN	H	H	240 280	360 445	
Poly-**BF3h**	CN	CH_3	H	240 280	400 sh[b] 435	
Poly-**BF3i**	CN	H	CH_3	240 280	360 455	
Poly-**BF3j**	C≡C-2-Pyr	H	H	240 346	395 470	
Poly-**BF3k**	$COOC_2H_5$	H	H	240 284	440	
Poly-**BF3l** (Poly-**BF1**)	$COOC_2H_5$	CH_3	H	240 285	445	380
Poly-2-MOEG-3-**BF1**	$COO(C_2H_4O)_3CH_3$	CH_3	H	240 288	445	
Poly-2-MOEG-9-**BF1**	$COO(C_2H_4O)_9CH_3$	CH_3	H	240 288	445	
Poly-6-MO-**BF3k**	$COOC_2H_5$	H	OCH_3	240 310	460 500 sh[b]	400 420 sh[b]
Poly-6-MOEG-2-**BF3k**	$COOC_2H_5$	H	$O(C_2H_4O)_2CH_3$	240 310	460 500 sh[b]	
Poly-6-MOEG-3-**BF3k**	$COOC_2H_5$	H	$O(C_2H_4O)_3CH_3$	240 310	460 500 sh[b]	
Poly-6-MOEG-9-**BF3k**	$COOC_2H_5$	H	$O(C_2H_4O)_9CH_3$	240 310	460 500 sh[b]	
Poly-6-MOEG-9-TM-**BF3k**	$COOC_2H_5$	H	$OCH_2(C_2HN_3)(C_2H_4O)_9CH_3$	240 310	460 500 sh[b]	
Poly-4,6-MO-**BF3k**				250 300 340	460 500 sh[b]	405 420 sh[b]
Poly-4,5,6-MO-**BF3k**				255 310	460 500 sh[b]	430
Poly-6-PO-**BF3k**	$COOC_2H_5$	H	OCH_2CCH	240 307	460 500 sh[b]	
Poly-6-AO-**BF3k**	$COOC_2H_5$	H	$OCH_2CH{=}CH_2$	240 310	460 500 sh[b]	

[a]Em MUM: this column reports the emission maxima of the appropriate monomeric unit models
[b]The value followed by sh refers to the shoulder observed in the spectrum

The fluorescence in solution of a macromolecule bearing aromatic side groups is usually formed by monomer and excimer emissions. In the emission spectrum of poly-**BF2** [15], the main component at 448 nm was assigned to the formation of an excimer resulting from chromophore stacking, while the secondary peak at 347 nm to the emission of the monomer in consideration of the fact that 3-phenylindene (the model of the monomeric unit) showed an emission peak at 345 nm. The emission spectra of poly-**BF1** and closely related poly-**3BFk** showed a broad peak centered at 440–445 nm, while both the models of poly-**BF1** monomeric units (**1,2-model** and **1,4-model**, see NMR section) showed maximum emissions at about 380 nm. On the basis of this observation and in agreement with the interpretation by Londergan and coworkers, we have attributed the emission around 350–380 nm in poly-**BF3** spectra to non-stacked monomeric units and the peaks at 440–470 nm to the stacked ones. This assumption led to the consideration that the degree of the chromophore stacking in poly-**BF3** could be highly variable. Thus, poly-**BF3a** (emitting at 350 and 450 nm) was assumed to be less stacked than both poly-**BF1** (emitting only at 445 nm) and poly-**BF3k** (emitting only at 440 nm), while poly-**BF3c–f** (emitting only at 390–395 nm) were even less stacked. As far as poly-**BF3g–j** are concerned, an intermediate situation has been hypothesized owing to the variable proportions between the two emission components.

MALDI-TOF Mass Spectrometry

MALDI-TOF MS has become an important tool for the analysis of chemical characterization of complex polymer samples as it is a highly sensitive, non-averaging technique that provides detailed structural information about the individual molecules contained in polymer systems [72–78]. It is a soft ionization technique that usually produces single charged quasi-molecular ions with negligible ion fragmentation and makes use of short pulses (3–10 ns duration) of laser light (i.e., N_2 at 337 nm) to induce the formation of intact gaseous ions. Analyte molecules are not directly exposed to laser light, but are homogeneously embedded in a large excess of "matrix," which consists of small organic molecules. The matrix perform two functions: (1) it absorbs photon energy from the laser beam and transfers it into excitation energy of the solid system (analyte/matrix) and (2) it serves to dilute the analyte molecules, so that their intermolecular forces are reduced and their aggregation is minimized. MALDI-TOF MS gives information on the mass of each individual macromolecule from which repeat units, end groups, the presence of rings, molar mass distributions, and other information can be derived, in particular in the mass range up to 20–30 kDa since the single species appears as mass resolved in the mass spectra. MALDI-TOF MS gives absolute determinations of average molar masses (M_w and M_n), in particular only for polymers with a narrow dispersity ($D = M_w/M_n < 1{,}1$) it succeeds independently of their structure, while fails for polymeric materials with a higher polydispersity ($D > 1{,}1$) [72, 73, 77], as well as for the polybenzofulvenes investigated [20, 22–24, 37, 52, 55]. It is known that in

MALDI-TOF MS analysis of polydisperse polymer low molecular mass species are more easily desorbed than high molecular mass ones, determining discrimination effect, and often macromolecules with molar mass higher than 10,000–15,000 g/mol were not revealed [72–78]. To overcome this failure the *off-line* SEC/MALDI-TOF MS coupling method has been used to evaluate the molar masses (M_w and M_n) of several polymer systems [72, 73, 77, 78]. The end groups of the oligomers revealed in the mass range lower than m/z 20,000 are representative of all the macromolecular species, leading information on polymerization reaction mechanism and on the experimental conditions; when the investigated polymers show high molar masses (M_w and M_n), they were not revealed by (^{1}H and ^{13}C)-NMR analysis because concentration of these end groups is below the sensibility of the NMR instruments used. Therefore, MALDI-TOF MS has been generally used in the study of our polybenzofulvene derivatives to confirm the molecular mass of the repeating unit and, in particular, to identify the end groups [20, 22–24, 37, 52, 55].

The first complete MALDI-TOF MS characterization of our polybenzofulvene derivatives was performed in the context of the structure property relationship study involving poly-**BF3k** derivatives (e.g., poly-6-MO-**BF3k**, poly-6-MOEG-2-**BF3k**, poly-6-MOEG-3-**BF3k**, and poly-6-MOEG-9-**BF3k**) [55]. These polymers were analyzed by MALDI-TOF MS using 2-(4-hydroxylphenylazo)benzoic acid (HABA) as a matrix, and the positive ion mass spectra recorded in linear mode show clusters of homologous peaks separated by a molecular mass corresponding to the repeating unit of the analyzed polymer. The mass spectrum of poly-6-MO-**BF3k** showed (in the range from 900 up to 8,800 g/mol) clusters of peak series corresponding to the oligomers with repeating units ranging from 3 (M_3) up to 28 (M_{28}). On the other hand, the MALDI-TOF spectra of the PEGylated poly-**BF3k** derivatives (poly-6-MOEG-2-**BF3k**, poly-6-MOEG-3-**BF3k**, and poly-6-MOEG-9-**BF3k**) showed peaks of oligomers formed by a lower number of repeating units as a result of the higher weight of the monomeric unit. MALDI-TOF spectra of the polymer fractions showing a narrow distribution (obtained by size exclusion chromatography, SEC fractionation) showed the same clusters of peaks in the mass range higher than 10,000 g/mol. Information on the polymer end groups and composition was derived from the analysis of the enlarged section of mass spectra reported in Fig. 37 [55]. All spectra showed a dominant series of peaks assigned to the protonated ions of the expected polymer chains terminated with hydrogen at both ends (H/H; species A_n). This peak was accompanied by two additional families of intense peaks, which were attributed to the macromolecular chains terminated with a benzofulvene unit bearing one aldehyde moiety (species B_n) or one carboxylic acid group (species C_n) in place of the methyl group of species A_n. The presence of the carbonyl groups belonging to the aldehyde and carboxyl acid end groups was confirmed by ^{13}C-NMR spectra (two peaks in the downfield region at 220.6 ppm and at 193.1 ppm) and was explained in terms of autoxidation of the methylene macro-radical species formed during the polymerization of the corresponding monomers. The result was considered to be an interesting piece of information in the development of a revised mechanistic hypothesis.

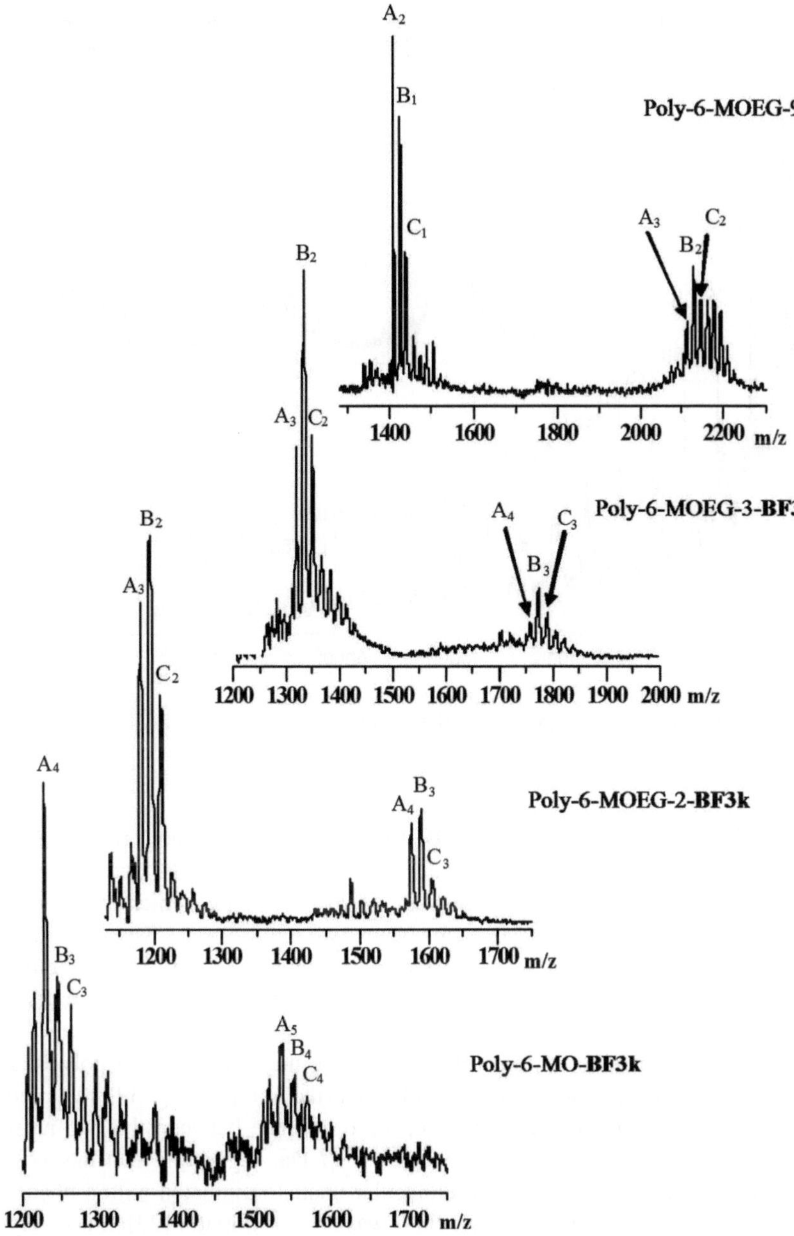

Fig. 37 Enlarged sections of MALDI-TOF mass spectra of poly-6-MO-**BF3k**, poly-6-MOEG-2-**BF3k**, poly-6-MOEG-3-**BF3k**, and poly-6-MOEG-9-**BF3k**. Reprinted with permission from Cappelli et al. [37]. Copyright 2010, Wiley Periodicals, Inc.

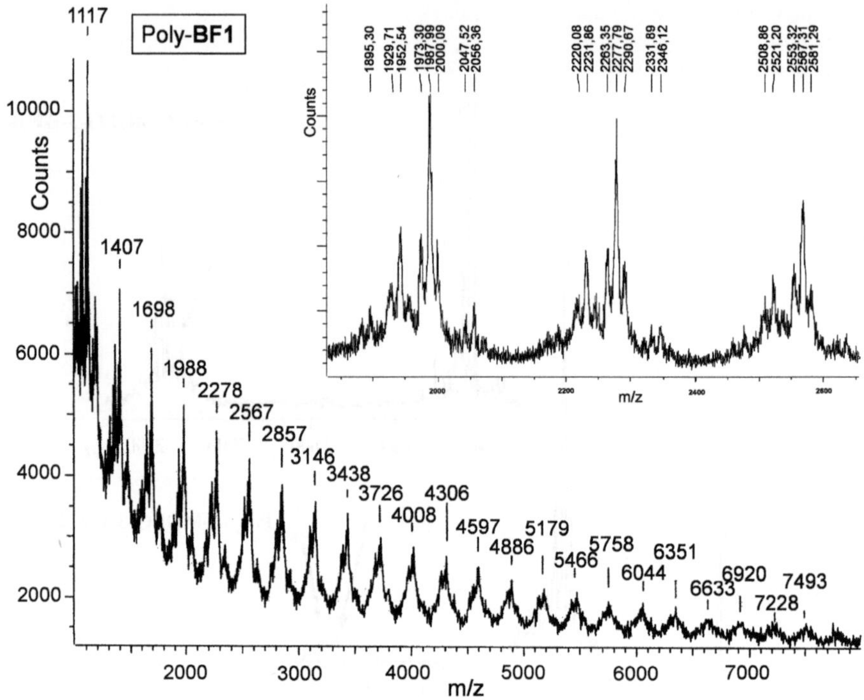

Fig. 38 MALDI-TOF mass spectrum of poly-**BF1**. Reprinted with permission from Cappelli et al. [20]. Copyright 2003, American Chemical Society

MALDI-TOF mass spectrum of poly-**BF1** obtained by spontaneous polymerization (Fig. 38), recorded in linear mode using dithranol (1,8,9-trihydroxyanthracene) as matrix, shows a series of peaks equidistant of 290.4 ± 0.3 Da, which correspond to the mass of **BF1** repeating unit. In each cluster, two intense peaks were spaced of about 44–45 Da, probably due to the loss of an acetaldehyde unit from the ethyl ester moiety. Moreover, each intense peak is also accompanied with two ancillary peaks at ± 14 Da [20].

The loss of 44–45 uma was also observed in the mass spectrum of poly-**BF1** synthesized by anionic polymerization with PhLi as the initiator (Fig. 39), recorded using the dithranol as matrix. In this case, the difference between the two most intense peaks belonging to each cluster is about 77 Da, suggesting that the most intense peaks correspond to the macromolecular chains terminated with a phenyl group (Ph) at one end, as expected since PhLi was used as initiator [23]. Since very similar mass spectra were recorded using other matrices, such as CHCA (α-cyano-4-hydroxycinnamic acid), DCTB [*trans*-2-[3-(4-*tert*-butylphenyl)-2-methyl-2-propenylidene]malononitrile], and HABA [2-(4-hydroxylphenylazo)benzoic acid], the question about the loss of a compound with a mass 44–45 requires appropriate investigations by means of post-source decay (PSD) and CID (collision induced dissociation) MALDI-TOF/TOF tandem mass spectrometry methods.

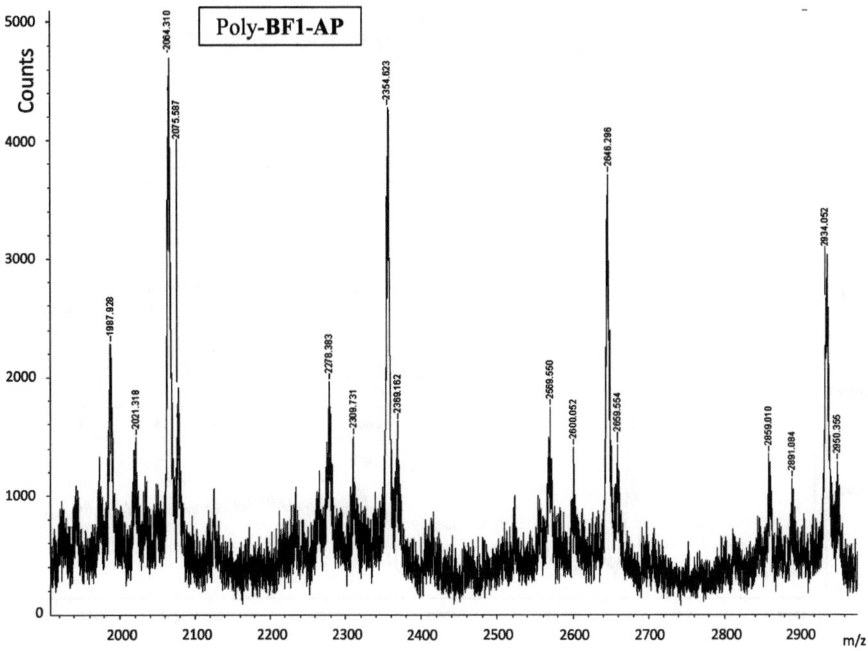

Fig. 39 MALDI-TOF mass spectrum of poly-**BF1-AP**

The MALDI-TOF characterization of other polymers belonging to the subfamily of the poly-**BF3k** derivatives confirmed a common behavior in the termination step, with some differences being observed in the composition (in the differently terminated macromolecules) of the polymer mixtures. For instance, the analysis of the biorelevant molecular brush poly-6-MOEG-9-TM-**BF3k** (Fig. 40, HABA as the matrix) suggested a binary composition in the end groups, with the H/H-terminated species (A_n) dominating on the one bearing an aldehyde group (species B_n) [52].

End group analysis by MALDI-TOF of poly-4,6-MO-**BF3k** and of poly-4,5,6-MO-BF3k samples revealed that the main peaks corresponded to the molecular ions of oligomers terminated with hydrogen at both ends [H/H; species A_n in Fig. 41, CHCA (α-cyano-4-hydroxycinnamic acid) as the matrix] [55]. The enlarged section of MALDI spectra of both polymers recorded in reflectron mode showed that the other two peaks (corresponding to oxidized benzofulvene oligomers bearing one an aldehyde group (CHO; species B_n) and the other one a carboxylic group (COOH; species C_n)) were less intense.

Finally, MALDI-TOF MS characterization of poly-6-AO-**BF3k** [DCTB (*trans*-2-[3-(4-*tert*-butylphenyl)-2-methyl-2-propenylidene]malononitrile) was found to be the best matrix] showed that the most intense peaks correspond to the molecular ions (M^+) of polybenzofulvene oligomers terminated with hydrogen at both ends (H/H, species A_n in Fig. 42) and the second intense peak series belongs to the molecular ions of macromolecules bearing an aldehyde moiety (H/CHO; species B_n),

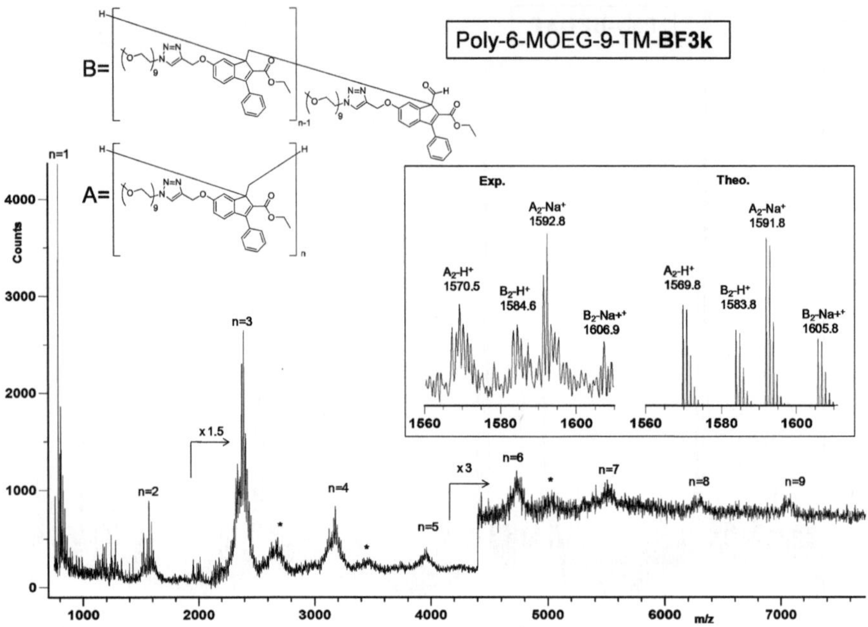

Fig. 40 MALDI-TOF mass spectrum of poly-6-MOEG-9-TM-**BF3k**. *Asterisk* symbols indicate the adduct with the matrix (HABA). Reprinted with permission from Cappelli et al. [52]. Copyright 2011, The Royal Society of Chemistry

whereas peaks corresponding to species terminated with one carboxylic group (H/COOH; C_n) appeared with weak intensity. Figure 43 shows that a similar trend was observed with poly-6-PO-**BF3k**.

The comparison of high-resolution mass spectra of poly-6-MOEG-9-TM-**BF3k** obtained by the grafting onto approach with those obtained with its macromolecular precursor poly-6-PO-**BF3k** (Fig. 44) revealed the existence of differences in the composition of the macromolecular mixtures.

In particular, the relative intensity of peaks corresponding to poly-6-PO-**BF3k** chains terminated with H/H, H/CHO, and H/COOH was about 54:39:7, while in the spectra of poly-6-MOEG-9-TM-**BF3k** (obtained by the grafting onto approach) the most intense peaks were assigned to chains terminated with H/CHO groups (species B_n in Fig. 44) so that the relative intensity of peaks belonging to chains terminated with H/H, H/CHO, and H/COOH groups was 30:42:28. In conclusion, the end group composition of the functionalized polymer was different from that of its precursor. This is probably related to the molecular weight decrease (and the polydispersity increase) indicating the occurrence of multiple ruptures in the polybenzofulvene backbone in the transformation of poly-6-PO-**BF3k** into poly-6-MOEG-9-TM-**BF3k**. The end group composition is probably related to the environment in which the macromolecular species are formed or broken with the formation of macro-radical intermediates.

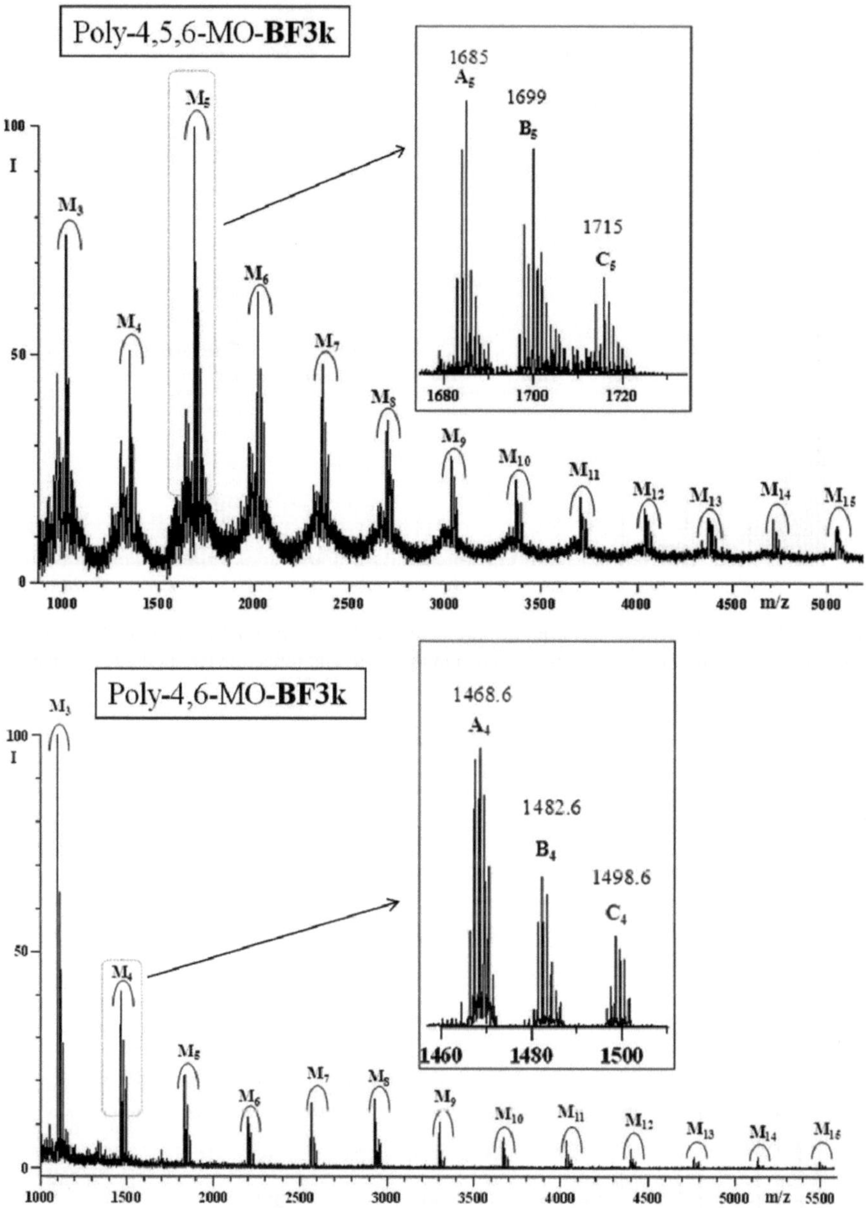

Fig. 41 MALDI-TOF mass spectra (in linear mode) of (*down*) poly-4,6-MO-**BF3k** and (*up*) of poly-4,5,6-MO-**BF3k** samples; the insets display the enlarged section of the corresponding mass spectra recorded in reflectron mode. Reprinted with permission from Cappelli et al. [55]. Copyright 2012, The Royal Society of Chemistry

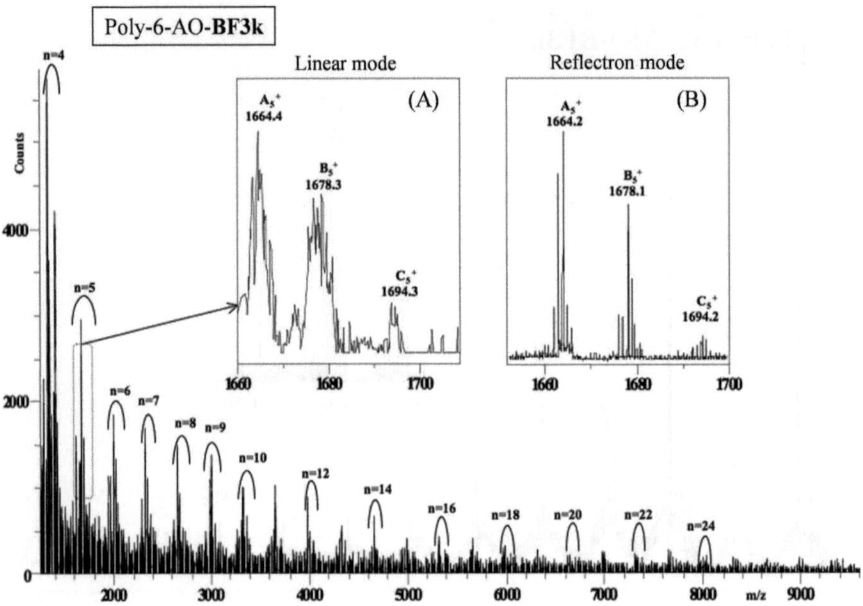

Fig. 42 MALDI-TOF mass spectrum of poly-6-AO-**BF3k** recorded in linear mode. Insets show enlarged sections of the mass spectra recorded in linear (**a**) and reflectron (**b**) mode, respectively

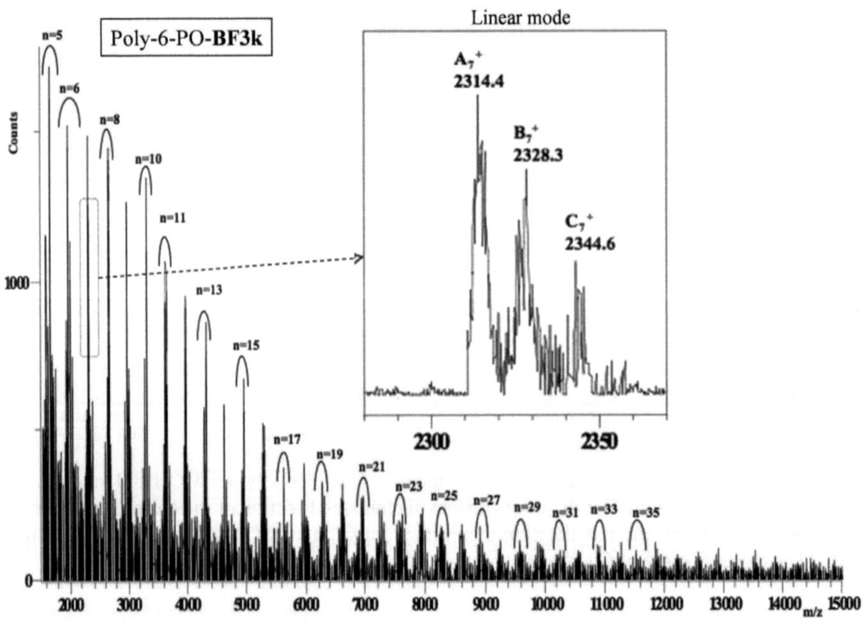

Fig. 43 MALDI-TOF mass spectrum of poly-6-PO-**BF3k** recorded in linear mode

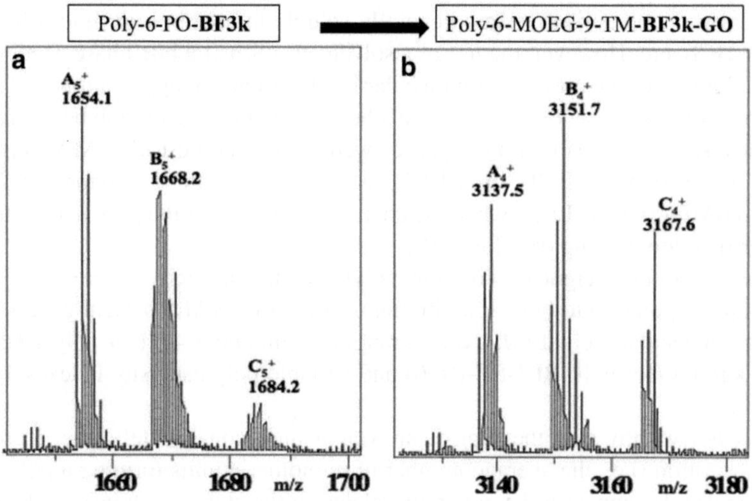

Fig. 44 Enlarged sections of MALDI-TOF mass spectra of (**a**) poly-6-PO-**BF3k** and (**b**) poly-6-MOEG-9-TM-**BF3k-GO**, recorded in reflectron mode

In conclusion, the mass spectra here presented unequivocally show the potential of the MALDI-TOF MS technique into the characterization of end groups of the polybenzofulvene derivatives and then on the understanding of the experimental conditions used to their syntheses.

Properties of Polybenzofulvenes

Molecular Weight Distribution

The molecular characterization of polybenzofulvene derivatives was performed by means of a multi-detector size exclusion chromatography (SEC or GPC) system. The macromolecular characterization consisted in the determination of both the MWD and the radius of gyration distribution (RGD), which is related to the size of macromolecules and to the conformation of the polymers. Owing to the different chemical structure, solubility, and aggregation liability, the experimental conditions (in particular the SEC fractionation) have been optimized carefully. The average molar masses of the polymers were measured by a multi-angle laser light scattering (MALS) detector online to the SEC system coupled with a differential refractometer (DRI) as a concentration detector. The molecular weight measured by means of an online MALS detector is absolute rather than relative to particular standards (polystyrene, e.g., in THF solvent). Sometimes an online differential viscometer (DV) has been also used for the determination of the intrinsic viscosity (η) and the conformation of the polymer by the Mark–Houwink–Sakurada (MHS) plot.

Polybenzofulvene derivatives are generally soluble in common organic solvent as THF, $CHCl_3$, etc. However, the use of a solvent as a SEC mobile phase needs complete solubility of the polymer at molecular level, absence of aggregation, and steric fractionation. Because of the different chemical structures of polybenzofulvene derivatives, different SEC mobile phases were used. The best SEC-MALS results were obtained with THF or $CHCl_3$, as well as with a mixture (8:2) of THF + (DMF + 0.01 M LiBr) when the chemical structure of the polymer contained a polar functional group (see Table 10).

The molecular weight of different polybenzofulvene derivatives ranges from very low (oligomers) to high and ultrahigh. This SEC-MALS system gave weight average molecular weight (M_w) values ranging from about 4.0 (for poly-**BF3a**) to 2,670 kg/mol (for poly-**BF1-SP**-P400) and variable polydispersity indexes M_w/M_n (Table 10).

It is noteworthy that the molecular weight and, consequently, the degree of polymerization (i.e., the average number of monomeric units forming a polymeric macromolecule) appear to be generally related to the steric hindrance in the close proximity of the 2-position of the indene nucleus. The comparison of the molecular weight values shown by the polymers with different substituents in position 2 (poly-**BF3a,c–g,k,j**) demonstrates that high values were related to the presence of a bulky group such as an ethyl ester (e.g., poly-**BF1** and poly-**BF3k**), whereas the values decreased with the diminishing of the dimension of the substituent, until it became minimum for poly-**BF3a** bearing a hydrogen atom. The presence of a hydrogen or a methyl group in para-position of the pendent phenyl group does not affect the molecular weight of these polymers (compare poly-**BF1-SP** with poly-**BF3k**). On the contrary, the introduction of methoxy groups at the 4-, 5-, and 6-position of indene nucleus of poly-**BF3k** monomeric unit causes the molecular weight of the resulting polymers (poly-6-MO-**BF3k**, poly-4,6-MO-**BF3k**, and poly-4,5,6-MO-BF3k) to be significantly reduced with respect to parent macromolecule poly-**BF3k**. Curiously, the presence at the 6-position of some substituents bulkier than the methoxy group (e.g., short MOEG chains, allyloxy or propargyloxy groups) is well tolerated and does not affect the very high molecular weight of these polybenzofulvene derivatives.

On the other hand, the presence of long MOEG chains at the 6-position (either directly attached as in poly-6-MOEG-9-**BF3k** or spaced by means of a triazolomethyl bridge as in poly-6-MOEG-9-TM-**BF3k**) is related to lower molecular weights. This effect could be explained in terms of steric hindrance generated by the high degree of MOEG-9 grafting on the polymer backbone. This hypothesis is based on evidence collected by molecular weight analysis of poly-6-MOEG-9-BF3k-GO synthesized by "grafting onto" approach. This polymer was synthesized from poly-6-PO-**BF3k** showing ultrahigh molecular weight values (M_w = 1,794 kg/mol) and relatively broad polydispersity indices (PDI = M_w/M_n = 2.2). When poly-6-PO-**BF3k** was reacted with MOEG-9-N_3 by CuAAC reaction to afford poly-6-MOEG-9-TM-**BF3k-GO**, a significant decrease in the molecular weight (M_W = 821 kg/mol) and an increase in polydispersity index (PDI = M_W/M_n = 8.9) were observed. These evidences have highlighted the possibility that the polybenzofulvene backbone undergoes ruptures, which were assumed to occur randomly

Table 10 Macromolecule features of polybenzofulvene derivatives

Polymer	R_1	R_2	R_3	Mobile phase	M_p (kg/mol)	M_w (kg/mol)	PDI	R_g (nm)	α
Poly-**BF1**-SP (P100) [21]	$COOC_2H_5$	CH_3	H	THF	955	867	2.0	30.8	–
Poly-**BF1**-SP (P200) [21]	$COOC_2H_5$	CH_3	H	THF	1,870	1,605	1.8	43.1	–
Poly-**BF1**-SP (P400) [21]	$COOC_2H_5$	CH_3	H	THF	2,950	2,670	1.6	60.2	–
Poly-**BF1**-SP (P800) [21]	$COOC_2H_5$	CH_3	H	THF	–	1,437	1.8	41.0	–
Poly-**BF1**-AP (entry I) [23]	$COOC_2H_5$	CH_3	H	THF	13.9	12	2.1	–	–
Poly-**BF1**-AP (entry J) [23]	$COOC_2H_5$	CH_3	H	THF	8.2	7.4	2.0	–	–
Poly-**BF3a** [22]	H	H	H	THF	–	4.0	1.3	–	–
Poly-**BF3c** [22]	F	H	H	THF	–	26	4.5	–	–
Poly-**BF3d** [22]	Cl	H	H	THF	–	48	2.1	–	–
Poly-**BF3e** [22]	Br	H	H	THF	–	109	3.1	–	–
Poly-**BF3f** [22]	CH_3	H	H	THF	–	531	4.7	43.0	–
Poly-**BF3g** [22]	CN	H	H	THF	–	392	6.3	32.5	–
Poly-**BF3h** [22]	CN	CH_3	H	THF	–	500	4.4	26.8	–
Poly-**BF3i** [22]	CN	H	CH_3	THF	–	476	4.3	24.3	–
Poly-**BF3j** [22]	C≡C-2-Pyr	H	H	THF	–	289	2.5	–	–
Poly-**BF3k** [55]	$COOC_2H_5$	H	H	THF	1,900	1,506	3.4	49.9	0.60
Poly-**BF3ma** [22]	COOH	CH_3	H	THF	–	695	9.6	–	–
Poly-2-MOEG-3-**BF1** [25]	COO $(C_2H_4O)_3CH_3$	CH_3	H	$CHCl_3$	–	108	2.4	–	–
Poly-2-MOEG-9-**BF1** [25]	COO $(C_2H_4O)_9CH_3$	CH_3	H	$CHCl_3$	–	175	3.0	–	–
Poly-6-MO-**BF3k** [37]	$COOC_2H_5$	H	OCH_3	THF	312	347	4.3	19.4	0.61
Poly-6-MOEG-2-**BF3k** [37]	$COOC_2H_5$	H	$O(C_2H_4O)_2CH_3$	THF	821	768	4.0	28.7	–
Poly-6-MOEG-3-**BF3k** [37]	$COOC_2H_5$	H	$O(C_2H_4O)_3CH_3$	THF	1,857	1,408	2.0	38.7	–
Poly-6-MOEG-9-**BF3k** [37]	$COOC_2H_5$	H	$O(C_2H_4O)_9CH_3$	THF	527	456	1.8	20.8	–
Poly-6-MOEG-9-TM-**BF3k-GT** [52]	$COOC_2H_5$	H	$OCH_2(C_2HN_3)(C_2H_4O)_9CH_3$	THF DMF[a]	193	358	3.9	15.5	–
Poly-6-MOEG-9-TM-**BF3k-GO**	$COOC_2H_5$	H	$OCH_2(C_2HN_3)(C_2H_4O)_9CH_3$	THF DMF[a]	484	821	8.9	33.6	0.50
Poly-6-PO-**BF3k**	$COOC_2H_5$	H	$OCH_2C{\equiv}CH$	THF	2,030	1,794	2.2	51.5	0.58
Poly-6-AO-**BF3k**	$COOC_2H_5$	H	$OCH_2CH{=}CH_2$	THF	1,864	1,497	2.6	47.7	0.52
Poly-4,6-MO-**BF3k** [55]	$COOC_2H_5$	H	Structure in "Absorption and Emission Spectroscopy" section	THF	185	207	1.8	11.8	0.94
Poly-4,5,6-MO-**BF3k** [55]	$COOC_2H_5$	H	Structure in " Absorption and Emission Spectroscopy" section	THF	252	245	1.8	12.9	0.92

[a]The mobile phase was a (8:2) mixture of THF+(DMF+0.01 M LiBr)

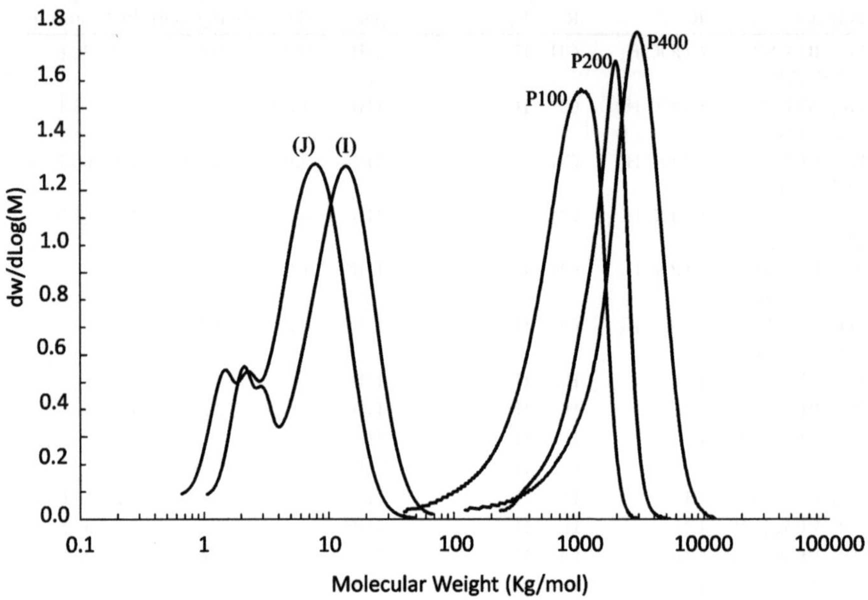

Fig. 45 MWD obtained by SEC-MALS, of two poly-**BF1-AP** samples (entries I and J of Table 10) and three poly-**BF1-SP** samples (P100, P200, and P400 of Table 10). Reprinted with permission from Cappelli et al. [23]. Copyright 2008, American Chemical Society

along the polymer backbone, thus providing a material with a high degree of MOEG-9 grafting and a molecular weight quite similar to that shown by the same polymer obtained by spontaneous polymerization of the macromonomer (poly-6-MOEG-9-TM-**BF3k-GT**).

Interesting results were provided by SEC-MALS analysis of poly-**BF1** obtained by anionic polymerization with phenyl lithium (PhLi) as the initiator. In fact, the average M_w value of poly-**BF1-AP** (Table 10) was found to be relatively low and very different from the corresponding values determined for poly-**BF1** obtained by spontaneous polymerization (Fig. 45). The average M_w values of two preparations of poly-**BF1-AP** (12.2 and 7.4 kg/mol) evidenced a little difference, which was assumed to be due to both the concentration variability of the commercial phenyl lithium solution and the experimental uncertainty. The spontaneous polymerization experiments performed with **BF1** (leading to different poly-**BF1-SP** samples labeled P100, P200, P400, and P800 in Table 10) suggested that the M_w average obtained by spontaneous polymerization was affected by the amount of the monomer in the starting solution. On the contrary, the average M_w of poly-**BF1-AP** samples significantly depends on the amount of the PhLi initiator used.

It is well known that the use of a MALS detector also provides the size of macromolecules R_g from the angular variation of the scattering intensity. Using an elastic or total intensity MALS detector online to a SEC chromatographic system the minimum R_g size measurable with a sufficient accuracy is about 8–10 nm.

Consequently, when the molecular weight and the size of the macromolecules were adequate, also the size distribution (RGD) and the polymer conformation can be studied by means of the conformation plot: $R_g = KM\alpha$.

Similarly to the molecular weight, the R_g value measured for polybenzofulvene derivatives ranges in a wide interval of sizes (i.e., from about 12 nm shown by poly-4,6-MO-**BF3k** to 59 nm shown by poly-**BF1**, P400). As expected, R_g values appeared to be strongly related to the degree of polymerization of polybenzofulvene derivatives.

The conformation plot is the scaling law between macromolecule size and molecular weight (i.e., the chain length) and it is very useful in studying the stiffness (flexible, stiff, or compact conformation) of polymers. Also branching, degree of functionalization, or derivatization could be studied using the conformation plot of the polymer. Therefore, it is very useful in the molecular characterization of polymers.

The conformation plot of several analyzed polybenzofulvene derivatives was almost always linear in the whole molecular weight range, and in general, these polymers did not show any meaningful degree of branching in particular of long-chain branching (LCB).

As far as the stiffness of macromolecules (conformation) is concerned, different results were obtained for the different polybenzofulvene derivatives. In general, the slope α of the conformation plot for the majority of polybenzofulvene derivatives ranged between 0.50 and 0.61. Such slope α values are typical of random coil polymers in an ideal theta solvent ($\alpha = 0.50$) or in a good solvent ($\alpha = 0.54$–0.58). Higher α values ($\alpha = 0.60$–0.61) substantially denote a more rigid conformation (semi-stiff). Some exceptions were represented by poly-4,6-MO-**BF3k** and poly-4,5,6-MO-BF3k, which showed slope α values higher than 0.90 indicating for these two polymers a very stiff (rod) conformation. Substantially, the stiffness of different polybenzofulvene derivatives strongly depends on the chemical structure of the monomeric unit and, obviously, on the solvent used (good, theta, poor). The derivatization of polybenzofulvene backbone provides a wide range of polymer conformations. For instance, Fig. 46 shows the conformation plot of some polybenzofulvene derivatives with the slope ranging from $\alpha = 0.52$ (flexible macromolecules close to the ideal theta condition) to $\alpha = 0.6$ (semi-stiff macromolecules) and to $\alpha = 0.92$ (stiff rod macromolecules).

Solubility and Aggregation

Poly-**BF1** and its most remarkable 4′-desmethyl derivative poly-**BF3k** are characterized by both a high solubility in the most common organic solvents ($CHCl_3$, CH_2Cl_2, THF, EtOAc, DMSO, etc.) and a weak interaction with water. In addition, they have been found to be strongly prone to aggregation. In fact, the significant change in the solubility of poly-**BF1** observed after its prolonged storage as a solid could be due to the possible formation of macromolecular aggregates derived from

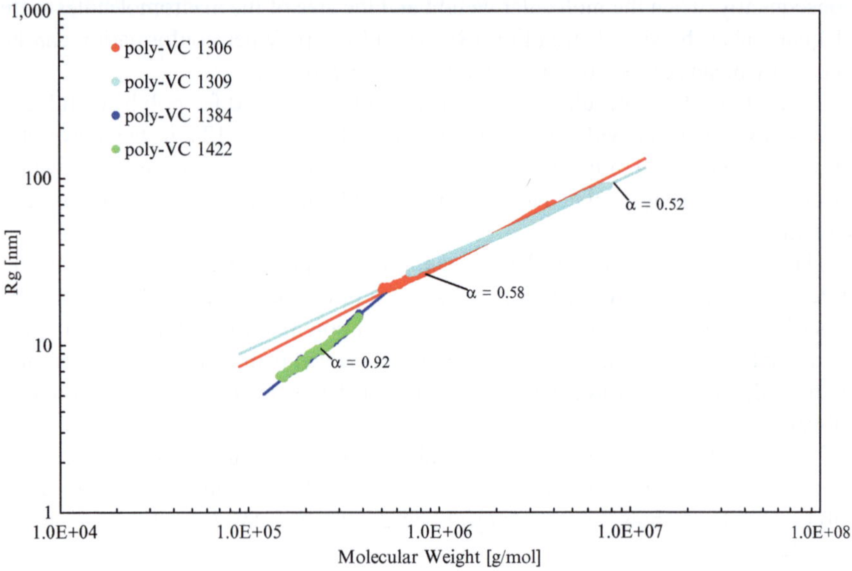

Fig. 46 Examples of conformation plot of some different polybenzofulvene derivatives: poly-6-PO-**BF3k** (*red*), poly-6-AO-**BF3k** (*cyan*), poly-4,6-MO-**BF3k** (*blue*), and poly-4,5,6-MO-**BF3k** (*green*)

specific combinations of the individual poly-**BF1** chains in the solid state [21]. This hypothesis has been tested by performing TEM studies on metal replicas that have shown the liability of this polymer to give nanostructured aggregates (Fig. 47) [21].

The observed architecture elements were believed to derive from the helical arrangement of this particular type of π-stacked polymer [21]. Poly-**BF1** aggregates were then studied by SEM, and the results obtained were comparable with the ones obtained by TEM.

Thus, systematic investigations on the aggregate structure of poly-**BF3** derivatives were performed by SEM and the results suggested that the aggregation properties (e.g., shape and dimensions of nanoaggregates) could be possibly related to MWD [22]. In particular, the polymers characterized by low molecular weight, such as poly-**BF3a**, were particularly inclined to form nanospheres with well-defined dimensions and morphology, whereas the polybenzofulvene derivatives showing a higher degree of polymerization (e.g., poly-**BF3f** and poly-**BF3i**) were found to form larger fractal-like structures (Fig. 48) [22]. Since the substituent in position 2 of monomeric unit indene nucleus was demonstrated to be capable of affecting MWD, as a consequence, it exerts its influence also on the aggregation behavior of the corresponding polybenzofulvene derivative by acting on MWD.

In order to support these findings further, the propensity towards aggregation shown by a lower molecular weight poly-**BF1** sample (M_w = 7.4 kg/mol) obtained by means of anionic polymerization (poly-**BF1-AP**) was compared with that shown by a higher molecular weight poly-**BF1** (M_w = 2,670.0 kg/mol) obtained by means of spontaneous polymerization (poly-**BF1-SP**) [23].

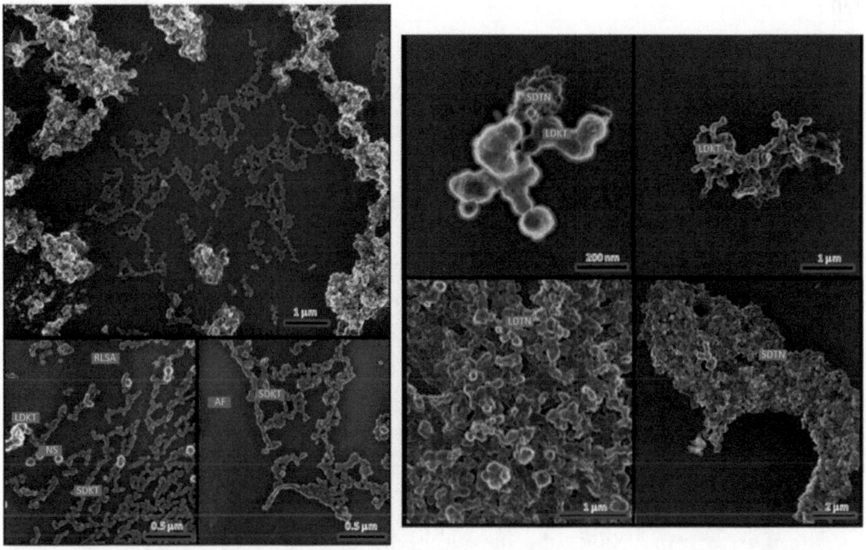

Fig. 47 Some of the most representative images obtained by TEM analysis performed on freshly prepared poly-**BF1** (*left panel*) and poly-**BF1** stored at room temperature for 6 months (*right panel*). The presence of several differently ordered nanostructures grown on an apparently amorphous film (**AF**) is shown in the images: (1) **RLSA** flat aggregates showing a raspberrylike structure; (2) **SDKT** small diameter (30 nm) knotted tubes; (3) **NS** nanospheres showing a diameter of about 75 nm; (4) **LDKT** large diameter (90–130 nm) knotted tubes; and (5) **SDTN** small diameter three-dimensional net which includes (6) (**LDTN**) a large diameter three-dimensional net. **SDTN** and **LDTN** are the main species detected in experiments performed with aged poly-**BF1**. To make the interpretation easier, the images are reported in negative contrast (e.g., dark shadowing). Reprinted with permission from Cappelli et al. [21]. Copyright 2005, Wiley Periodicals, Inc.

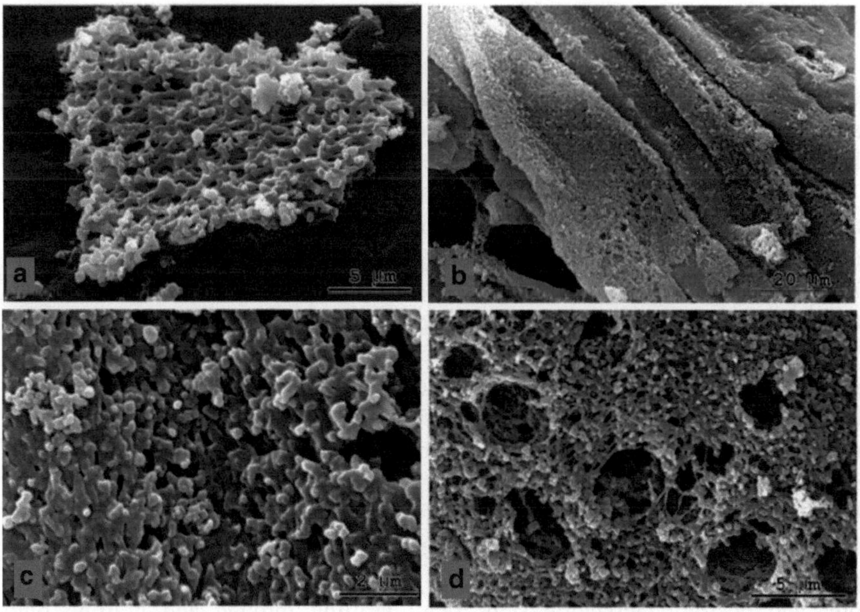

Fig. 48 Scanning electron microscopy (SEM) micrographs of freshly prepared poly-**BF3i** (**a**) and poly-**BF3f** (**b–d**). Reprinted with permission from Cappelli et al. [22]. Copyright 2007, American Chemical Society

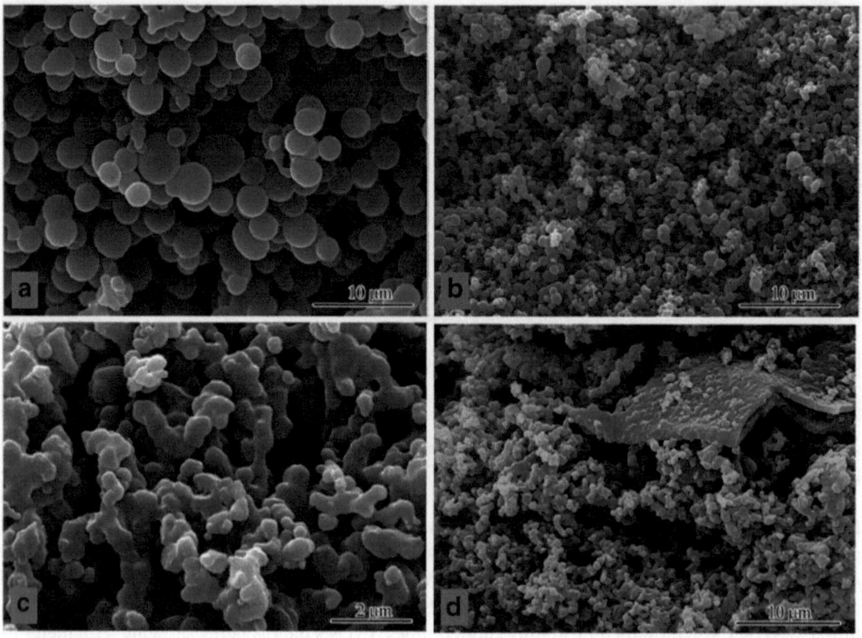

Fig. 49 Molecular aggregate structure found by SEM analysis of poly-**BF1** precipitated from a dichloromethane solution with ethanol. (**a**) Microspheres obtained with poly-**BF1**-**AP**; (**b**) nanoparticles obtained with poly-**BF1**-**AP**; and (**c**) and (**d**) more complex aggregates obtained with poly-**BF1**-**SP** shown at different magnification ratios. Reprinted with permission from Cappelli et al. [23]. Copyright 2008, American Chemical Society

SEM studies confirmed a plausible relationship between molecular weight and aggregation behavior. In fact, poly-**BF1**-**AP** appeared to be liable to give nanospheres and microspheres showing favorable shapes and dimensions, while poly-BF1-SP has the tendency to generate larger fractal-like aggregates (Fig. 49) [23].

In conclusion, the abovementioned aggregation properties could be explained in terms of aromatic stacking interactions among the hydrophobic polybenzofulvene backbones of individual polymer chains. The intermolecular collapse appeared to be dependent on molecular weight and it could be modulated by introducing suitable substituents on 3-phenylindene moiety.

The weak interaction with water and the pronounced tendency to the aggregation shown by polybenzofulvene derivatives were tentatively exploited in the formation of a nanocomposite material potentially useful in drug release [24], but, at the same time, were considered to be limiting factors hampering their potential development in biological applications. Thus, by virtue of the susceptibility to molecular manipulation of the benzofulvene scaffold, these features have been improved by the insertion of monomethyl nona(ethylene glycol) (MOEG-9) side chains onto polybenzofulvene hydrophobic backbones using suitable coupling reactions both in pre- and post-polymerization steps (e.g., grafting through and grafting onto

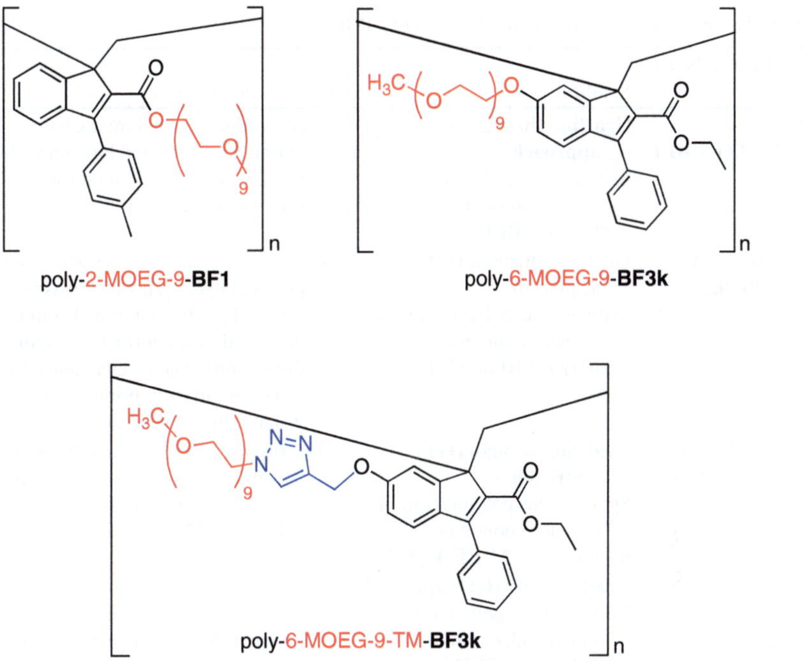

Fig. 50 Structures of PEGylated polybenzofulvene brushes (PPBBs). For the sake of clarity, the structural moieties used to define the polymer names are colored differently. MOEG-9 side chains are reported in *red*, while triazolomethyl spacer is shown in *blue*. Finally, polybenzofulvene backbones of the parent macromolecules poly-**BF1** and poly-**BF3k** are depicted in *black*

synthetic strategies). The insertion of MOEG-9 fragments in specific positions (e.g., 2 and 6) of the benzofulvene scaffold has led to three PEGylated polybenzofulvene brushes (PPBBs), namely, poly-2-MOEG-9-**BF1** [25], poly-6-MOEG-9-**BF3k** [37], and poly-6-MOEG-9-TM-**BF3k** [52] (Fig. 50).

On the basis of similarities existing among benzofulvene, methacrylate, and cyanoacrylate derivatives, the polymerization of preformed PEG macromonomers such as OEGMAs has inspired the covalent conjugation of PEG oligomers (permanent PEGylation) with polybenzofulvene backbones.

The presence and the location of MOEG 9 side chains have been demonstrated to significantly modulate the hydrophilic character and the aggregation properties of these amphiphilic macromolecules belonging to a new family of polymer brushes. In fact, they showed different behaviors in interacting with the water environment ranging from the formation of a compact physical hydrogel to the water solubility of isolated polybenzofulvene macromolecules (Table 11) [25, 37, 52].

In particular, poly-2-MOEG-9-**BF1** showed an amphiphilic character and interacted with a large variety of organic solvents (e.g., $CHCl_3$, THF, DMF, DMA, DMSO, ethanol, methanol) to originate transparent and dissoluble gel aggregates and with water to give a more compact physical hydrogel, stabilized by weak non-covalent interactions and stable when exposed to ultrasounds [25, 36]. On the other hand,

Table 11 Synthetic and hydrophilic features of PPBBs

Polybenzofulvene brushes	Synthesis	Interaction with water
Poly-2-MOEG-9-**BF1**	**Grafting through (GT) approach** Spontaneous polymerization of macromonomer 2-MOEG-9-**BF1** [25]	After swelling, the polymer generated a strong, compact, and biocompatible physical hydrogel, which was stable to ultrasounds [25, 36]
Poly-6-MOEG-9-**BF3k**	**Grafting through (GT) approach** Spontaneous polymerization of macromonomer 6-MOEG-9-**BF3k** [37]	After swelling, the polymer generated a physical hydrogel, which was rapidly broken by ultrasounds and could be dissolved in water (thermoresponsive dispersion), where isolated macromolecules are in equilibrium with nanosized water-soluble aggregates [37]
Poly-6-MOEG-9-TM-**BF3k**	**Grafting through (GT) approach** Spontaneous polymerization of macromonomer 6-MOEG-9-TM-**BF3k** [52]	The polymer formed clear water solutions becoming slightly opalescent when heated at 37 °C (thermoresponsive behavior) [52]
	Grafting onto (GO) approach CuAAC[a] post-polymerization functionalization of poly-6-PO-**BF3k**[b] with MOEG-9-N_3 in the presence of DIPEA and CuBr[c]	The polymer appeared to be molecularly dissolved in water at room temperature giving a thermoresponsive colloidal dispersion

[a]CuAAC = Cu(I)-catalyzed azide–alkyne cycloaddition
[b]Poly-6-PO-**BF3k** was the corresponding precursor polymer bearing one clickable propynyloxy groups (PO) in position 6
[c]The synthetic **GO** approach was performed in THF by means of microwave irradiation for 40 min (4 × 10 min, T = 60 °C, W = 150)

poly-6-MOEG-9-**BF3k** appeared to be more readily soluble in organic solvents and less liable to aggregation in aqueous medium than poly-2-MOEG-9-**BF1**. In fact, the former polymer generated a physical hydrogel stabilized by weaker non-covalent interactions, which can be weakened by means of ultrasounds to give a milky water suspension leading to the separation of a transparent hydrogel. Interestingly, the clear supernatant solution has been demonstrated (by ^{1}H NMR spectroscopy) to contain a substantial amount of poly-6-MOEG-9-**BF3k** and became cloudy when gently warmed at 35–37 °C, suggesting thermoresponsive features [37].

Thus, variations in the temperature were able to affect the aggregation ability of this PPBB in the aqueous environment. Poly-6-MOEG-9-**BF3k** macromolecular structure in water was investigated also by TEM analysis that suggested the coexistence of different conformations and the liability of this brush to generate water-soluble nanoparticles showing heterogeneous shapes and dimensions around 30–50 nm [37]. The preponderance observed for globular architectures together with the presence of aggregates in the analyzed samples led to assume for poly-6-MOEG-9-**BF3k** the tendency to collapse both intramolecularly and intermolecularly [37].

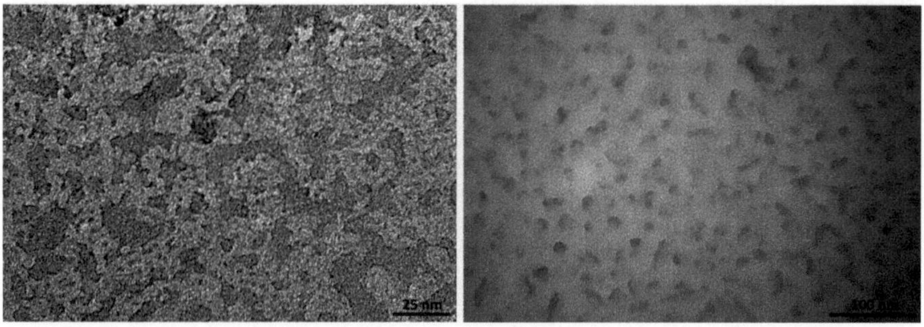

Fig. 51 Structure of macromolecular entities present in the solutions of poly-6-MOEG-9-TM-**BF3k-GT** (*left panel*) and poly-6-MOEG-9-TM-**BF3k-GO** (*right panel*)

The insertion of relatively long MOEG-9 side chains in poly-**BF1** and poly-**BF3k** macromolecules leading to poly-2-MOEG-9-**BF1** and poly-6-MOEG-9-**BF3k** allowed these amphiphilic polybenzofulvene derivatives to interact with water, but the length of side chains was found to be inadequate to warrant a complete dissolution in the aqueous environment. Thus, in order to enhance the hydrophilic character of polybenzofulvene derivatives by increasing side chain protrusion, a triazolo methyl (TM) spacer was inserted between the hydrophobic polybenzofulvene backbone and each MOEG-9 side chain. The resulting polybenzofulvene brush carrying triazolomethyl spacers (poly-6-MOEG-9-TM-**BF3k**) was found to be promptly soluble in water, confirming that the elongation of side chains increases the efficiency of the interaction with aqueous medium by separating the hydrophilic domain (MOEG-9 fragments) from the hydrophobic one (polybenzofulvene backbone) [52]. In addition, it appeared to be less prone to aggregation in water because of its improved solvation shell and because the greater steric hindrance of the side chains makes the hydrophobic interactions leading to aggregation more difficult to be established.

However, TEM studies performed on water solutions of the polymer synthesized by the "grafting through" approach (poly-6-MOEG-9-TM-**BF3k-GT**) revealed the presence of objects (Fig. 51) showing dimensions larger than those determined by SEC-MALS analysis for the isolated macromolecules in a good solvent (radius of gyration around 15 nm).

As confirmed by DLS measurements, the objects observed by TEM were macromolecular aggregates showing average dimensions of 245 nm at 25 °C and of 302 nm at 37 °C (Fig. 52). This result suggested thermoresponsive colloidal properties for poly-6-MOEG-9-TM-**BF3k-GT** water dispersion [52].

Poly-6-MOEG-9-TM-**BF3k** prepared by the "grafting onto" approach (poly-6-MOEG-9-TM-**BF3k-GO**) showed solubility features consistent with that observed with the abovementioned poly-6-MOEG-9-TM-**BF3k-GT**. In short, also the transparent water solutions of poly-6-MOEG-9-TM-**BF3k-GO** showed the character of thermoresponsive colloidal dispersions with the difference that the polymer brush synthesized by the "grafting onto" approach appeared to be molecularly dissolved at room temperature, probably because of its higher Z potential value (−7.23 mV with respect to −1.24 mV shown by poly-6-MOEG-9-TM-**BF3k-GT**). DLS

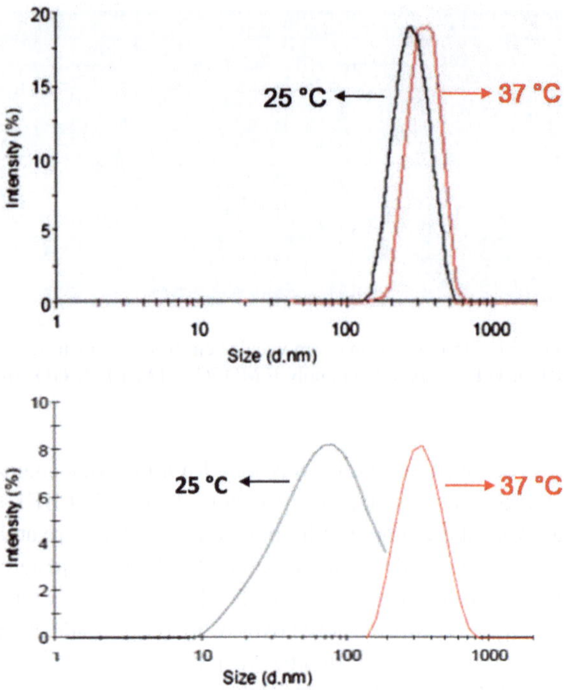

Fig. 52 Size distribution curves of poly-6-MOEG-9-TM-**BF3k-GT** (*top*, sample concentration 0.1 mg/mL) and –6-MOEG-9-TM-**BF3k**-GO (*bottom*, sample concentration 0.5 mg/mL) recorded at 25 and 37 °C by DLS

measurements suggested that the small objects (average dimensions of 57 nm) observed in water solution at room temperature increased their dimensions with the rising of the temperature (290 nm at 37 °C) (Fig. 52). The dimensions observed at room temperature are significantly smaller than those detected with the same polymer brush obtained by spontaneous polymerization of macromonomer 6-MOEG-9-TM-**BF3k** ("grafting through" approach) and agree with the radius of gyration as measured by SEC-MALS analysis (≈34 nm).

The DLS data were in agreement with the images obtained by TEM studies performed at room temperature. In fact, Fig. 51 shows the presence of wormlike objects, which were assumed to be isolated macromolecules, together with relatively abundant globular species presumably resulting from intramolecular collapses, and rare small aggregates confirming the low propensity towards aggregation shown by poly-6-MOEG-9-TM-**BF3k-GO**.

The above-described features have allowed us to propose a possible aggregation mechanism for amphiphilic PPBBs in water (Fig. 53). By segregating and hiding the hydrophobic domain in core structure, a homogeneous three-dimensional distribution of the MOEG-9 side chains represents the essential requirement to obtain a complete solvation of isolated macromolecules. However, these systems are not motionless and the molecular dynamics may induce variations leading to expose the lipophilic moieties of polybenzofulvene backbone. Thus, their incompatibility with the aqueous environment may cause both a hydrophobic intramolecular collapse (backbone

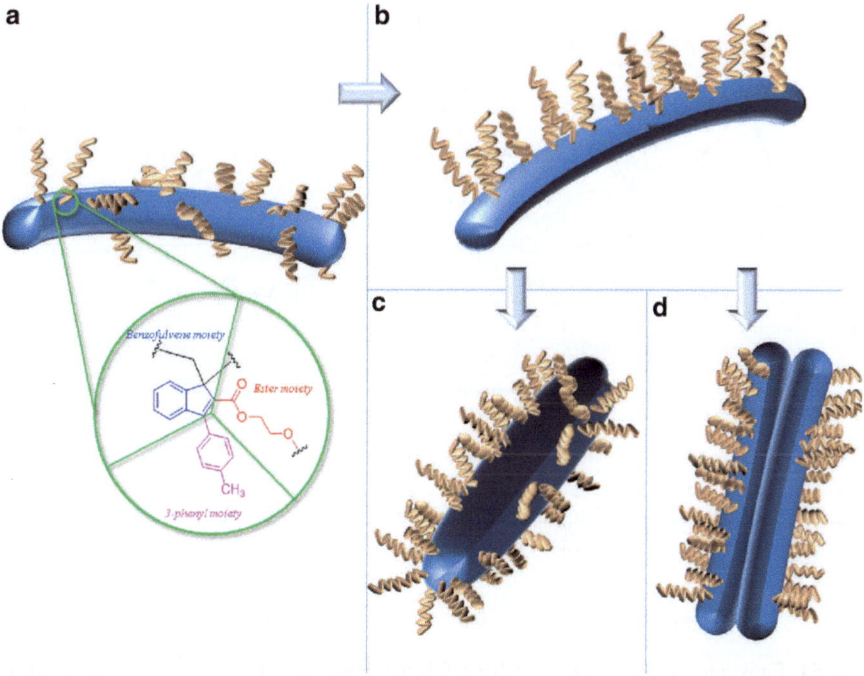

Fig. 53 Hypothesis proposed for the aggregation mechanism of PEGylated polybenzofulvene brushes in water. By way of example, poly-2-MOEG-9-**BF1** is reported (see the magnified insert). The *blue* tubes represent the polybenzofulvene backbones, whereas the *wavy golden tubes* represent the MOEG-9 side chains. (**a**) PPBB interacting with aqueous environment shows a regular distribution of the MOEG-9 side chains. (**b**) When MOEG-9 chains modify their regular distribution around the polybenzofulvene backbone, hydrophobic portions of the polybenzofulvene backbone become partially exposed and available for hydrophobic interactions both (**c**) intramolecularly (backbone folding) and (**d**) intermolecularly. Intermolecular collapse leading to aggregation by means of hydrophobic non-covalent interactions (e.g., π-stacking) among lipophilic backbones

folding by virtue of its flexibility) and an intermolecular one (aggregation). The steric hindrance due to grafting density of MOEG-9 side chains, together with their location, was assumed to explain the progressively decreasing tendency to give water aggregates observed for PPBBs.

Photophysical Properties

The analysis of the absorption spectra of polybenzofulvene derivatives (see Table 9 of "Absorption and Emission Spectroscopy" section) showed the existence of a common absorption component centered around 240 nm attributed to the 3-phenylindene moiety. Moreover, a second component around 285 nm was

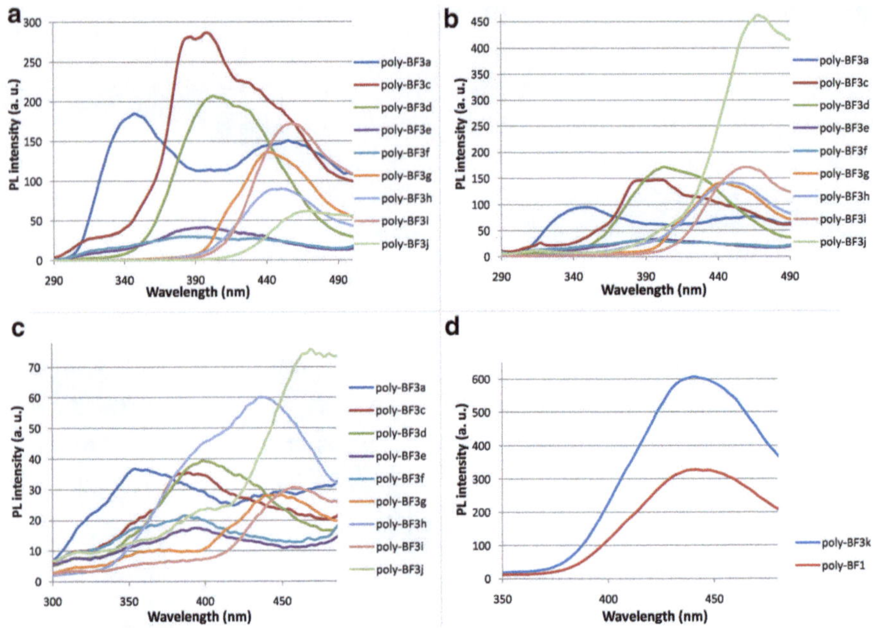

Fig. 54 Emission spectra of some polybenzofulvene derivatives. The spectra were obtained in dichloromethane at room temperature and the solutions were excited at 260 nm. The concentration in monomeric units was about 1.8×10^{-4} M for the spectra in panel (**a**), 1.8×10^{-5} M for the spectra in panel (**b**), and 1.8×10^{-6} M for the spectra in panel (**c**) and (**d**)

particularly evident in the spectra of the polymers bearing an ester group at the 2-position of the indene moiety (e.g., poly-**BF1**, poly-**BF3k**). This second absorption component is probably related to the presence of an extended π-conjugation (for instance, an ethyl cinnamate fragment can be recognized in the monomeric unit of poly-**BF3k**) and found to be sensitive to the presence of auxochromes such as methoxy groups in the methoxy derivatives of poly-**BF3k**. On the other hand, the second absorption component was less evident in the polymers bearing a halogen atom, a methyl group, or a hydrogen atom at the 2-position of the indene nucleus of the monomeric unit. Finally, a component centered at about 346 nm was present in poly-**BF3j** spectrum owing to its even more extended conjugation. In conclusion, the absorption features of polybenzofulvene derivatives are modulated by the nature of the substituents present in the indene nucleus of the monomeric unit. A good correspondence has been noticed between the absorption spectra of the polymers and those of the model compounds.

The polybenzofulvene derivatives bearing an ester group in position 2 of the indene moiety (e.g., poly-**BF1**, poly-**BF3k**) have been found to show intense emission bands around 440–460 nm (see Table 9 of "Absorption and Emission Spectroscopy" section), whereas the presence of substituents different from the COOR was found to decrease significantly the emission intensity (Fig. 54). This result complicated the study and the eventual exploitation of the emission properties of poly-**BF3** derivatives

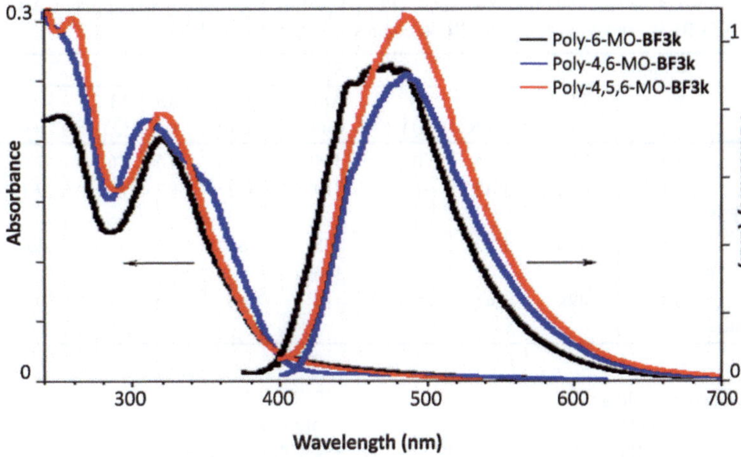

Fig. 55 Comparison of absorption and emission spectra of poly-6-MO-**BF3k** (*black*), poly-4,6-MO-BF3k (*blue*), and poly-4,5,6-MO-**BF3k** (*red*) thin films (absorption) and powders (emission). Reprinted with permission from Cappelli et al. [55]. Copyright 2012, The Royal Society of Chemistry

having a halogen atom, a methyl group, a cyano group, or a hydrogen atom at the 2-position of the indene nucleus of the monomeric unit and was one of the reasons leading us to develop mainly the chemistry of the poly-**BF3k** derivatives.

One of the most interesting developments in the chemistry of the poly-**BF3k** derivatives was the design of polybenzofulvene molecular brush poly-6-MOEG-9-TM-BF3k, its preparation by both "grafting onto" and "grafting through" approaches, and the evidence that poly-6-MOEG-9-TM-**BF3k**-GO was molecularly dissolved at room temperature. Interestingly, the absorption/emission features of poly-6-MOEG-9-TM-**BF3k** were found to be compatible with its detection by means of fluorescence microscopy equipments with standard filters commonly available for the use with fluorescence stain DAPI (4′-6-diamidino-2-phenylindole). This offered the rational basis for the design of nanocarriers, which can be monitored in their movements in the biological environment by fluorescence microscopy.

Owing to the favorable absorption/emission features shown by poly-6-MO-BF3k, poly-4,6-MO-**BF3k**, and poly-4,5,6-MO-**BF3k** in solution, the optical and electrical properties of these polymers were characterized in the solid state (Fig. 55) in the aim of evaluating their use as active materials in electronic devices.

No significant variations were observed in the absorption and emission spectra passing from diluted solutions to films for the three polybenzofulvene derivatives. In the case of PVK, Stokes shift ($\Delta\nu$) was found to double in the solid state indicating stronger carbazole interactions, while for the three poly-**BF3k** derivatives, $\Delta\nu$ was already quite high in diluted solutions and did not increase further in the solid state. This result once more supported the assumption of a tight chromophore stacking in diluted solutions that induces excimer emission via intramolecular interactions. The photoluminescence (PL) quantum yield (QY) in the solid state was similar and even

Table 12 Absorption (abs), photoluminescence (PL) properties, and mobility (μ) of poly-6-MO-BF3k, poly-4,6-MO-**BF3k,** and poly-4,5,6-MO-**BF3k** compared with those of PVK

	Solution				Solid state				
Polymer	Abs λ_{max} (nm)	PL λ_{max} (nm)	$\Delta\nu$ (eV)	PL QY (%)	Abs λ_{max} (nm)	PL λ_{max} (nm)	$\Delta\nu$ (eV)	PL QY (%)	μ (cm²/Vs)
PVK	344	368	0.23	10	344	412	0.59	16	1.67×10^{-8}
Poly-4,5,6-MO-**BF3k**	314	490	1.42	4	319	487	1.34	8	7.12×10^{-9}
Poly-4,6-MO-**BF3k**	305 348sh	488	1.52	5	308 346sh	485	1.47	7	1.07×10^{-9}
Poly-6-MO-**BF3k**	314	488	1.41	6	318	465	1.23	7	9.31×10^{-10}

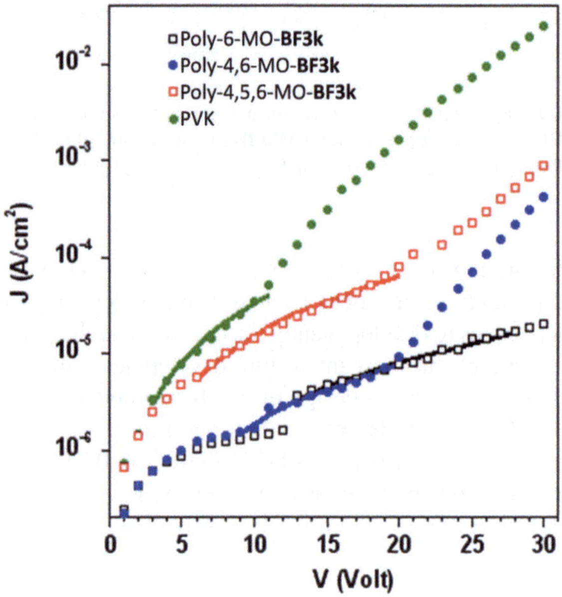

Fig. 56 Current–voltage characteristics of ITO/PEDOT/Polymer/MbO_3/Al devices obtained with either PVK**,** poly-4,5,6-MO-**BF3k**, poly-4,6-MO-**BF3k,** or poly-6-MO-**BF3k**. Reprinted with permission from Cappelli et al. [55]. Copyright 2012, The Royal Society of Chemistry

slightly higher than in solution (Table 12) suggesting for these polymers a potential utility in optoelectronic applications. It is noteworthy that the decreased emission efficiency in the solid state with respect to the solution observed in most conducting polymers represents a limit for their use in light-emitting devices.

The hole-transporting features of the poly-**BF3k** derivatives were evaluated in a comparative study involving poly-6-MO-**BF3k**, poly-4,6-MO-**BF3k**, poly-4,5,6-MO-BF3k**,** and **PVK**, which possesses good hole-transporting properties and is therefore largely used in optoelectronic applications. The hole mobility in the three polybenzofulvene derivatives was evaluated by using a simple hole only device structure (ITO/PEDOT/Polymer/MbO_3/Al) that has been shown to give good ohmic contacts. The J–V curves of the polymers showed ohmic behavior at low fields (Fig. 56), space charge-limited current (SCL) at intermediate fields, and trap fill-limited current at high fields. The mobility values were obtained by fitting the data at intermediate voltages with the Mott–Gurney law and found to be in the range of 10^{-8}–10^{-9} cm^2/V/s, about one order of magnitude lower than that previously reported for PVK measured with an ITO/PEDOT/PVK/Al device. Very intriguingly, the hole mobility appeared to increase in a parallel way with the increase in the number of methoxy substituents up to the corresponding value shown by poly-4,5,6-MO-BF3k, which showed a mobility value only 2.3-fold lower than that of PVK measured in the same conditions (see Table 12). Therefore, the results demonstrated that the polybenzofulvene derivatives possess interesting optical and electrical properties. Furthermore, the propensity shown by benzofulvene scaffold towards its structural manipulation by the introduction of suitable substituents together with the good solubility in the most common organic solvents and the consequent processability (e.g., filmability) pave the way in the development of advanced materials for optoelectronics.

Biological Properties

Our basic assumption in studying the biological properties of polybenzofulvene derivatives has been that the interaction of a polymer with the biological environment requires suitable hydrophilic properties. Therefore, PPBBs have been synthesized and demonstrated to be lacking of cytotoxic effects in the experimental conditions used, suggesting a potential biocompatibility for these materials.

In particular, the potential cytotoxicity and the effects of poly-2-MOEG-9-**BF1** on cell viability were evaluated by LDH (lactate dehydrogenase) and MTT [3-(4,5-dimethylthiazol-2-yl)-2,5-diphenyltetrazolium bromide] tests, respectively, in human hepatocellular carcinoma cells JHH6 (undifferentiated) and HepG2 (hepatocyte-like) incubated at 37 °C in the presence of the solubilized polymer (50 μm) for 1 or 2 days. On the basis of the results, poly-2-MOEG-9-**BF1** was demonstrated to possess a negligible biological impact on human hepatocellular carcinoma cells HepG2 and JHH6, characterized by different phenotypes [25].

Subsequently, a further cell viability study was performed on MCF-7 cell line (human breast cancer cells) and the results confirmed the absence of toxic effect of poly-2-MOEG-9-**BF1** at concentrations of 50, 100, and 200 μM in monomer units for either 24 or 72 h [36].

Similarly, the potential biocompatibility of poly-6-MOEG-9-**BF3k** has been assessed on a healthy human bronchial epithelial (16-HBE) cell line in the same environmental conditions applied for MCF-7 cells used in the abovementioned test. More specifically, cell viability was evaluated by MTS [3-(4,5-dimethylthiazol-2-yl)-5-(3-carboxymethoxyphenyl)-2-(4-sulfophenyl)-2H-tetrazolium] assay on 16-HBE cells, incubated at 37 °C in the presence of polymer aggregate dispersions in Dulbecco's modified Eagle's medium (DMEM, at concentrations of 0.1, 0.25, and 0.5 mg/mL, highly superior to the critical aggregation concentration) for 4 and 24 h. The results provided sound evidence that no cytotoxic effects were produced by this polybenzofulvene derivative, even at the highest polymer concentration tested still after 24 h incubation.

The potential biocompatibility of poly-6-MOEG-9-**BF3k** was evaluated also in vivo by injecting a polymer dispersion (concentration of 0.5 mg/mL) subcutaneously to rats. Then, the animal healthy parameters were monitored and, after 1 and 7 days, histological analyses of tissue cross sections were carried out on the epidermis, dermis, and subcutis. The image analysis of the scrutinized cross sections suggested that no inflammatory process and stromal reaction were generated both after 1 and 7 days. Indeed, the lack of necrosis areas or neutrophilic granulocyte infiltrations indicated that poly-6-MOEG-9-**BF3k** did not stimulate inflammatory response and could be considered to be a highly biocompatible and bioreabsorbable material [79].

The favorable absorption/emission features of poly-6-MOEG-9-TM-**BF3k** allowed fluorescence microscopy experiments to be easily performed with standard equipments. The results of the studies performed with poly-6-MOEG-9-TM-**BF3k-GT** demonstrated that the polymer obtained by the "grafting through" approach was unable to enter HEK293T live cells (human embryonic kidney 293 T cells) and to alter their morphology, probably because of its tendency to form aggregates showing dimensions in the range of the hundreds of nanometers in aqueous medium [52].

Owing to the relatively smaller dimensions of the macromolecular entities generated in aqueous medium by poly-6-MOEG-9-TM-**BF3k** prepared by the "grafting onto approach," its potential cytotoxicity was assessed in two different cell lines (i.e., immortalized mouse fibroblasts NIH3T3 and human neuroblastoma IMR-32 cells). Poly-6-MOEG-9-TM-**BF3k-GO** showed low cytotoxicity, which was well related with its low propensity to interact and cross the cytoplasmic membrane as suggested by the weak fluorescence uptake observed in both the different cell lines used, even after a prolonged incubation time.

In conclusion, the PPBBs are devoid of cytotoxicity and did not stimulate inflammatory response. The interesting biorelevant features make these biocompatible and bioreabsorbable polymers potentially useful in a broad range of biological applications.

Applications of Polybenzofulvene Derivatives

On the basis of their interesting properties (which can be suitably modulated by structure manipulation) intriguing applications for polybenzofulvene derivatives can be envisioned in synthesis, pharmaceutics, optoelectronics, and in several other fields of materials science.

Firstly, polybenzofulvene derivatives can be synthesized completely free from by-products, impurities, or harmful substances because their preparation does not require the use of catalysts or initiators. Secondly, the absence of any relevant toxic effects and cell viability impairments allows PPBBs to be potentially functional in a wide range of biological, biomedical, and biotechnological applications, such as drug delivery, molecular recognition, biosensing, protein and DNA separation, micro- and nanofluidics, as well as tissue engineering.

By means of PEGylation, also the potentially limiting aggregation features shown by polybenzofulvene derivatives could be modulated in order to develop new macromolecular nanocarriers suitable for drug loading and release, and the nanometer dimensions shown by the macromolecular entities present in the water solutions of poly-6-MOEG-9-TM-**BF3k-GO** make this biocompatible polybenzofulvene brush a perfect candidate for this aim.

It is important to recall that valuable features of PEG (water solubility, biocompatibility, lack of toxicity, shielding properties preventing opsonization) allow this inexpensive FDA-approved material to be used in pharmaceutical field to confer a better solubility in the physiological environment and/or to prevent the interaction with plasma proteins and cells.

The interesting physical hydrogel generated by poly-2-MOEG-9-**BF1** showed the typical elastic properties of strong gels, as confirmed by rheological studies. The combination of rheology and NMR transverse relaxation measurements provided information about the hydrogel network structure and the probable pore size distribution [25]. The picture obtained from these studies was that the hydrogel could be considered as a highly concentrated suspension of homogeneous gel particles, in which small intra-particle pores are accompanied by larger channels formed by the imperfect contact among gel particles. In the light of the suggested mean mesh size (≈3–7 nm), poly-2-MOEG-9-**BF1** hydrogel appeared to be capable of acting as a controlled delivery matrix in which small molecules, such as theophylline (van der Waals radius = 0.37 nm) and vitamin B12 (van der Waals radius = 0.85 nm), can migrate at different rates, while larger molecular systems (e.g., proteins, enzymes, dendrimers) can be entrapped inside gel network or confined onto polymer surface [25]. Thus, poly-2-MOEG-9-**BF1** has been investigated as hydrogel-forming material able to complex and release HIgG at physiological mimicking conditions [36]. Release studies have provided experimental evidence about the ability of the hydrogel to release intact the functional form of the loaded model protein HIgG as a function of pH. Biphasic and prolonged HIgG release profiles (Fig. 57) were detected at both the two different pH values with a relatively low first burst release (probably due to the release of the protein adsorbed onto hydrogel surface) and a second slower release stage controlled by the diffusion of the protein through the large hydrogel channels. The pH dependence of the release kinetics was assumed to be related to the different protein solubility at the two pH values taken into consideration.

On the basis of the protein dimensions (height = 14.5 nm, width = 8.5 nm, thickness = 40 nm) and the mesh size distribution, the antibody was indeed assumed to be hosted outside the polymer network on the surface of gel channels pervading poly-2-MOEG-9-**BF1** matrix, as supported by the analysis of SEM images (Fig. 58).

Similarly, the interesting features (e.g., aggregation properties, biocompatibility, and bioreabsorbability) shown by second-generation PPBB poly-6-MOEG-9-**BF3k** stimulated its evaluation as a material for the production of injectable thermoresponsive microgels for the release of low molecular weight peptide drugs. The physical hydrogel generated by poly-6-MOEG-9-**BF3k** was found to be relatively unstable to ultrasound exposure and formed a dispersion leading to thermoresponsive microgels. Indeed, the dispersion was transparent (as a solution) at temperature

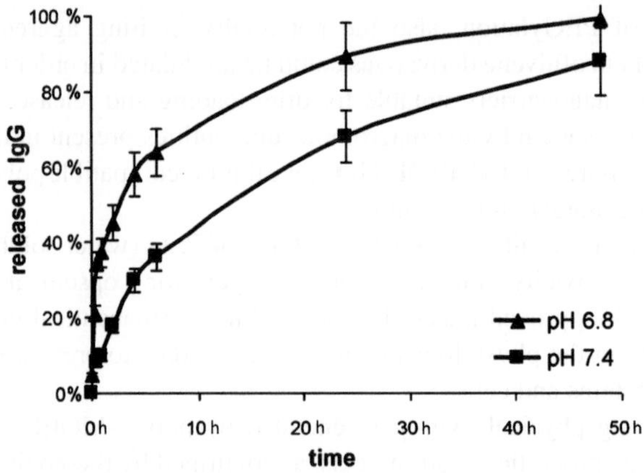

Fig. 57 pH-dependent release profile of HIgG from the complex poly-2-MOEG-9-**BF1**/antibody. A biphasic and prolonged protein release can be observed at both pH values (6.8 and 7.4). Reprinted with permission from Licciardi et al. [36]. Copyright 2010, Elsevier

Fig. 58 SEM images of lyophilized samples of poly-2-MOEG-9-**BF1** loaded with HIgG. Magnification values: 1,000× (*left*) and 10,000× (*right*). HIgG aggregates are present on the polymer surface and in the gel fractures probably produced by cryogenic freezing. This particular distribution of the carried guest can explain the biphasic profile of HIgG release from hydrogel. Reprinted with permission from Licciardi et al. [36]. Copyright 2010, Elsevier

values below 25 °C, while it became opalescent when temperature increased to provide a homogeneous microgel population near 37 °C (at this temperature the critical aggregation temperature was estimated to be 0.019 mg/mL, and the average diameter of the aggregate was 459 nm). The thermoresponsive nature of the system was exploited in developing a drug delivery system potentially capable of protecting peptide drugs from degradation and regulating the drug release rate in function of the temperature. Leuprolide, the peptide drug used in the treatment of hormone-responsive tumors such as breast cancer, was selected as a model drug because it contains in its structure aromatic amino acids (tryptophan, tyrosine, and histidine),

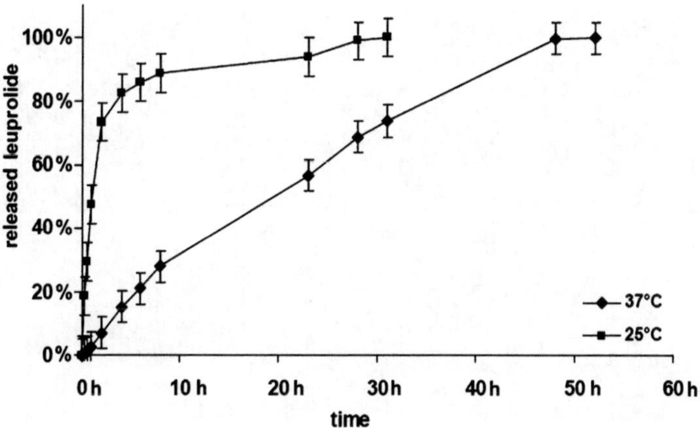

Fig. 59 Release profile of leuprolide from the complex with poly-6-MOEG-9-**BF3k**. Reprinted with permission from Licciardi et al. [79]. Copyright 2012, Elsevier

which were assumed to establish aromatic interactions with poly-6-MOEG-9-**BF3k** aromatic rings. These interactions were thought to be the basis for the suitable loading and delayed diffusion of the peptide drug from microgel aggregates. Leuprolide was easily and effectively loaded in poly-6-MOEG-9-**BF3k**. Then, the release studies showed temperature-dependent release profiles (Fig. 59) with a prolonged, constant, and complete release at 37 °C and a prominent burst effect (50 % of leuprolide was released after 2 h) at 25 °C in agreement with the thermoresponsive behavior shown by this PPBB [77].

The establishment of aromatic interactions was the driving idea also in the design of the nanocomposite material constituted by poly-**BF3a** nanoparticles (filler) and the drug model **CR3124** (matrix). We assumed that the aromatic interactions between the polymer and the drug molecule (on the nanoparticle surface) could stabilize the amorphous form of the latter despite of its pronounce tendency to crystallization. Two different nanoprecipitation/desolvation methods (procedure A, gradual excess of bad solvent; procedure B, continuous excess of bad solvent) were evaluated either in the absence or in the presence of PEG1000 as wetting agent to obtain various nanocomposite materials, which were characterized in terms of loading, morphology, and release. The formation of nanoparticles showing favorable shape and dimensions was related to the procedure used and to the presence of PEG that played a beneficial role in affecting nanoparticle morphology (Fig. 60). The inclusion of the drug molecule in the formulation produced nanoparticle aggregation and the interaction of the resulting materials with the aqueous environment was found to depend on the hydrophilicity of the matrix (**CR3124** and PEG when present) [24].

Similarly, the release profile was dramatically affected by matrix composition. The high wettability of the binary matrix composed by **CR3124** and PEG produced very rapid releases, whereas in the absence of PEG, the lower wettability of **CR3124** caused the dissolution kinetics to slow down (Fig. 61).

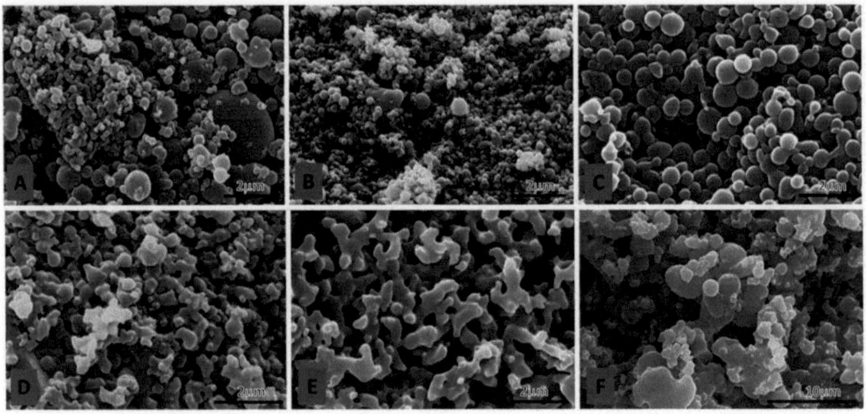

Fig. 60 SEM micrographs of poly-**BF3a** nanoparticles (*top* panels (**a**)–(**c**)) and **CR3124**-loaded poly-**BF3a** nanoparticles (*bottom* panels (**d**)–(**f**)) obtained in different conditions. Panel (**a**): poly-BF3a 100 %, procedure A. Panel (**b**): poly-**BF3a** 100 %, procedure B. Panel (**c**): poly-**BF3a** 70 %, PEG1000 30 %, procedure B. Panel (**d**): poly-**BF3a** 80 %, **CR3124** 20 %, procedure A. Panel (**e**): poly-**BF3a** 90 %, **CR3124** 10 %, procedure B. Panel (**f**): poly-**BF3a** 60 %, PEG1000 30 %, **CR3124** 10 %, procedure B. Reprinted with permission from Cappelli et al. [24]. Copyright 2010, Springer

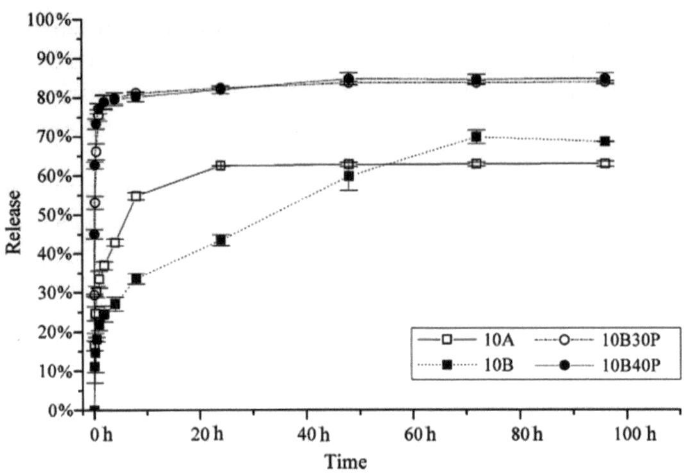

Fig. 61 **CR3124** release profiles of different poly-**BF3a** nanoparticle formulations. **Legend.** Preparation 10A: poly-**BF3a** 90 %, **CR3124** 10 %, procedure A. Preparation 10B: poly-**BF3a** 90 %, **CR3124** 10 %, procedure B. Preparation 10B30P: poly-**BF3a** 60 %, PEG1000 30 %, **CR3124** 10 %, procedure B. Preparation 10B40P: poly-**BF3a** 50 %, PEG1000 40 %, **CR3124** 10 %, procedure B. Reprinted with permission from Cappelli et al. [24]. Copyright 2010, Springer

In conclusion, polybenzofulvene nanoparticles demonstrated to be useful as substrates for the stabilization of the amorphous form of drug molecules showing pronounced tendency to the crystallization and for the modulation of their dissolution

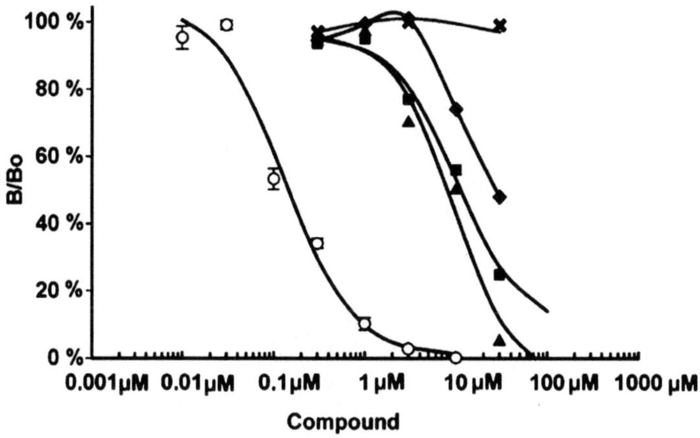

Fig. 62 AT_1 receptor binding: competition curves of poly-**BF-AT$_1$**, its monomer **BF-AT$_1$** and PTSA. Stock solutions of **BF-AT$_1$** and poly-**BF-AT$_1$** were freshly prepared and stored at 37 °C for 3 or 6 days. PTSA (*cross*) was considered an impurity of poly-**BF-AT$_1$** and it was assayed to evidence its inactivity. The assays performed on freshly prepared solutions of poly-**BF-AT$_1$** (*filled diamond*) revealed a dose-dependent inhibition of the radioligand-specific binding AT_1 receptors. The inhibition curve of poly-**BF-AT$_1$** is significantly shifted to the left when the stock solution was used after stored in buffer at 37 °C for the first 3 days (*filled square*), but does not undergo significant changes after stored for additional 3 days (*filled triangle*). On the contrary, monomer **BF-AT$_1$** (*open circle*) tested in parallel in each assay was insensitive to the storage conditions; values indicated the mean ± standard error of three replicates. Reprinted with permission from Cappelli et al. [18]. Copyright 2006, American Chemical Society

kinetics. Thus, polybenzofulvene derivatives could find applications as inactive components of different drug delivery systems (e.g., hydrogels, microgels, and nanoparticles). However, the thermoreversible polymerization of AT_1 receptor ligand **BF-AT1** opens the perspective of applying polybenzofulvene derivatives as active components of drug delivery systems such as polymeric prodrugs. The angiotensin II AT_1 receptor affinity shown by **BF-AT1** (binding IC_{50} = 60 nM) was apparently lost as a consequence of the spontaneous polymerization, but was partially recovered when the poly-**BF-AT1** was incubated in buffer at pH 7.4 and 37 °C for three days (Fig. 62) [18].

The appreciable depolymerization of poly-**BF-AT1** observed in buffer at room temperature increased its extent during the storage of freshly prepared solutions at 37 °C up to establishment of an apparent equilibrium within about 3 days. The IC_{50} values measured for the polymer and its corresponding monomer appeared to be different in terms of 2 orders of magnitude, letting us to assume that in these conditions poly-**BF-AT1** was able to release one molecule of monomer **BF-AT1** each about 100 monomeric units [18].

Although the design of new polymeric prodrugs could be considered a fascinating research theme, we have soon realized the complexity of the development matter and the results of several different attempts performed in our laboratories lay still unpublished over our desks.

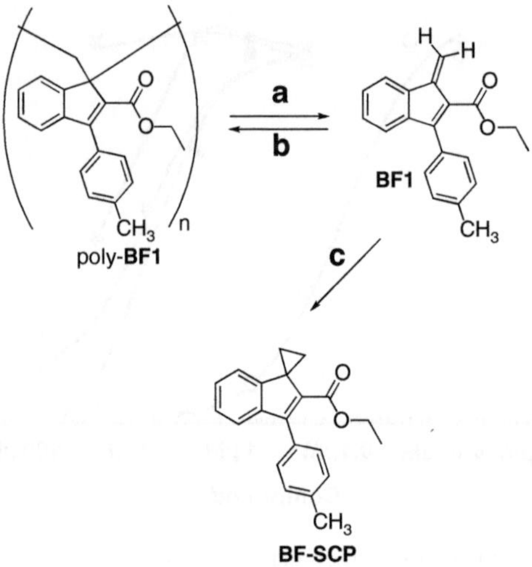

Scheme 19 Synthesis of spirocyclopropane derivative **BF-SCP** from poly-**BF1**. Reagents: (*a*) solvent, heating (*open triangle*); (*b*) solvent removal; and (*c*) CH_2N_2, $(CH_3COO)_2Pd$, $CHCl_3$

One of the first applications evaluated for polybenzofulvene derivatives was their use as stock forms of the corresponding monomers in synthetic approaches. For instance, *trans*-diene **BF1** generated from poly-**BF1** has been found to be synthetically useful to obtain a spirocyclopropane derivative **BF-SCP** by means of a palladium-catalyzed cyclopropanation with diazomethane (Scheme 19) [20].

Finally, the promising photophysical properties of π-stacked poly-4,6-MO-**BF3k** and poly-4,5,6-MO-**BF3k**, such as the interesting hole mobility shown by the latter polybenzofulvene derivative, can pave the way in the development of advanced polybenzofulvene materials for optoelectronic applications. A suitable chemical design based on the apparent key role of the methoxy substituents in controlling the electrical features of this new class of polymers (i.e., enhancing effect on charge mobility) might allow to achieve further improvements in order to obtain newsworthy alternatives to the host polymers commonly used in electronic devices [55].

Conclusions and Perspectives

Since its discovery in 2002, the spontaneous polymerization of benzofulvene derivatives has attracted our attention because of the particular features shown, and we felt committed to bringing them to the attention of the scientific community. After a decade of research work in our laboratories, around 50 new polymers belonging to this family have been synthesized and the relevant studies have been published in the most important journals dealing with polymer science or in different ones depending on the application explored for the newly synthesized polybenzofulvene derivatives. Thus, the spontaneous polymerization of benzofulvene monomers and the special features of the corresponding polymers appear to become a well-established topic. However, many things remain to do. The exploration of the chemistry is still ongoing and provides everyday new evidence. At a first sight, the chemical space around benzofulvene derivatives could appear limited, but we are sure that a portion of the infinity is still too big for our limited possibilities. Moreover, new applications are under evaluation for these polymers as new advanced functional materials.

References

1. Day JH (1953) Chem Rev 53:167
2. Bergmann ED (1968) Chem Rev 68:41
3. Möllerstedt H, Piqueras MC, Crespo R, Ottosson H (2004) J Am Chem Soc 126:13938
4. Stępień BT, Krygowsky TM, Cyrański M (2002) J Org Chem 67:5987
5. Abelt CJ, Roth HD (1985) J Am Chem Soc 107:6814
6. Nakano T, Takewaki K, Yade T, Okamoto Y (2001) J Am Chem Soc 123:9182
7. Nakano T, Yade T (2003) J Am Chem Soc 125:15474
8. Nakano T, Yade T, Fukuda Y, Yamaguchi T, Okumura S (2005) Macromolecules 38:8140
9. Nakano T, Nakagawa O, Tsuji M, Tanikawa M, Yade T, Okamoto Y (2004) Chem Commun 21:144
10. Nakano T (2003) Heat-decomposable polymer. WO03095523
11. Nakano T (2002) Polymer having unique optical property and polymerizable monomer thereof. WO2002088202
12. Nakano T (2003) Optically active high-molecular compounds. WO2003102039
13. Nakano T, Yade T, Yokoyama M, Nagayama N (2004) Chem Lett 33:296
14. Nakano T (2010) Polym J 42:103
15. Londergan TM, Teng CJ, Weber WP (1999) Macromolecules 32:1111
16. The story of serendipity, http://undisci.berkeley.edu/article/serendipity
17. Cappelli A, Pericot Mohr G, Gallelli A, Rizzo M, Anzini M, Vomero S, Mennuni L, Ferrari F, Makovec F, Menziani MC, De Benedetti PG, Giorgi G (2004) J Med Chem 47:2574
18. Cappelli A, Pericot Mohr G, Giuliani G, Galeazzi S, Anzini M, Mennuni L, Ferrari F, Makovec F, Kleinrath EM, Langer T, Valoti M, Giorgi G, Vomero S (2006) J Med Chem 49:6451
19. Cappelli A, Nannicini C, Gallelli A, Giuliani G, Valenti S, Pericot Mohr G, Anzini M, Mennuni L, Ferrari F, Caselli G, Giordani A, Peris W, Makovec F, Giorgi G, Vomero S (2008) J Med Chem 51:2137
20. Cappelli A, Pericot Mohr G, Anzini M, Vomero S, Donati A, Casolaro M, Mendichi R, Giorgi G, Makovec F (2003) J Org Chem 68:9473
21. Cappelli A, Anzini M, Vomero S, Donati A, Zetta L, Mendichi R, Casolaro M, Lupetti P, Salvatici P, Giorgi G (2005) J Polym Sci A Polym Chem 43:3289
22. Cappelli A, Galeazzi S, Giuliani G, Anzini M, Donati A, Zetta L, Mendichi R, Aggravi M, Giorgi G, Paccagnini E, Vomero S (2007) Macromolecules 40:3005
23. Cappelli A, Galeazzi S, Giuliani G, Anzini M, Aggravi M, Donati A, Zetta L, Boccia AC, Mendichi R, Giorgi G, Paccagnini E, Vomero S (2008) Macromolecules 41:2324
24. Cappelli A, Galeazzi S, Zanardi I, Travagli V, Anzini M, Mendichi R, Petralito S, Memoli A, Paccagnini E, Peris W, Giordani A, Makovec F, Fresta M, Vomero S (2010) J Nanopart Res 12:895

25. Cappelli A, Galeazzi S, Giuliani G, Anzini M, Grassi M, Lapasin R, Grassi G, Farra R, Dapas B, Aggravi M, Donati A, Zetta L, Boccia AC, Bertini F, Samperi F, Vomero S (2009) Macromolecules 42:2368
26. Duncan R (2003) Nat Rev Drug Discov 2:347
27. Langer R (1990) Science 249:1527
28. Lutz J-F, Andrieu J, Uzgiin S, Rudolph C, Agarwal S (2007) Macromolecules 40:8540
29. Lutz J-F, Hoth A (2006) Macromolecules 39:893
30. Oh JK, Min K, Matyjaszewski K (2006) Macromolecules 39:3161
31. Tao L, Mantovani G, Lecolley F, Haddleton DM (2004) J Am Chem Soc 126:13220
32. Ishizone T, Seki A, Hagiwara M, Han S, Yokoyama H, Oyane A, Deffieux A, Carlotti S (2008) Macromolecules 41:2963
33. Wang X-S, Armes SP (2000) Macromolecules 33:6640
34. Wang X-S, Lascelles SF, Jackson RA, Armes SP (1999) Chem Commun 18:1817
35. Zhang M, Müller AHE (2005) J Polym Sci A Polym Chem 43:3461
36. Licciardi M, Grassi M, Di Stefano M, Feruglio L, Giuliani G, Valenti S, Cappelli A, Giammona G (2010) Int J Pharm 390:183
37. Cappelli A, Paolino M, Anzini P, Giuliani G, Valenti S, Aggravi M, Donati A, Mendichi R, Zetta L, Boccia AC, Bertini F, Samperi F, Battiato S, Paccagnini E, Vomero S (2010) J Polym Sci A Polym Chem 48:2446
38. Kolb HC, Finn MG, Sharpless KB (2001) Angew Chem Int Ed 40:2004
39. Meldal M, Tornøe CW (2008) Chem Rev 108:2952
40. Dondoni A (2008) Angew Chem Int Ed 47:8995
41. Kade MJ, Burke DJ, Hawker CJ (2010) J Polym Sci A Polym Chem 48:743
42. Lodge TP (2009) Macromolecules 42:3827
43. Gao H, Matyjaszewski K (2007) J Am Chem Soc 129:6633
44. Zhang X, Lian X, Liu L, Zhang J, Zhao H (2008) Macromolecules 41:7863
45. Campos LM, Killops KL, Sakai R, Paulusse JMJ, Damiron D, Drockenmuller E, Messmore BW, Hawker CJ (2008) Macromolecules 41:7063
46. Goldmann AS, Walther A, Nebhani L, Joso R, Ernst D, Loos K, Barner-Kowollik C, Barner L, Müller AHE (2009) Macromolecules 42:3707
47. Fleischmann S, Komber H, Voit B (2008) Macromolecules 41:5255
48. Munteanu M, Choi SW, Ritter H (2008) Macromolecules 41:9619
49. Kempe K, Krieg A, Becer R, Schubert US (2012) Chem Soc Rev 41:176
50. van der Ende AE, Harrell J, Sathiyakumar V, Meschievitz M, Katz J, Adcock K, Harth E (2010) Macromolecules 43:5665
51. Golas PL, Matyjaszewski K (2010) Chem Soc Rev 39:1338
52. Cappelli A, Paolino M, Grisci G, Giuliani G, Donati A, Mendichi R, Boccia AC, Samperi F, Battiato S, Paccagnini E, Giacomello E, Sorrentino V, Licciardi M, Giammona G, Vomero S (2011) Polym Chem 2:2518
53. Morisaki Y, Fernandes JA, Chujo Y (2009) Macromol Rapid Commun 30:2107
54. Xie L-H, Deng X-Y, Chen L, Liu R-R, Hou X-Y, Wong K-Y, Ling Q-D, Huang W (2009) J Polym Sci A Polym Chem 47:5221
55. Cappelli A, Paolino M, Grisci G, Giuliani G, Donati A, Mendichi R, Boccia AC, Botta C, Mróz W, Samperi F, Scamporrino A, Giorgi G, Vomero S (2012) J Mater Chem 22:9611
56. Grignard V, Courtot C (1915) Compt Rend 160:500
57. Neuenschwander M, Vogeli R, Fahrni H-P, Lehmann H, Ruder J-P (1977) Helv Chim Acta 60:1073
58. Larock RC, Doty MJ, Cacchi S (1993) J Org Chem 58:4579
59. Marsili A (1961) Ann Chim 51:237
60. Koelsch CF (1960) J Org Chem 25:2088
61. Carpino LA (1987) Acc Chem Res 20:401
62. Carpino LA, Mansour EM, Knapczyk J (1983) J Org Chem 48:666
63. Bordwell FG, Satish AV (1992) J Am Chem Soc 114:10173

64. Hall HK, Padias AB (1990) Acc Chem Res 23:3
65. Hall HK, Padias AB (1997) Acc Chem Res 30:322
66. Walling C (1949) J Am Chem Soc 71:1930
67. Kazantsev OA, Shirshin KV (2004) Polymer 45:5021
68. Pepper DC (1978) J Polym Sci Polym Symp 62:65
69. Greer SC (1998) J Phys Chem B 102:5413
70. Ryan B, McCann G (1996) Macromol Rapid Commun 17:217
71. DeVoe H (1969) Ann N Y Acad Sci 158:298
72. Montaudo G, Montaudo MS, Samperi F (2002) Martix-assisted laser desorption/ionization mass spectrometry of polymers (MALDI-MS). In: Montaudo G, Lattimer RP (eds) Mass spectrometry of polymers, vol 10. CRC, Boca Raton, p 419
73. Montaudo G, Samperi F, Montaudo MS (2006) Prog Polym Sci 31:277
74. Räder HJ, Schrepp W (1998) Acta Polym 49:272
75. Pasch H, Schrepp W (2003) MALDI-TOF mass spectrometry of polymers. Springer, Berlin, p 298
76. Nielen MWF (1999) Mass Spectrom Rev 18:309
77. Montaudo G, Garozzo D, Montaudo MS, Puglisi C, Samperi F (1995) Macromolecules 28:7983
78. Samperi F, Battiato S, Puglisi C, Giovanella U, Mendichi R, Destri S (2012) Macromolecules 45:1811
79. Licciardi M, Amato G, Cappelli A, Paolino M, Giuliani G, Belmonte B, Guarnotta C, Pitarresi G, Giammona G (2012) Int J Pharm 438:279

Cyclophane-Based π-Stacked Polymers

Yasuhiro Morisaki and Yoshiki Chujo

Introduction

Cyclophane is a general term for a cyclic compound containing at least one aromatic ring in the cyclic skeleton. [2.2]Paracyclophane, which plays an important role in the discussion in this chapter, is a typical cyclophane compound consisting of two benzene rings bridged at the *para*-positions by two ethylene chains. It was first synthesized by Brown and Farthing in 1949 as a pyrolysis product of *p*-xylene [1]. In 1951, Cram and Steinberg reported its direct synthesis by the Wurtz-type intramolecular cyclization of 1,4-bis(bromomethyl)benzene [2]. Since then, several synthetic routes to [2.2]paracyclophane have been developed as a result of the notable progress made in the last 60 years in organic chemistry [3, 4]. At present, [2.2]paracyclophane is commercially available and readily obtained on the gram scale.

[*m.n*]Cyclophanes (in particular, $m \le 3$ and $n \le 3$, where *m* and *n* are the number of carbon atoms in the alkyl chains), consisting of two stacked benzene rings and two bridging alkyl chains, have unique structures and characteristic π-electron systems. Considerable attention has been paid to their synthesis, reactivities, optoelectronic properties, and theoretical features in the field of organic chemistry [3, 4]. However, despite the large number of studies on cyclophane chemistry, conjugated polymers containing cyclophane skeletons in the polymer main chain have been largely ignored [5–10]. In addition, the few cyclophane-containing polymers reported were insoluble in common organic solvents and did not have film-forming abilities [11–16]; hence, their structures and properties were not elucidated in detail until recently.

Y. Morisaki (✉) • Y. Chujo (✉)
Department of Polymer Chemistry, Graduate School of Engineering, Kyoto University, Katsura, Nishikyo-ku, Kyoto, 615-8510, Japan
e-mail: ymo@chujo.synchem.kyoto-u.ac.jp; chujo@chujo.synchem.kyoto-u.ac.jp

T. Nakano (ed.), *π-Stacked Polymers and Molecules: Theory, Synthesis, and Properties*, DOI 10.1007/978-4-431-54129-5_3,

Over the last decade, we have focused our attention on the unique structures of [2.2]paracyclophane and related compounds. In particular, we studied the synthesis of novel conjugated polymers that included [2.2]paracyclophane as a repeating unit in their main chains [5, 6, 8–10]. Incorporation of a [2.2]paracyclophane skeleton into a π-conjugated polymer backbone leads to a π-stacked structure, in which π-electron systems are partly π-stacked in proximity in a polymer chain. In this chapter, we will focus on unprecedented π-stacked polymers based on [2.2]paracyclophane. Because of the limited number of studies on such π-stacked polymers, we will primarily introduce our recent results, together with important examples of cyclophane-containing π-stacked polymers developed by other research groups.

π-Stacked Polymer Based on [2.2]Metacyclophane

The first example of a π-stacked polymer containing a cyclophane skeleton as a repeating unit in the polymer chain was reported in 1985 [11, 12]. In this case, 8,16-dihydroxy[2.2]metacyclophane **1** was incorporated as a key component, as shown in Scheme 1. Metacyclophane **1** is an electron-rich compound, which led to oxidative coupling with air (oxygen) under sunlight to yield a metacyclophane dimer **2** in the form of bright reddish-orange whiskers. Compound **2** was used as a monomer and heated at 300–400 °C to yield poly[2.2]metacyclophane **3**. The obtained polymer **3** was insoluble in common organic solvents; the reaction was monitored and the structure was confirmed by FTIR spectroscopy. The reported estimate of the electrical conductivity of polymer **3** was of the order of 10^{-9} S cm^{-1}, and it increased dramatically to 0.25 S cm^{-1} by chemical doping with H_2SO_4 vapor.

The synthesis of π-stacked polymer **4**, in which diphenoquinone units were stacked and layered, was reported by the same group in 1985 [11]. As shown in Scheme 1, the oxidative coupling polymerization of **1** in the presence of $FeCl_3$ afforded diphenoquinone-stacked polymer **4** quantitatively. This polymer was also

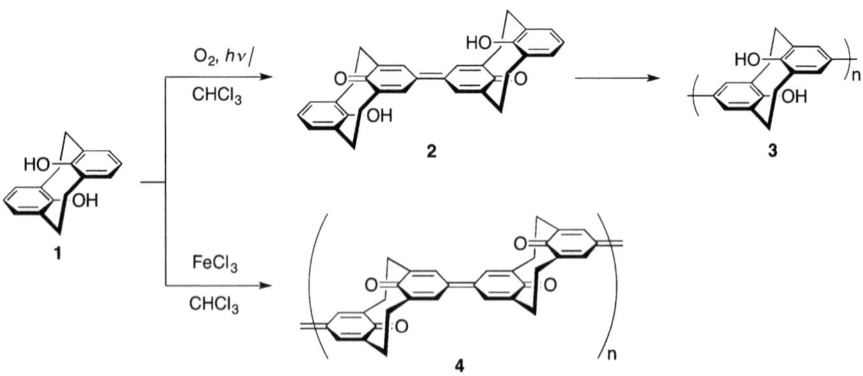

Scheme 1 Synthesis of polymers **3** and **4**

insoluble in any solvents and was therefore characterized by FTIR spectroscopy as well as elemental analysis. The electrical conductivity of **4** was measured. The value for the pristine sample was 10^{-9} S cm^{-1}, and chemical doping with H_2SO_4 vapor provided a conductivity of 10^{-4} S cm^{-1}.

π-Stacked Polymers Based on Pseudo-*para*-Disubstituted [2.2] Paracyclophane

After the report of poly[2.2]metacyclophane **3** described above, there were no reports on the synthesis of cyclophane-based π-stacked polymers for over 15 years. In 2001 and 2002, polymerizations of thiophene-substituted [2.2]paracyclophanes were reported independently by two research groups [13–16]. Both groups synthesized π-stacked polymers by electrochemical polymerization using cyclic voltammetry. For example, Audebert and coworkers reported that the electrochemical polymerization of monomer **5** deposited π-stacked polymer **6** on a glassy carbon or indium–tin oxide electrode (Scheme 2) [13, 14].

Collard and coworkers synthesized polymer **8**, which was obtained by electrochemical oxidation as well as chemical oxidation using $FeCl_3$ in $CHCl_3$ (Scheme 2) [15, 16]. The octyl groups in the thiophene rings of polymer **8** enhanced solubility in organic solvents such as hexane and $CHCl_3$. The molecular weight of the polymer could be measured by gel permeation chromatography (GPC) in $CHCl_3$, and the weight-average molecular weight (M_w) of the $CHCl_3$-soluble portion of **8** was calculated to be 50,000 [16].

Palladium-catalyzed cross-coupling is a powerful tool for the synthesis of various π-conjugated molecules and polymers through sp^2 and sp carbon–carbon bond formation [17–20]. Scheme 3 shows the synthetic routes for poly(*p*-arylene–ethynylene)-type (PAE-type) π-stacked polymers by Sonogashira–Hagihara cross-coupling [21, 22]. Polymer **13** was obtained from the monomers pseudo-*para*-dibromo[2.2]

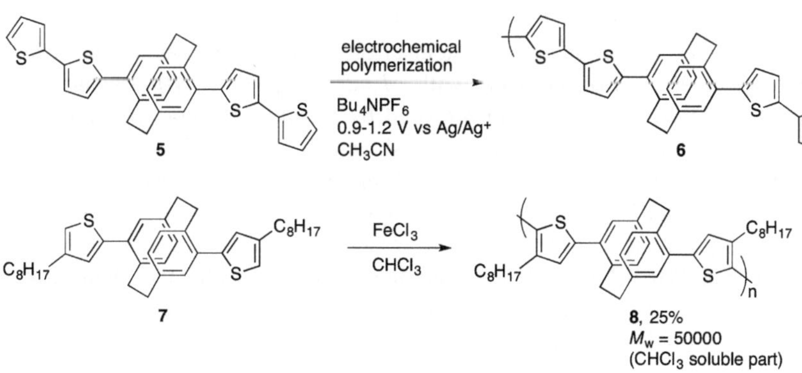

Scheme 2 Synthesis of polymers **6** and **8**

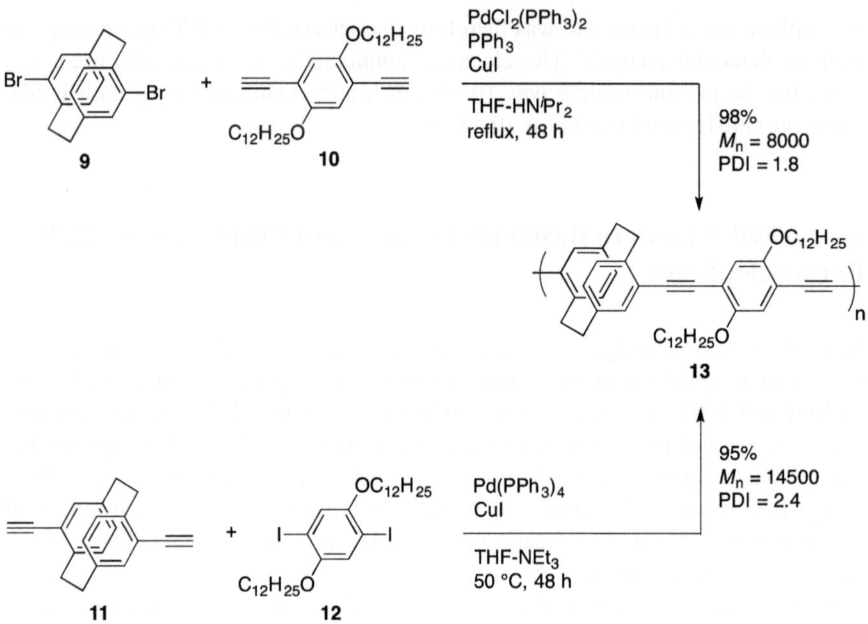

Scheme 3 Synthesis of PAE-type polymer **13**

paracyclophane **9** and diethynylbenzene **10** in the presence of a catalytic amount of $PdCl_2(PPh_3)_2$ [23]. The polymerization of pseudo-*para*-diethynyl[2.2]paracyclophane **11** with diiodobenzene **12** afforded the same π-stacked polymer **13** with a higher number-average molecular weight (M_n = 14,500) than that of the polymer obtained from **9** to **10** (M_n = 8,000). The dibromo[2.2]paracyclophane **9** was considered to be relatively inert to oxidative addition by palladium because of steric hindrance and the electron-rich nature of the paracyclophane moiety.

Scheme 4 shows the synthesis of poly(*p*-arylenevinylene)-type (PAV-type) π-stacked polymer by Heck–Mizoroki cross-coupling [24, 25]. The treatment of pseudo-*para*-divinyl[2.2]paracyclophane **14** with diiodofluorene **15** using a $Pd(OAc)_2$/P(o-Tol)$_3$/NBu_3 catalytic system afforded the corresponding π-stacked polymer **16** in 96 % isolated yield with M_n = 5,200 [26]. Poly(*p*-arylene)-type (PA-type) π-stacked polymers were prepared by Suzuki–Miyaura cross-coupling [27, 28], and a representative example is shown in Scheme 5. The $Pd(PPh_3)_4$-catalyzed polymerization of monomer **9** with bis(boronic acid ester) monomer **17** afforded the polymer **18** in 89 % isolated yield (M_n = 2,600) [29].

All the polymers described above were reasonably soluble in common organic solvents such as THF, $CHCl_3$, CH_2Cl_2, and toluene, which enabled the characterization in their solutions by NMR, UV–vis absorption, and fluorescence spectroscopy. The polymers exhibited sufficiently good film-forming abilities, and the thin films were readily formed by casting and spin coating from the corresponding toluene solutions. They also exhibited good thermal stability; for example, polymer **18** was

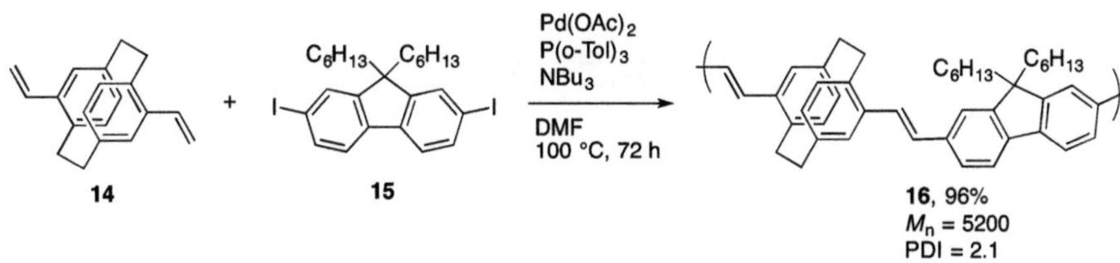

Scheme 4 Synthesis of PAV-type polymer **16**

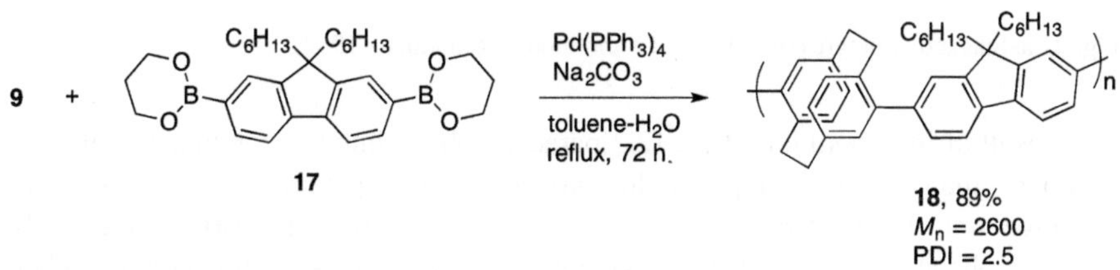

Scheme 5 Synthesis of PA-type polymer **18**

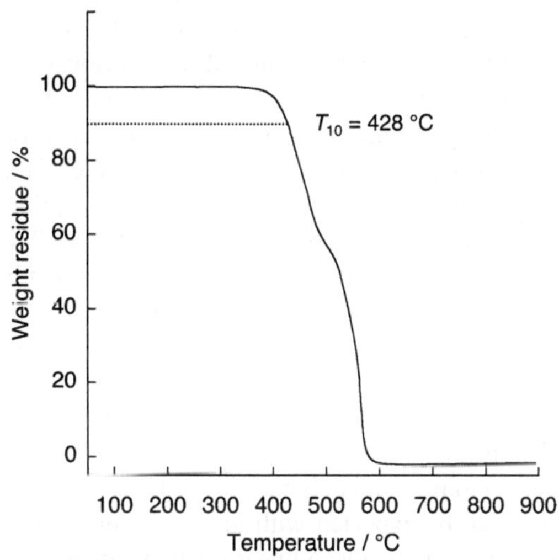

Fig. 1 Thermogravimetric analysis of polymer **18** under air (10 °C min^{-1})

found to have a 10 % weight loss temperature (T_{10}) of 428 °C, as determined by thermogravimetric analysis (Fig. 1) [29].

One of the most characteristic features of [2.2]paracyclophane-based polymers is the unique π-stacked structure of π-electron systems. Here, we will focus on the structure of PAE-type polymer **13**. As illustrated in Fig. 2, we can see that polymer **13** consists of π-electron systems corresponding to compound **19**. In a single polymer chain, the π-electron systems are partly stacked and layered one-dimensionally.

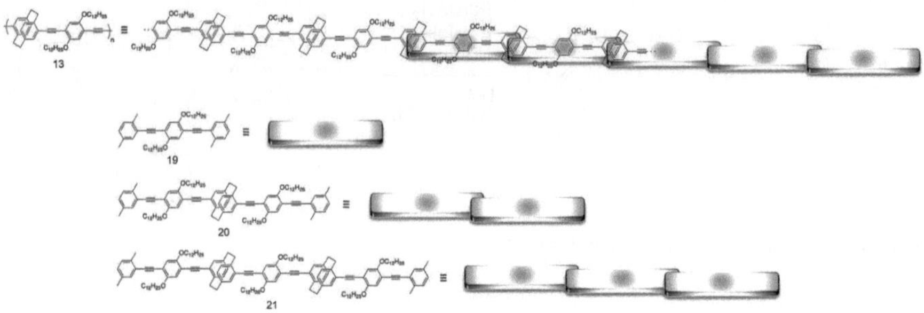

Fig. 2 π-Stacked structures of polymer **13** and model compounds **19–21**

As a result of incorporation of the [2.2]paracyclophane into the conjugated polymer backbone, the [2.2]paracyclophane does not act as a [2.2]paracyclophane unit but as a bridge unit linking the π-electron systems. The principal part of the [2.2]paracyclophane-based π-stacked polymers is the stacked π-electron system rather than the [2.2]paracyclophane; therefore, their opto-electrical properties depend on the properties of the stacked π-electron systems, regardless of [2.2]paracyclophane. In this case, the properties of compound **19** are strongly reflected in the properties of **13**.

To obtain an insight into the conjugation system of [2.2]paracyclophane-based π-stacked polymers [30], compounds **20** and **21** were prepared (Fig. 2), in which two and three π-electron systems were π-stacked, respectively. In addition, π-stacked polymer **22** containing xylyl groups at the polymer chain ends was synthesized; **22** was separated into four fractions **22a–d** by size-exclusion chromatography (SEC), as shown in Scheme 6. Figure 3a shows the UV–vis absorption spectra of compounds **19–21**, polymers **22a–d**, and polymer **13** in dilute $CHCl_3$ (1.0×10^{-5} M). All the compounds and polymers exhibited a typical π–π* transition band derived from the arylene–ethynylene skeleton. The absorption maxima (λ_{max}) of **19–21** were observed at 371, 378, and 381 nm, respectively. Their λ_{max} and absorption edges exhibited a clear redshift with an increase in the number of the stacked π-electron systems. As would be expected, such a redshift in the spectra of polymers **22a–d** was observed with an increase in molecular weight.

For the conjugated oligomers and polymers, systematic changes in properties such as UV–vis absorption can be expected with increasing numbers of repeating units [31–36]. For example, it is known that a linear relationship exists between the band-gap energy (E_g/cm^{-1}), estimated from the λ_{max} and the reciprocal of the number of repeating units (1/*n*). We applied this relationship to the π-stacked polymers and investigated whether this relationship held for the present system [30]. Figure 3b shows the plot of E_g against 1/*n* for compounds **19–21** and polymers **22a–d**. A good linear relationship ($R^2 = 0.99$) was obtained, which indicated the existence of so-called through-space conjugation. On the basis of this relationship, the λ_{max} of PAE-type π-stacked polymers ($n = \infty$) was predicted to be 388 nm, which was in good agreement with the observed λ_{max} (387 nm) of polymer **13** ($M_n = 14{,}500$, $n = 21$). According to the empirical power-law relationship ($E = 1/\lambda_{max} \sim n^{-x}$) and the data pertaining to compounds **19–21** and polymers **22a–d**, the effective conjugation length, i.e., the repeating unit number (*n*), of the π-stacked polymer was calculated to be 11.4.

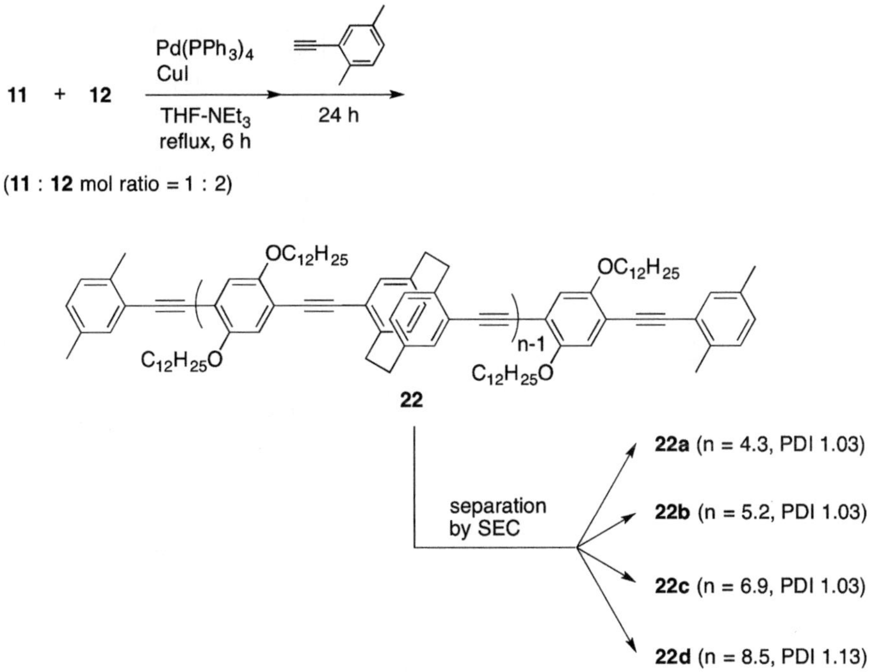

Scheme 6 Synthesis and separation of polymer **22**

On the other hand, it was reported that the electronic interactions among the π-stacked aromatic rings in poly(dibenzofulvene) [37–44] extended to five repeating units. In addition, the effective conjugation repeating unit of poly(phenylnorbornane) [45–47] was also studied and was estimated to be 4–5. Those effective conjugation repeating units were smaller than that of the [2.2]paracyclophane-based π-stacked polymer. In both of poly(dibenzofulvene) and poly(phenylnorbornane), π-stacked and unstacked conformations are in equilibrium in a single polymer chain, as shown in Fig. 4; therefore, 4–5 aromatic rings comprise the π-stacked sequence uniformly in solution. In the [2.2]paracyclophane-based π-stacked polymer, π-electron systems are forced to be π-stacked with each other by proximity, leading to a conformationally stable π-stacked structure. Thus, electronic interactions extend through at least ten π-electron systems via the through-space interaction in the ground state.

The emission behaviors of [2.2]paracyclophane-based π-stacked polymers were investigated in detail. Almost all the polymers exhibited photoluminescence in solution with high photoluminescence quantum efficiencies (Φ_{PL}). For example, the relative Φ_{PL} values of polymers **13** and **16** were calculated to be 0.82 and 0.70, respectively. The photoluminescence spectra of **13** and the stacked model compound **19** in $CHCl_3$ (1.0×10^{-5} M) were obtained upon excitation at the absorption maximum (Fig. 5). In polymer **13**, the emission spectrum was observed in the visible blue region with a peak top wavelength of 430 nm. It is noteworthy that the

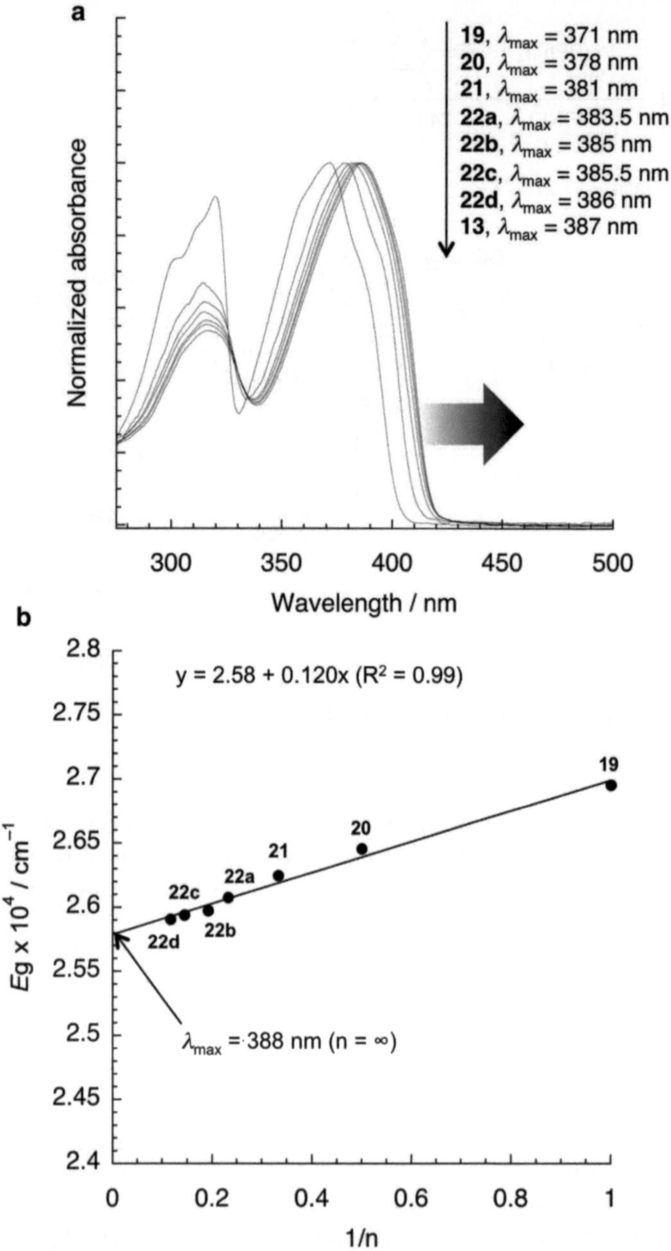

Fig. 3 (**a**) UV–vis absorption spectra of **19–21**, polymers **22a–d** and **13** in $CHCl_3$ (1.0×10^{-5} M). (**b**) Correlation between λ_{max} and $1/n$

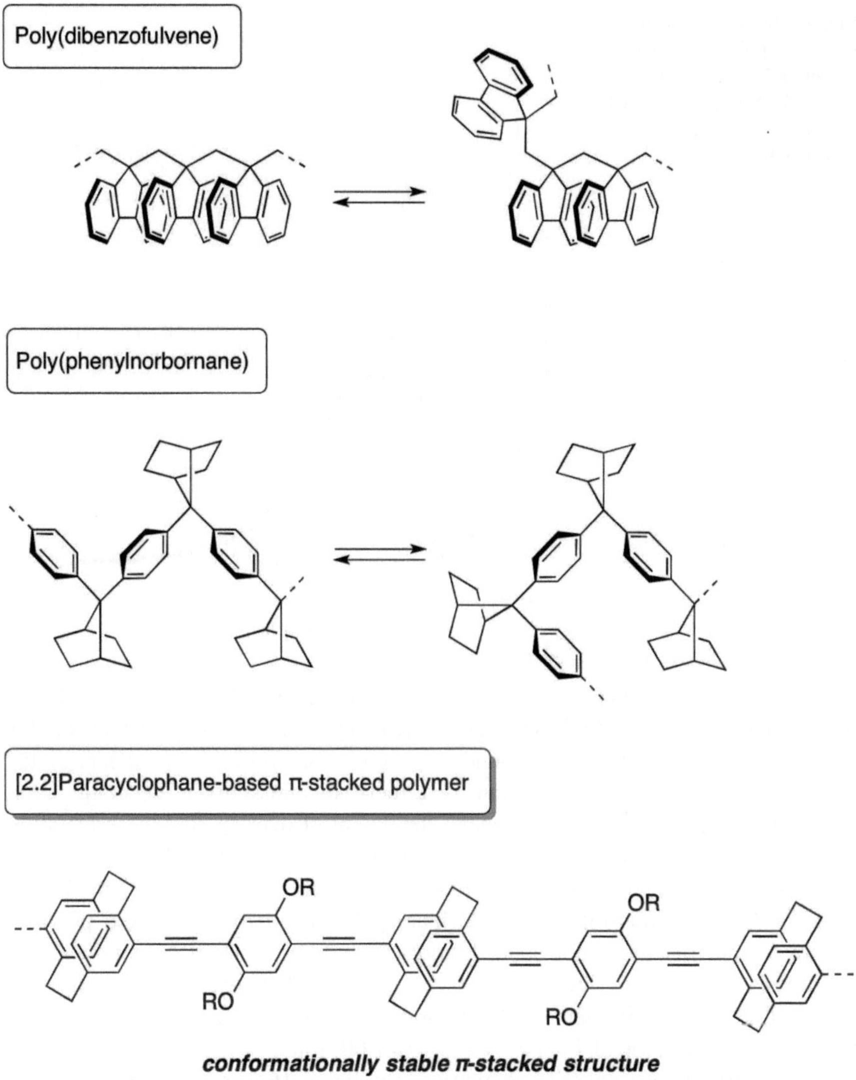

Fig. 4 Conformations of π-stacked polymers: poly(dibenzofulvene), poly(phenylnorbornane), and [2.2]paracyclophane-based polymer

spectrum of **13** exhibited clear vibrational structures as well as a high Φ_{PL} of 0.82. The spectrum and Φ_{PL} of **13** were similar to those of **19**. Thus, the π-stacked polymers based on [2.2]paracyclophane did not exhibit broad spectra with a low Φ_{PL}, like an excimer emission, although two dozen π-electron systems were stacked in proximity in a single polymer main chain.

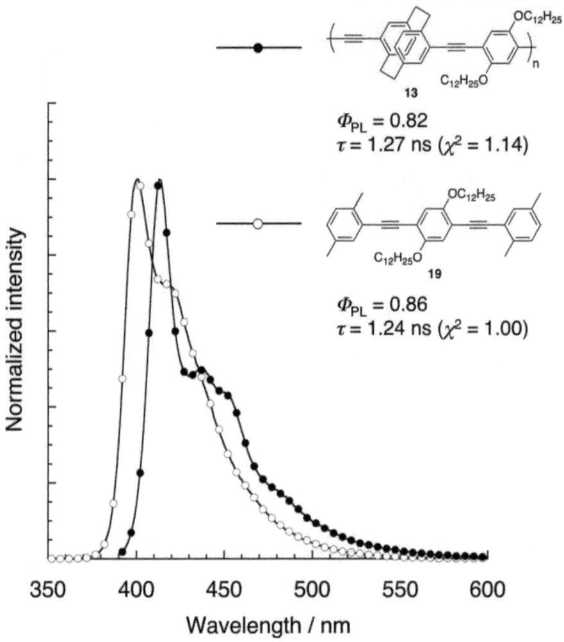

Fig. 5 Photoluminescence spectra of polymer **13** and compound **19** in $CHCl_3$ (1.0×10^{-5} M) excited at each absorption maximum wavelength

Bazan and coworkers investigated the emission behavior of [2.2]paracyclophane-based compounds [48–55]. They reported two types of emission mechanisms for [2.2]paracyclophane-based compounds, i.e., emission from "the monomer state" and emission from "the phane state (excimer-like emission)." The conjugation length of the stacked π-electron system, the extent of overlap, and the orientation between the stacked π-electron systems determine the mechanism. According to the photoluminescence spectra of **13** and **19** (Fig. 5) and their high Φ_{PL} (0.82 for **13** and 0.86 for **19**), the emission of the [2.2]paracyclophane-based π-stacked polymer occurred from the monomer state. Fluorescence lifetime studies supported this hypothesis. Both fluorescence decay curves of **13** and **19** were a single exponential, and the fluorescence lifetime (τ) of the polymer was 1.27 ns ($\chi^2 = 1.14$), which was identical to the lifetime of **19** ($\tau = 1.24$ ns, $\chi^2 = 1.00$) [30].

The electroluminescent (EL) properties of polymer **16** were examined by fabricating a standard EL device, ITO/PEDOT-PSS/**16**/BCP/Alq_3/LiF/Al (ITO = indium–tin oxide, PEDOT = poly(3,4-ethylenedioxythiophene), PSS = poly(4-styrenesulfonic acid), BCP = 2,9-dimethyl-4,7-diphenyl-1,10-phenanthroline, and Alq_3 = tris(8-hydroxyquinolinato)aluminum [26]. The EL spectrum exhibited a maximum peak at around 490 nm with a turn-on voltage of 13 V. The luminescence brightness was found to be 150 cd m^{-2} at a drive voltage of 19 V with a current density of 546 A m^{-2}. The external quantum efficiency was estimated to be 0.88 % at 38 cd m^{-2} under a bias of 14 V. Thus, we expect the application and development of [2.2]paracyclophane-based π-stacked polymers in optoelectronic devices.

After the polymerization of monomers **11** and **12** (mol ratio = 10:11) for 24 h using the Sonogashira–Hagihara protocol, ethynylxylene was added to afford the

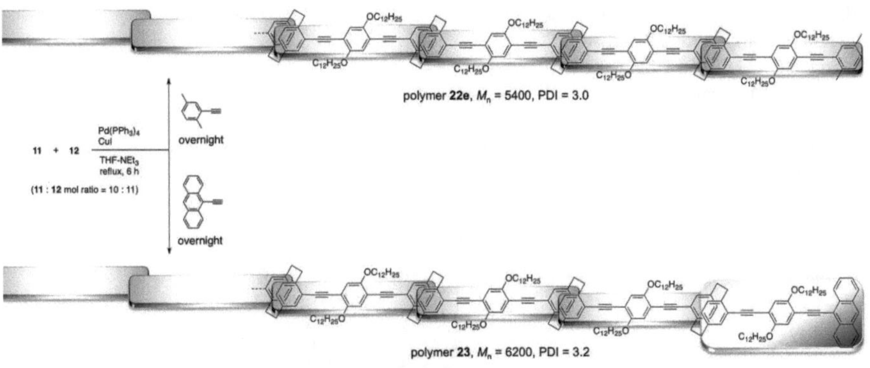

Scheme 7 Synthesis of end-capped polymers **22e** and **23**

corresponding polymer **22e** with xylenes as the end-capping groups in 94 % isolated yield (M_n=5,400), as shown in Scheme 7 [30]. In addition, polymer **23** end-capped with anthracenes (55 %, M_n=6,200) was synthesized using the same procedure (Scheme 7). The M_n values indicated that **22e** consisted of an average of seven π-electron systems and that **23** consisted of an average of six π-electron systems with two anthracene-containing π-electron systems stacked at the polymer chain ends.

Figure 6a shows the UV–vis absorption spectra of compound **19** and the anthracene-containing π-electron system **24** in dilute $CHCl_3$ (1.0×10^{-5} M). In addition, Fig. 6b shows the UV–vis absorption spectra of polymers **22e** and **23** in $CHCl_3$ (1.0×10^{-5} M). The absorption spectrum of **19** exhibited π–π* transition bands corresponding to the typical arylene–ethynylene moiety at around 320 and 370 nm, whereas the spectrum of **24** exhibited a unique absorption band corresponding to the anthracene ring at 270 nm as well as at around 430 nm. In the spectrum of polymer **23** (Fig. 6b), absorption bands derived from both compounds **19** and **24** were observed; namely, the broad band at around 450 nm in polymer **23** was assigned to the transition band of the anthracene-containing π-electron systems at the polymer chain ends.

When compounds **19** and **24** were excited at each absorption maximum in dilute $CHCl_3$ solutions (1.0×10^{-5} M), they emitted with peak maxima at around 400 and 470 nm, respectively, as shown in Fig. 7a. Figure 7b shows the photoluminescence spectra of polymers **22e** and **23** in $CHCl_3$ solutions (1.0×10^{-5} M). Polymer **22e** emitted with a photoluminescence peak maximum at around 410 nm upon excited at 385 nm, suggesting that the emission originated from the stacked π-electron systems. In the case of polymer **23**, upon photoexcitation at 385 nm, the spectrum exhibited a small emission peak at around 410 nm and a main peak at around 470 nm (Fig. 7b). Comparison of the spectra of **19** and **24** implied that the peak at 470 nm was attributed to the emission from the anthracene-containing π-electron systems at the polymer chain ends. Time-resolved photoluminescence spectra of **23** were obtained, as shown in Fig. 8. The intensity of the emission peak from the stacked π-electron systems decreased, whereas that of the emission peak from the anthracene-containing π-electron systems increased. The good overlap between the emission peak of the stacked π-electron systems (around 410 nm) and the absorption peak of

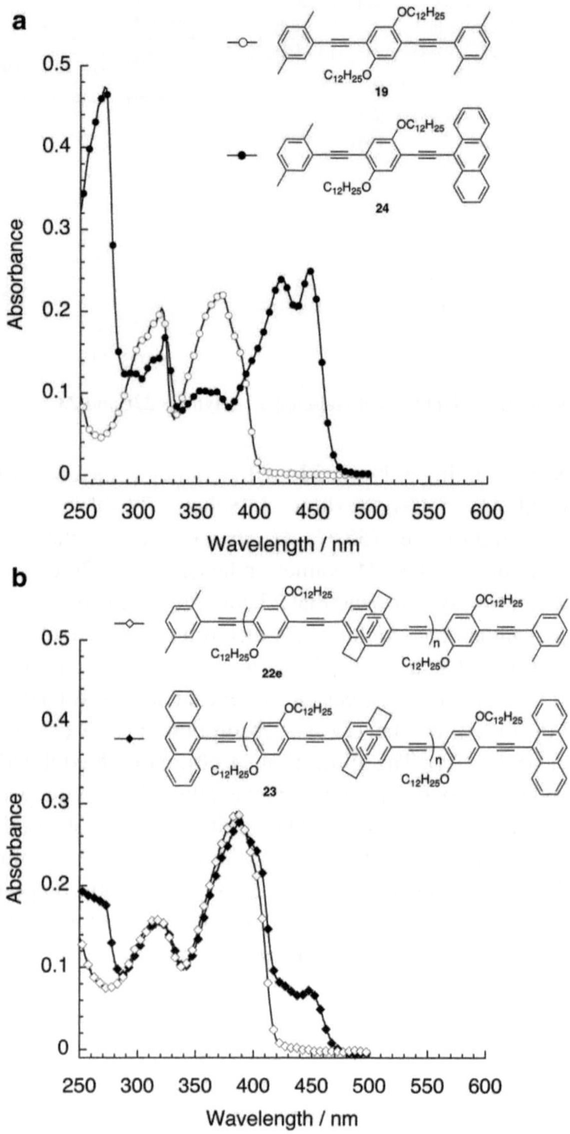

Fig. 6 (**a**) UV–vis absorption spectra of compounds **19** and **24** in $CHCl_3$ (1.0×10^{-5} M). (**b**) UV–vis absorption spectra of polymers **22e** and **23** in $CHCl_3$ (1.0×10^{-5} M)

the anthracene-containing π-electron systems (390–470 nm) caused fluorescence resonance energy transfer (FRET) [56] from the stacked π-electron systems to the terminal anthracene-containing π-electron systems in polymer **23**.

The electrical conductivity of the π-stacked polymer was investigated by the flash-photolysis time-resolved microwave conductivity (FP-TRMC) method [57–68]. We could observe the charge mobility in **22e** through the stacked π-electron systems in a single polymer chain using this noncontact measurement technique [30]. Polymer **22e** was dispersed in a polystyrene film without any dopants, and the cast film on quartz was excited by a laser pulse at 355 nm. The kinetic traces of microwave

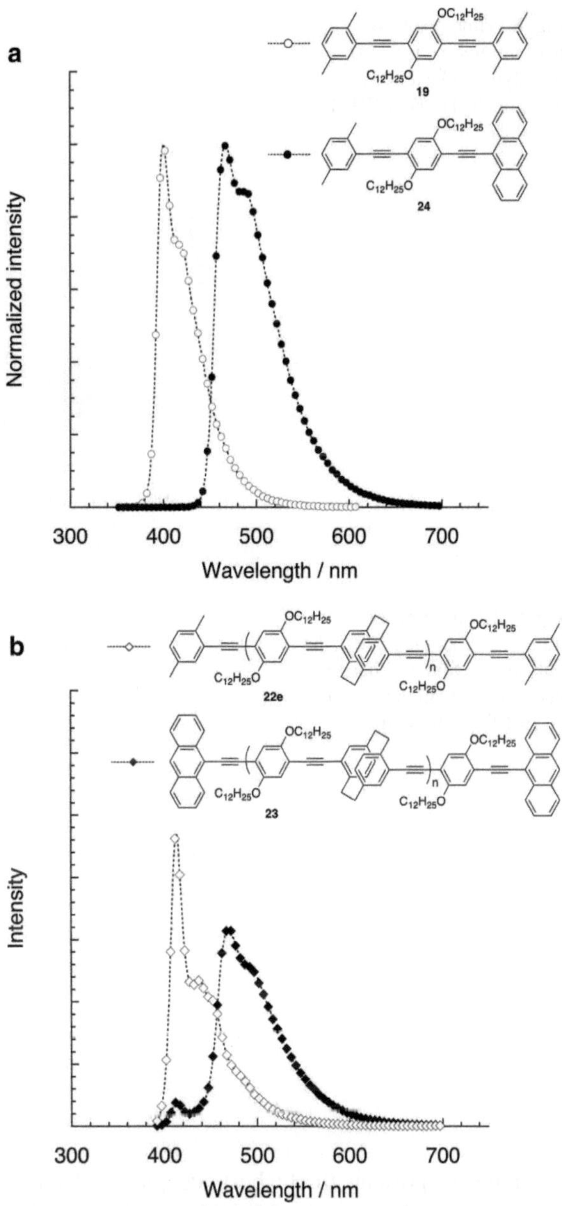

Fig. 7 (**a**) Photoluminescence spectra of compound **19** in $CHCl_3$ (1.0×10^{-5} M) excited at 365 nm and compound **24** in $CHCl_3$ (1.0×10^{-5} M) excited at 468 nm. (**b**) Photoluminescence spectra of polymers **22e** and **23** in $CHCl_3$ (1.0×10^{-5} M) excited at 385 nm

conductivity transients are shown in Fig. 9. The product of the photo carrier generation yield (ϕ) and the sum of charge mobilities ($\Sigma\mu$) was estimated to be $\phi\Sigma\mu = 1.9\times10^{-5}$ cm^2 V^{-1} s^{-1}. Studies of the triplet states by photoexcitation and direct photocurrent measurements revealed that intramolecular charge carrier mobility (μ) of **22e** (through-space system) was of the order of 10^{-2} cm^2 V^{-1} s^{-1} and a few tens of percentage higher than the through-bond system.

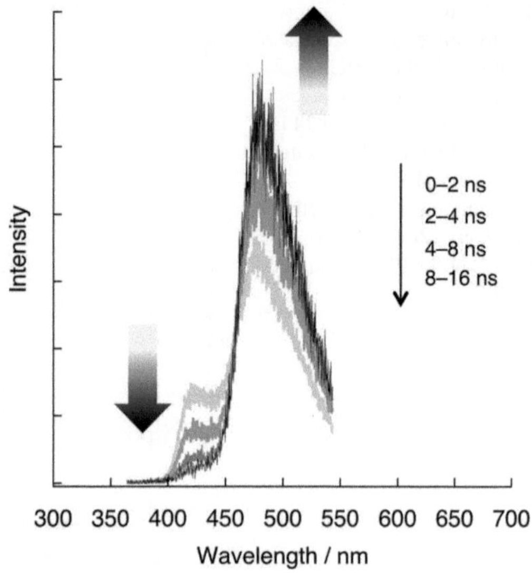

Fig. 8 Time-resolved photoluminescence spectra of polymer **23** excited at 337 nm (N_2 laser) in $CHCl_3$ (1.0×10^{-6} M)

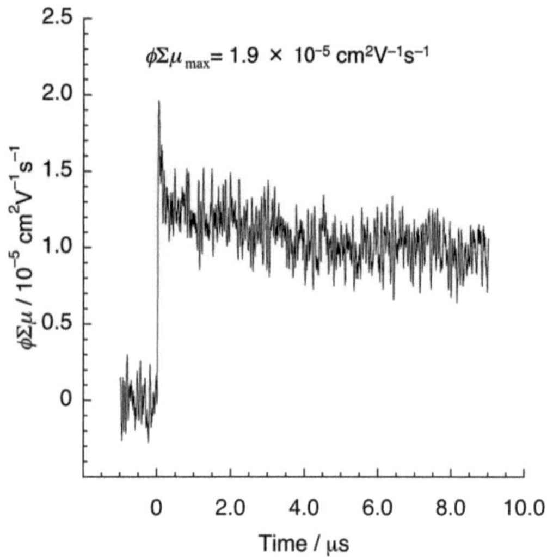

Fig. 9 TRMC conductivity transient of polymer **22e** in the polystyrene film on a quartz under the excitation at 355 nm (1.8×10^{16} photons cm^{-2})

Figure 10 presents the highest occupied molecular orbital (HOMO) and lowest unoccupied molecular orbital (LUMO) of the neutral state of the dimer model of the π-stacked polymer obtained by the density functional theory (DFT) calculations. The molecule had the delocalized HOMO and LUMO over the two repeating π-electron systems. Radical cations of the singly occupied molecular orbital (SOMO) were also delocalized in a similar manner. In addition, the stacked π-electron systems were observed to be highly planar. The structure of the dimer model did not change significantly between the neutral and cationic states, and the SOMO was obviously delocalized through the space, not only at the central cyclophane unit but

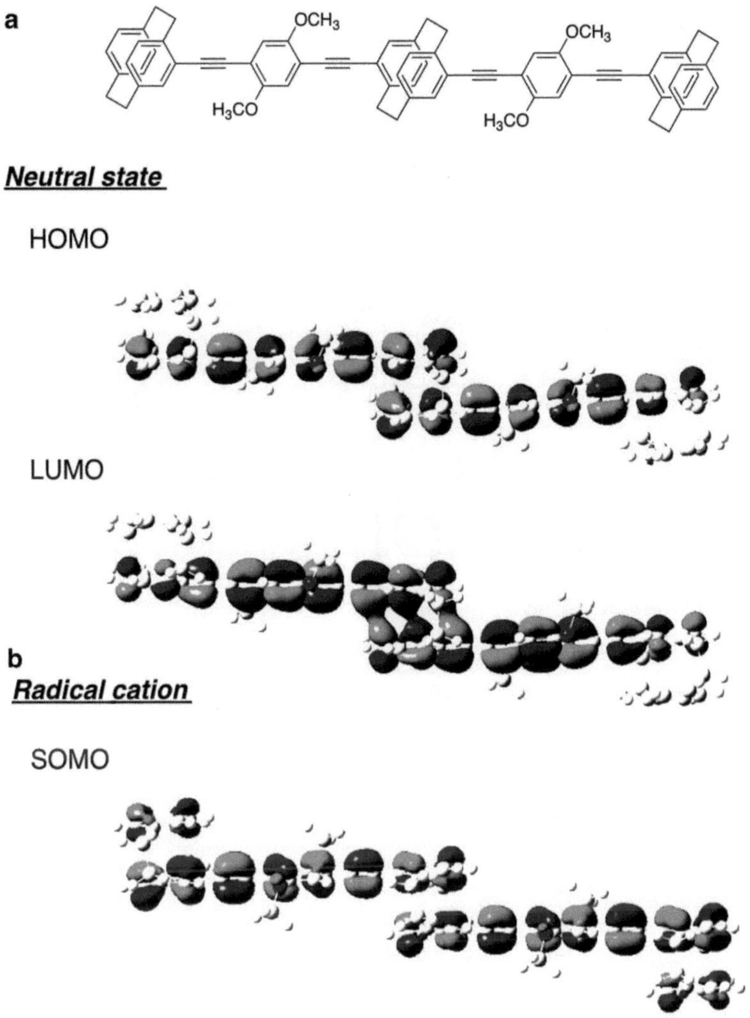

Fig. 10 DFT calculation using B3LYP/6-31G(d,p) of dimer model compound: (**a**) HOMO and LUMO of neutral states and (**b**) SOMO of radical cations

also at the end cyclophane units. This result indicates that the rigidity of the main chain of polymer **22e** can decrease the reorganization energy for charge transport and can contribute to the increase in intramolecular through-space charge carrier mobility [68–70]. In addition to the results on energy transfer, [2.2]paracyclophane-based π-stacked polymers are promising candidates as the single molecular wires that can carry flows of energy and/or charge through stacked π-electron systems.

The synthesis of conjugated polymers, including [2.2]paracyclophane-based π-stacked polymers, has been accomplished using several useful reactions. Scheme 8 shows the synthesis of PAV-type π-stacked polymers [71, 72] by Knoevenagel

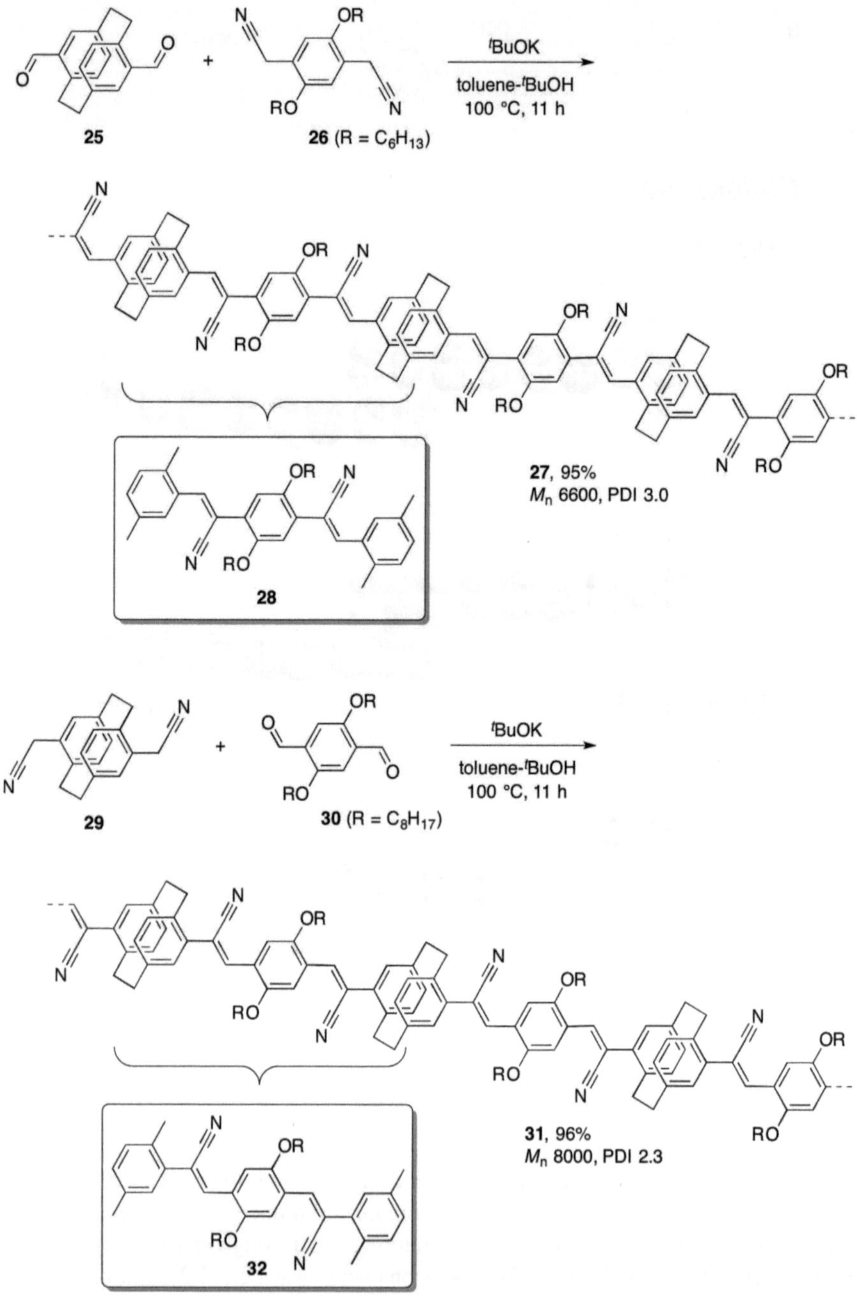

Scheme 8 Synthesis of PAV-type polymers **27** and **31**

reaction [73], which provided an electron-withdrawing substituent at the olefin moiety. The treatment of pseudo-*para*-diformyl[2.2]paracyclophane **25** with bis(cyanomethyl)benzene monomer **26** in the presence of ^{t}BuOK afforded cyano-substituted polymer **27** in 95 % isolated yield (M_n=6,600, PDI=3.0). Polymer **31**, which also contained cyano substituents at different positions, was obtained in 96 % isolated yield (M_n=8,000, PDI=2.3) by the same polymerization of pseudo-*para*-bis(cyanomethyl)[2.2]paracyclophane **29** and diformylbenzene monomer **30**. As shown in Scheme 8, polymers **27** and **31** consist of repeating units of compounds **28** and **32**, respectively. The difference between polymers **27** and **31** was the substitution position of the cyano groups in the vinylene moieties, i.e., they were substituted at the α- and β-positions from the dialkoxyphenylene units in **27** and **31**, respectively.

The optical properties of polymers **27** and **31**, as well as compounds **28** and **32**, were compared. UV–vis absorption spectra and photoluminescence spectra are shown in Fig. 11, and the data are summarized in Table 1. The λ_{max} of **27** and **31** were observed at 380 and 436 nm, respectively, which were redshifted in comparison with those of **28** (λ_{max}=370 nm) and **32** (λ_{max}=413 nm), as a result of the through-space conjugation among the stacked π-electron systems.

Photoluminescence spectra of polymers **27** and **31** were obtained in sufficiently diluted $CHCl_3$ solution, and peak maxima were observed at around 477 and 510 nm, respectively. It is noted that their Φ_{PL} values were different from each other. The Φ_{PL} values of the polymers were calculated to be less than 0.01 and 0.32, respectively. The Φ_{PL} values of compounds **28** and **32** were also estimated to be less than 0.01 and 0.30, respectively. Polymer **27** and compound **28** were poor emitters, which was attributed to the substitution position of the cyano groups. As shown in Fig. 12, π-conjugation between the dialkoxyphenylene unit and the cyano groups at the α-positions of the vinylene moieties in **27** and **28** was apparently broken. On the other hand, for **31** and **32**, the π-conjugation lengths were extended (Fig. 12), and there was a donor–acceptor resonance effect. X-ray crystallography supported the optical behaviors; it was reported that the center phenylene units were twisted, depending on the substitution position of the cyano groups. This case of the cyano-substituted π-stacked polymers is a good example of how the properties of the [2.2]paracyclophane-based π-stacked polymers reflect those of the stacked π-electron systems.

A donor–acceptor-type π-stacked polymer, in which a donor π-electron system and an acceptor π-electron system were alternately stacked, was synthesized (Scheme 9) [74]. The treatment of cyclophane-containing monomer **33** with **34** in the presence of a catalytic amount of palladium complex and a bulky phosphine [75] afforded the corresponding donor–acceptor π-stacked polymer **35** in good isolated yield with the M_n of 17,000 (PDI=2.9). The structure of polymer **35** comprised compounds **36** and **37**, in which fluorene and benzothiadiazole existed as donor and acceptor units, respectively.

In Fig. 13, the UV–vis absorption spectra (1.0×10^{-5} M in $CHCl_3$) and photoluminescence spectra (1.0×10^{-6} M in $CHCl_3$) of polymer **35**, donor compound **36**, and acceptor compound **37** are shown. As shown in Fig. 13a, the absorption band attributable to the π–π* transition of the fluorene moiety in **36** was observed at around 300 nm, and the maximum emission peak appeared at 364 nm with a vibrational

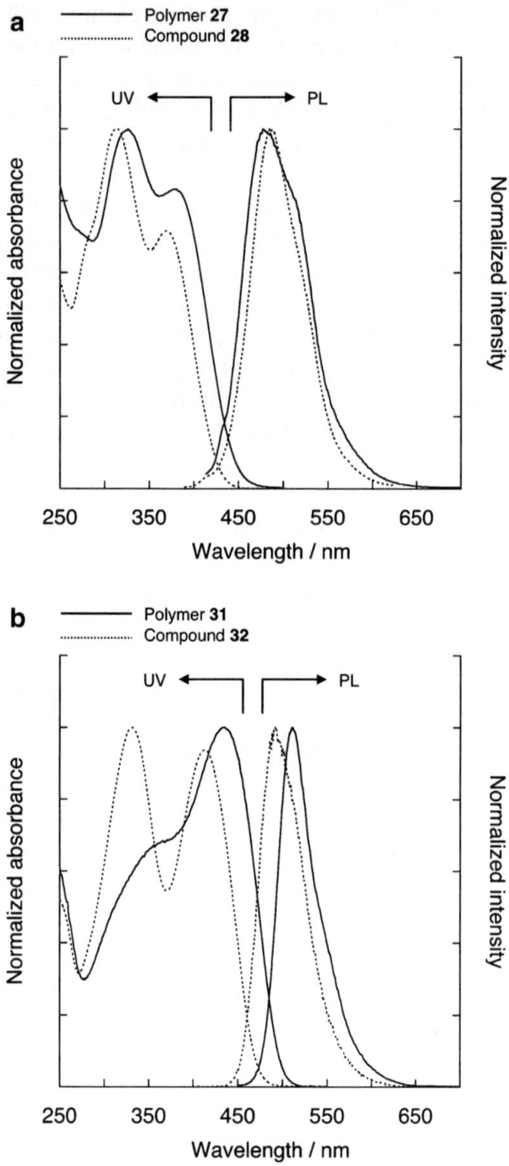

Fig. 11 (**a**) UV–vis absorption spectra (3.0×10^{-5} M in $CHCl_3$) and photoluminescence spectra (3.0×10^{-6} M in $CHCl_3$) of polymer **27** and compound **28**. (**b**) UV–vis absorption spectra (2.0×10^{-5} M in $CHCl_3$) and photoluminescence spectra (2.6×10^{-6} M in $CHCl_3$) of polymer **31** and compound **32**

Table 1 Optical properties of **27**, **28**, **31**, and **32**

	$\lambda_{Abs,max}$/nm [a]	$\lambda_{PL,max}$/nm [b]	Φ_{PL} [c]
Polymer **27**	327, 380	477	<0.01
Compound **28**	315, 370	484	<0.01
Polymer **31**	360, 436	510	0.32
Compound **32**	330, 413	491	0.30

[a] $\lambda_{Abs, Max}$ of UV–vis absorption spectra in $CHCl_3$
[b] $\lambda_{PL,max}$ of photoluminescence spectra in $CHCl_3$
[c] Relative photoluminescence quantum efficiencies calculated by using 9-anthracenecarboxylic acid in CH_2Cl_2 as a standard

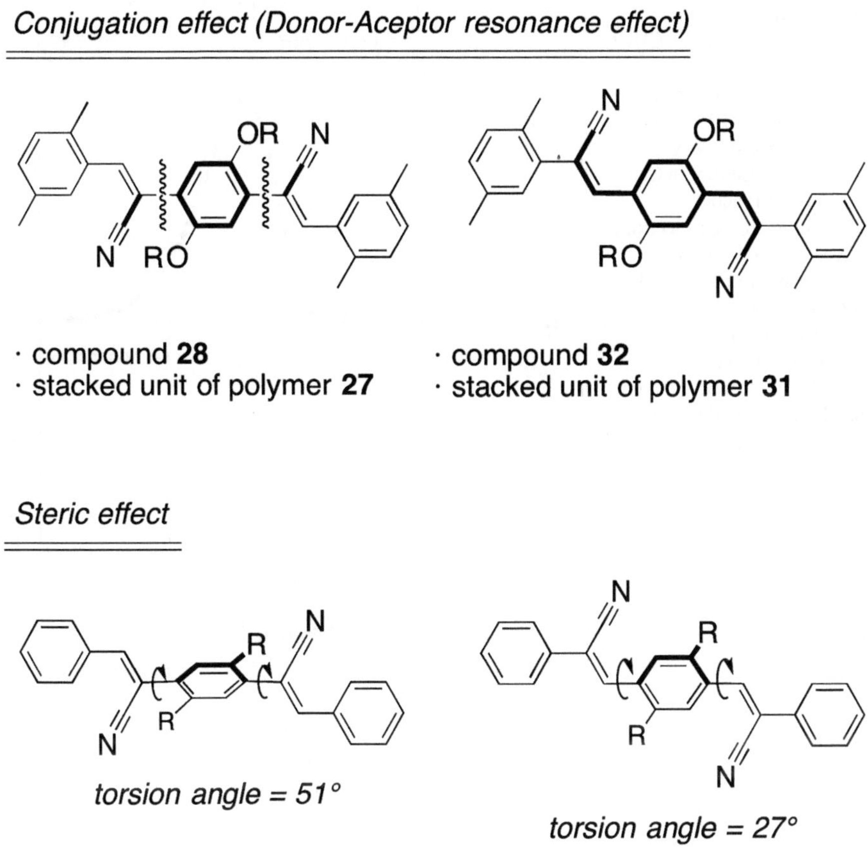

Fig. 12 Structures of compounds **28** and **32**. Schematic figures for their conjugation and steric effects

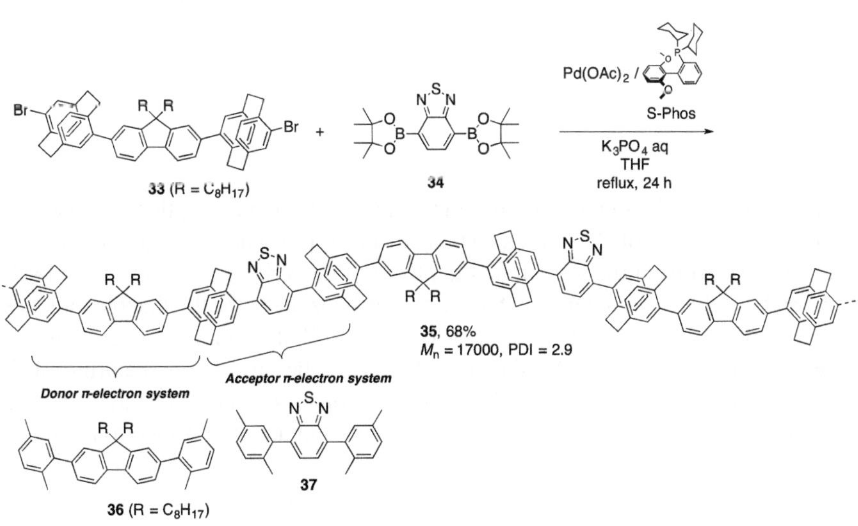

Scheme 9 Synthesis of donor-acceptor-type polymer **35**

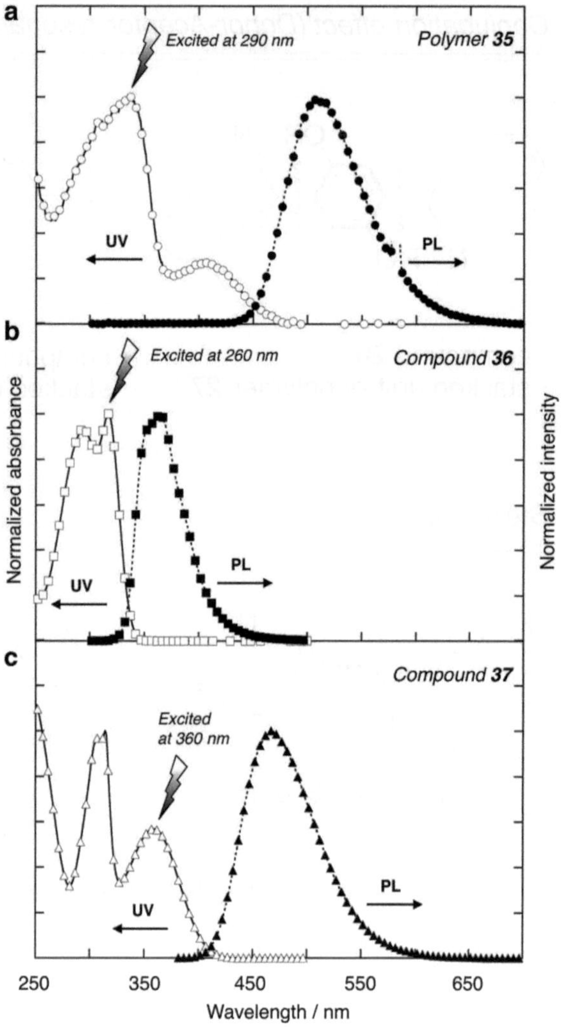

Fig. 13 (**a**) UV–vis absorption spectrum in $CHCl_3$ (1.0×10^{-5} M) and photoluminescence spectrum in $CHCl_3$ (1.0×10^{-6} M) of polymer **35**. (**b**) UV–vis absorption spectrum in $CHCl_3$ (1.0×10^{-5} M) and photoluminescence spectrum in $CHCl_3$ (1.0×10^{-6} M) of donor π-electron system **36**. (**c**) UV–vis absorption spectrum in $CHCl_3$ (1.0×10^{-5} M) and photoluminescence spectrum in $CHCl_3$ (1.0×10^{-6} M) of acceptor π-electron system **37**

structure upon excitation at 260 nm. Figure 13b shows the absorption and photoluminescence spectra of **37**, which exhibited absorption bands at around 310 and 360 nm derived from the π–π* transition band of the xylyl–phenylene–xylyl backbone and the charge-transfer (CT) band from the benzene to the thiadiazole moieties, respectively. By the excitation of the CT band of **37** at 360 nm, a green emission peak appeared at 467 nm. As shown in Fig. 13c, the absorption spectrum of **35** exhibited broad peaks at around 325 nm and around 400 nm, indicating that the spectrum comprises the spectra of both the donor and acceptor π-electron systems. On the other hand, when the donor π-electron system in **35** was excited at 290 nm, only a green emission at 475 nm was observed. The complete overlap of the photoluminescence spectrum of the donor **36** with the absorption spectrum of the CT band of the acceptor **37** resulted in effective intramolecular FRET from the donor π-electron systems to the acceptor π-electron systems in the single polymer chain.

π-Stacked Polymers Based on Pseudo-*ortho*-Disubstituted [2.2]Paracyclophane

There are several stereoisomers in disubstituted [2.2]paracyclophane, such as *ortho*-, *meta*-, *para*-, pseudo-*ortho*-, pseudo-*meta*-, pseudo-*para*-, and pseudo-*geminal*-disubstituted [2.2]paracyclophanes. Iron-catalyzed dibromination of commercially available [2.2]paracyclophane yields the mixtures of the dibromo[2.2]paracyclophane isomers, and pseudo-*para*-dibromo[2.2]paracyclophane **9** can be obtained by recrystallization as it has the poorest solubility in common organic solvents [76]. Thermal isomerization of **9** in triglyme afforded the pseudo-*ortho* isomer **38**, as shown in Scheme 10. This transformation is an equilibrium reaction; when the reaction was allowed to cool to room temperature, pseudo-*para* isomer **9** was precipitated as a result of its low solubility and was readily separated by simple filtration to obtain pseudo-*ortho* isomer **38** in moderate isolated yield [76]. Pseudo-*ortho*-dibromo[2.2]paracyclophane **38** was converted into pseudo-*ortho*-diethynyl[2.2]paracyclophane **40** via pseudo-*ortho*-diformyl[2.2]paracyclophane **39** (Scheme 10) [77]. Sonogashira–Hagihara coupling polymerization of **40** with **12** afforded the corresponding π-stacked polymer **41** in 23 % isolated yield with the M_n of 3,800 [78].

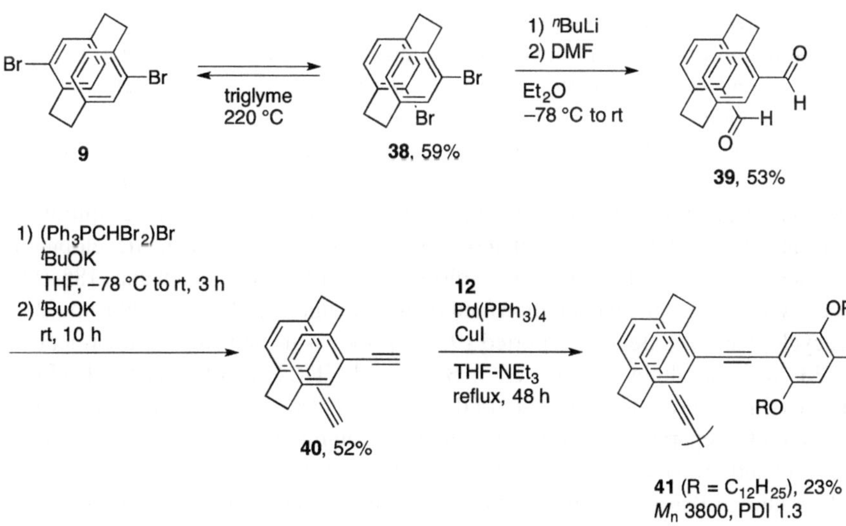

Scheme 10 Synthesis of polymer **41**

Table 2 Optical properties of **13**, **41**, and **9**

	$\lambda_{Abs,max}$/nm [a]	$\lambda_{PL,max}$/nm [b]	Φ_{PL} [c]
Pseudo-*para*-linked polymer **13**	319, 386	411	0.82
Pseudo-*para*-linked polymer **41**	319, 377	407	0.86
Compound **19**	320, 369	398	0.86

[a] $\lambda_{Abs, Max}$ of UV–vis absorption spectra in $CHCl_3$ (1.0×10^{-5} M)
[b] $\lambda_{PL,max}$ of photoluminescence spectra in $CHCl_3$ (1.0×10^{-7} M)
[c] Relative photoluminescence quantum efficiencies calculated by using 9-anthracenecarboxylic acid in CH_2Cl_2 as a standard

The optical profiles of pseudo-*ortho*-linked polymer **41** were almost identical to those of pseudo-*para*-linked polymer **13** [78]. The optical property data of compound **19**, pseudo-*para*-linked polymer **13**, and pseudo-*ortho*-linked polymer **41** are summarized in Table 2. Both polymers **13** and **41** consisted of the same π-electron systems corresponding to **19**, which was π-stacked straight in polymer **13** and π-stacked with a zigzag structure in polymer **41**. Therefore, there was no substantial difference between the optical properties of **13** and **41**. The absorption spectrum of **41** was consistent with that of **13**, and the maximum of the absorption peak of **41** was observed at 377 nm (Table 2). This λ_{max} value was smaller than that of **13** (Table 2) because of the shorter chain length of **41** (M_n = 3,800) relative to that of **13** (M_n = 14,500). The photoluminescence spectra of **13** and **41** were also similar to the spectrum of **19**, showing a clear vibrational structure with peak maxima at around 400 nm. In addition, the Φ_{PL} values of **13**, **41**, and **19** were estimated to be 0.82, 0.86, and 0.86, respectively (Table 2). This result suggests that the pseudo-*ortho*-linked π-stacked polymer **41** would emit bright blue light from the monomer state rather than the phane state in a similar manner to pseudo-*para*-linked polymer **13**.

Substituted [2.2]paracyclophane compounds exhibit planar chirality [3, 4]. In contrast to conformationally flexible chiral molecules with an axis of rotation and a rotatable bond around a stereogenic center, the planar chirality of [2.2]paracyclophane compounds possesses a conformationally stable chiral space because of the fixed aromatic rings. From this structural viewpoint, various methods for the optical resolution of planar chiral [2.2]paracyclophanes have been developed since the 1990s [79–92]. In comparison with the establishment of the optical resolution of monosubstituted [2.2]paracyclophanes, there was more scope for the development of optical resolution of pseudo-*ortho*-disubstituted [2.2]paracyclophanes [85–92]. Recently, a simple optical resolution method for racemic pseudo-*ortho*-dibromo[2.2]paracyclophane *rac*-**38** was reported [93]. As shown in Scheme 11, treatment of *rac*-**38** with a slight excess of *n*-BuLi and (1*R*,2*S*,5*R*)-(−)-menthyl (*S*)-*p*-toluenesulfinate provided a mixture of diastereomers (R_p,*S*)-**42** and (S_p,*S*)-**42**. These were readily separated by conventional SiO_2 column chromatography to afford (R_p,*S*)-**42** and (S_p,*S*)-**42** (each in 39 % isolated yield).

It is known that the sulfinyl–lithium exchange reaction readily proceeds upon treatment with alkyllithium reagents [81, 94–96]. The reaction of (R_p,*S*)-**42** with excess *t*-BuLi afforded the (R_p)-pseudo-*ortho*-dilithio[2.2]paracyclophane (R_p)-**43**.

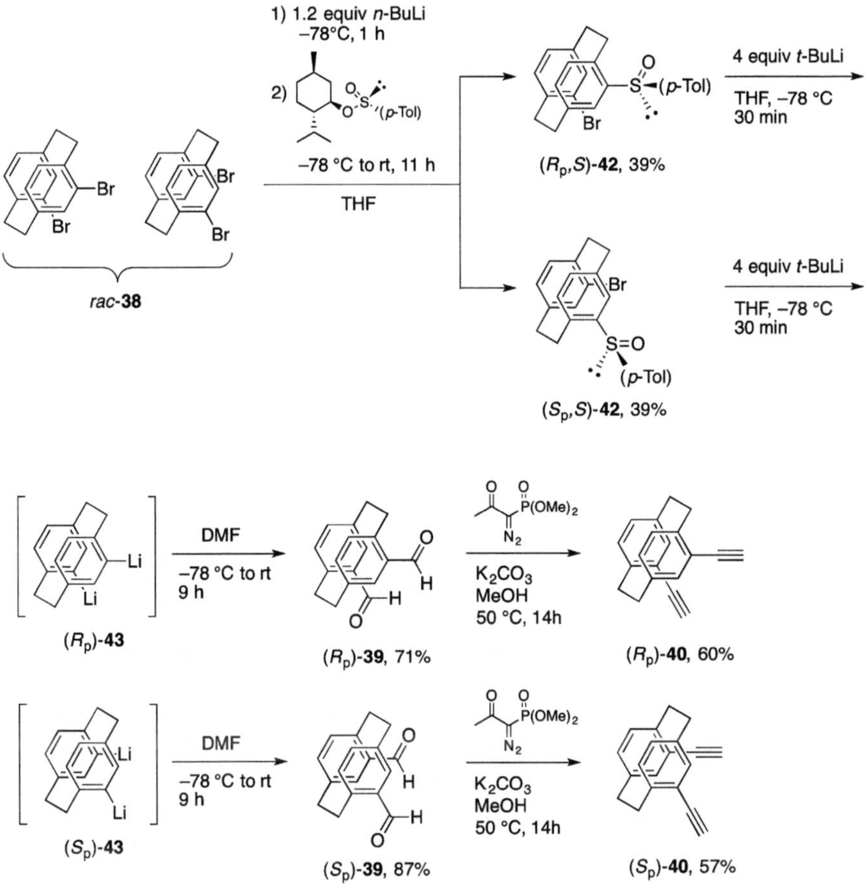

Scheme 11 Optical resolution of racemic pseudo-*ortho*-disubstituted cyclophane and synthesis of optically active diethynyl monomers (R_p)-**40** and (S_p)-**40**

This intermediate was allowed to react with electrophiles such as DMF to obtain (R_p)-pseudo-*ortho*-diformyl[2.2]paracyclophane (R_p)-**39** in 71 % isolated yield (Scheme 11). Enantiomer (S_p)-**39** was obtained in 87 % isolated yield from (S_p,*S*)-**2** through the same route. The formyl groups of (R_p)-**39** were converted into the ethynyl groups using the Ohira–Bestmann reagent to give enantiopure (R_p)-pseudo-*ortho*-diethynyl[2.2]paracyclophane (R_p)-**40** in 60 % isolated yield; the enantiomer (S_p)-**40** was similarly obtained in 57 % yield.

As shown in Scheme 12, π-stacked polymers (R_p)-**41** and (S_p)-**41**, in which the π-electron systems partially overlapped to form chiral zigzag structures, were synthesized from (R_p)-**40** and (S_p)-**40** [97], respectively, by Sonogashira–Hagihara polymerization. The appropriate combination of a palladium complex and ligand was important for successful polymerization. In this case, the $Pd_2(dba)_3/P(t\text{-}Bu)_3$ (dba = dibenzylideneacetone) catalytic system gave the best result, and the polymers were obtained in better yields and with higher M_n than those obtained through

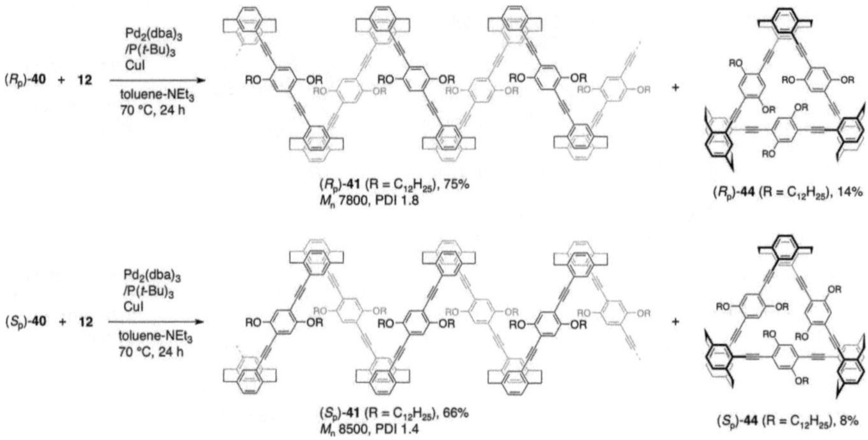

Scheme 12 Synthesis of optically active polymers (R_p)-**41** and (S_p)-**41** and cyclic compounds (R_p)-**44** and (S_p)-**44**

$Pd(PPh_3)_4$ catalysis (Scheme 10). In addition to the target polymer, enantiopure trimers (R_p)-**44** and (S_p)-**44** were obtained in 14 % and 8 % isolated yields, respectively.

The UV–vis absorption spectra of (R_p)-**40** and (S_p)-**40** in dilute $CHCl_3$ solution (1.0×10^{-5} M) are shown in Fig. 14. The spectra exhibited typical π–π* transition bands attributable to the stacked π-electron systems and peak maxima at 378 nm. Figure 14 includes the circular dichroism (CD) spectra of (R_p)-**40** and (S_p)-**40** solutions (1.0×10^{-5} M in $CHCl_3$). The spectra exhibited strong Cotton effects in the π–π* transition regions of the π-electron systems, and they were complete mirror images. The absolute value of their molar ellipticity [θ] was found to be approximately 0.8×10^5 deg cm^2 dmol^{-1}. This large [θ] value was caused by the stable chiral zigzag orientation of the π-electron systems, which was created by the planar chiral pseudo-*ortho*-disubstituted [2.2]paracyclophane moieties.

As mentioned above, polymer **40** emitted bright blue photoluminescence in dilute $CHCl_3$ solution, and, of course, (R_p)-**40** and (S_p)-**40** showed identical photoluminescence (Fig. 15). Polymers (R_p)-**40** and (S_p)-**40** exhibited intense circularly polarized luminescence (CPL), and their CPL spectra in dilute $CHCl_3$ (1.0×10^{-5} M, excitation wavelength: 320 nm) are shown in Fig. 15. The mirror-image CPL signals in the observed photoluminescence range appeared, and the CPL *g*-factors of (R_p)-**40** and (S_p)-**40** at the photoluminescence peak maxima were calculated to be 2.3×10^{-3} and -2.0×10^{-3}, respectively. The CPL observed in several π-conjugated polymers was induced only by their ordered structures in the film [98–103], partly aggregated states [104, 105], chiral network structures [106], and liquid crystalline states [107, 108]. In addition, it was reported that chirality of the optically active solvent such as limonene and pinene was transferred to π-conjugated polymer aggregates [109, 110]. Because of the conformationally stable structure of the planar chiral [2.2]paracyclophane, (R_p)-**40** and (S_p)-**40** showed intense CPL with a large CPL *g*-factor in such dilute solutions (1.0×10^{-5} M).

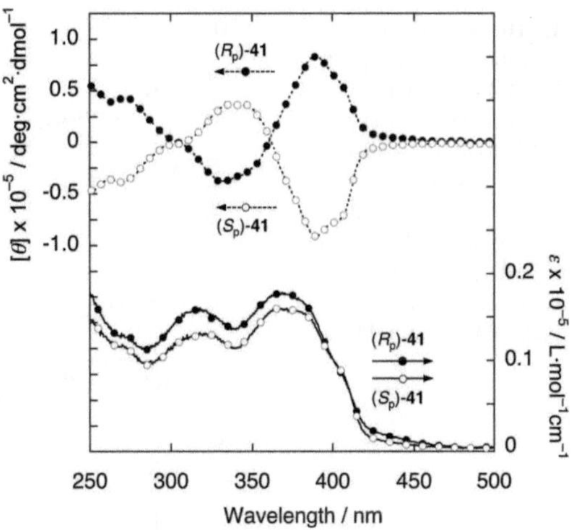

Fig. 14 UV–vis absorption and CD spectra of polymers (R_p)-**41** and (S_p)-**41** in $CHCl_3$ (1.0×10^{-5} M)

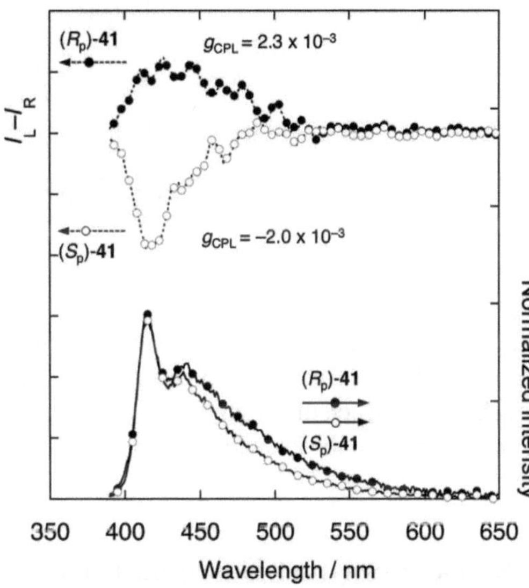

Fig. 15 Photoluminescence spectra (in $CHCl_3$, 1.0×10^{-6} M, excited at 370 nm) and CPL spectra (in $CHCl_3$, 1.0×10^{-5} M, excited at 320 nm) of polymers (R_p)-**41** and (S_p)-**41**

The UV–vis, CD, photoluminescence, and CPL spectra of cyclic trimers (R_p)-**44** and (S_p)-**44** are shown in Fig. 16. The absolute [θ] values were reached at approximately 2.0×10^5 deg cm^2 dmol^{-1} (Fig. 16a), which were greater than the values for the polymers (0.8×10^5 deg cm^2 dmol^{-1}) because of the completely fixed chiral triangular structure. As shown in Fig. 16b, CPL was clearly observed for (R_p)-**44** and (S_p)-**44** in the observed photoluminescence range; however, the absolute CPL *g-factor* values were smaller than those of the polymers. Although the reason for that remains unclear, the following may be considered. The polymer adopts a second-ordered structure such as an optically active zigzag or one-handed helix in the

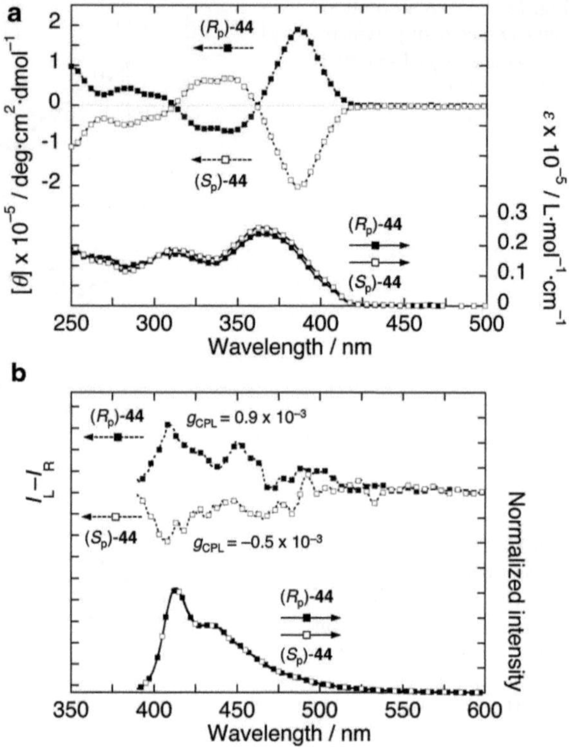

Fig. 16 (**a**) UV–vis absorption and CD spectra of (R_p)-**44** and (S_p)-**44** in $CHCl_3$ (1.0×10^{-5} M). (**b**) Photoluminescence spectra (in $CHCl_3$, 1.0×10^{-6} M, excited at 370 nm) and CPL spectra (in $CHCl_3$, 1.0×10^{-5} M, excited at 320 nm) of (R_p)-**44** and (S_p)-**44**

excited state, whereas the structure in the ground state is random with local chirality around the planar chiral cyclophane moieties. In the cyclic trimer, the structure does not change between the ground state and excited state. The higher-ordered structure of the polymer in the excited state would result in the larger CPL *g*-factor value.

π-Stacked Polymer Based on Pseudo-*geminal*-Disubstituted [2.2]Paracyclophane

Recently, the synthesis of π-stacked polymer **45** consisting of a pseudo-*geminal*-disubstituted [2.2]paracyclophane skeleton, in which the π-electron systems are fully stacked in the polymer main chain (Fig. 17), was reported [111, 112]. The UV–vis absorption spectrum exhibited considerable tailing. In addition, a large Stokes shift was observed, and the photoluminescence spectrum was broad and featureless, indicating that the emission from **45** arises from the phane state. Synthetic details and a further discussion of the π-stacked polymers based on pseudo-*geminal*-linked [2.2]paracyclophane are reviewed in "π-Stacked Oligomers as Models for Semiconducting Conjugated Organic Materials" in this volume.

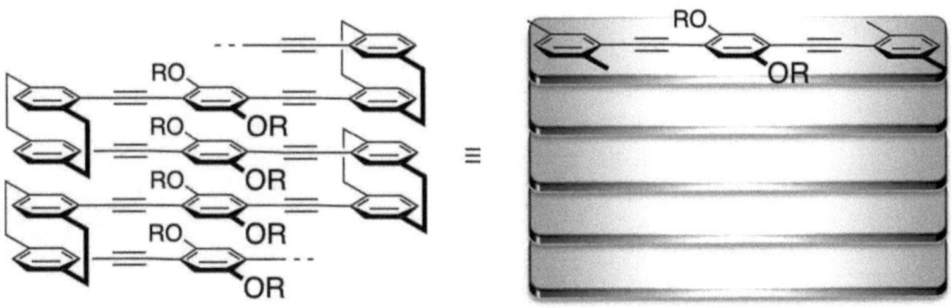

45 (R = C_9H_{19})

Fig. 17 π-Stacked polymer **45** consisting of pseudo-*geminal*-disubstituted [2.2]paracyclophane

Conjugated Microporous Polymers Based on [2.2] Paracyclophane

Network polymers constructed of organic molecules have attracted considerable attention in polymer chemistry as well as in materials chemistry [113–127]; those composed of π-conjugated frameworks have been extensively studied over the past 5 years. These compounds, known as conjugated microporous polymers (CMPs), constitute a new class of polymers [128–137]. They have received considerable attention as new materials because of their merits such as facile synthesis, high chemical stability, rigid micropores, and delocalization of π-electrons throughout their frameworks. Recently, CMPs containing [2.2]paracyclophane units in their conjugated main chains were synthesized and characterized [138, 139]. As shown in Scheme 13, cyclophane monomer **11** and 1,3,5-triiodobenzene **46** were used as bifunctional and trifunctional monomers, respectively, to construct the rigid network. Polymerization was carried out by Sonogashira–Hagihara polymerization to yield the CMP **47** containing pseudo-*para*-disubstituted [2.2]paracyclophanes, and CMP **48** without the cyclophane skeletons was also prepared for comparison (Scheme 13). In addition, 4,7,13,16-tetrasubstituted [2.2]paracyclophane **49** was also polymerized with diiodobenzene **50** to obtain CMP **51** (Scheme 14), in which a step crisscross structure of tetrasubstituted [2.2]paracyclophane existed as a network junction in the CMP.

The CMPs were precipitated during the polymerization and were completely insoluble in any solvents; therefore, their purification was carried out by washing repeatedly with organic solvents and water using a Soxhlet extractor. The structures were confirmed by the solid-state ^{13}C NMR and FTIR spectra. The networks were characterized by the N_2 adsorption measurement, and the Brunauer–Emmett–Teller (BET) surface areas (S_{BET}) of CMPs **47** and **51** were found to be 501 and 640 $m^2 g^{-1}$, respectively. The S_{BET} values decreased in comparison with that of **48** (822 $m^2 g^{-1}$) by the incorporation of the [2.2]paracyclophane skeletons; however, they possessed large surface areas over 500 $m^2 g^{-1}$. It is considered that a three-dimensional network structure was formed effectively in **48** because of the bending of the struts. On

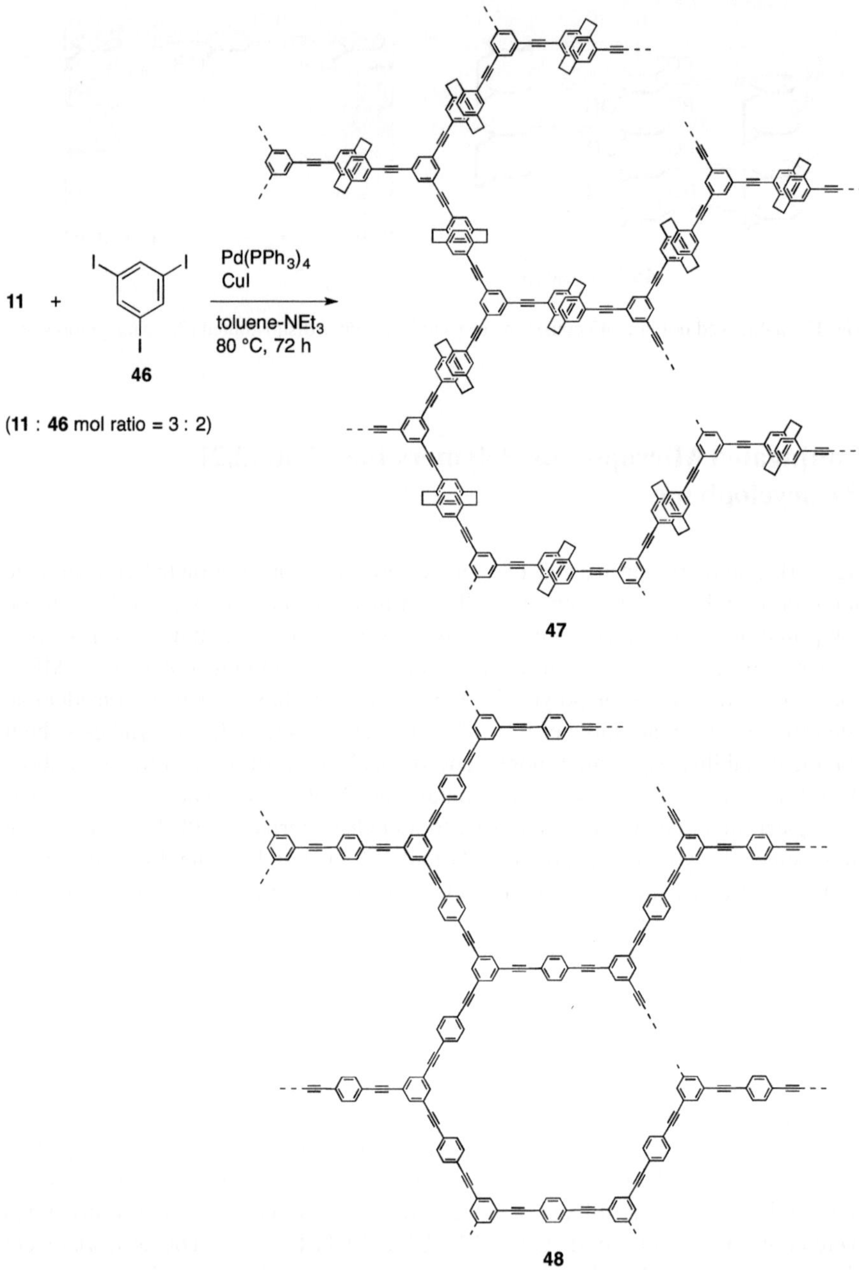

Scheme 13 Synthesis of CMP **47** consisting of pseudo-*para*-disubstituted [2.2]paracyclophane and reference CMP **48**

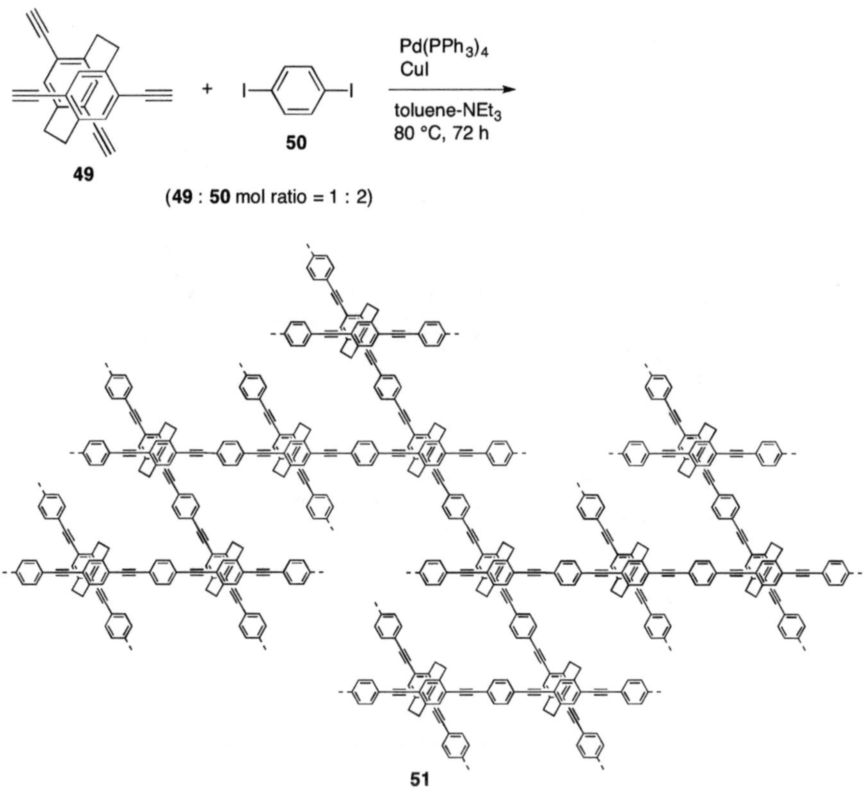

Scheme 14 Synthesis of CMP **51** consisting of tetrasubstituted [2.2]paracyclophane

the other hand, π-conjugated frameworks were offset by the [2.2]paracyclophane units to form a partial two-dimensional stacked structures in **47** and **51**, which resulted in the covered nanopores and a smaller surface area. Figure 18 shows the N_2 adsorption–desorption isotherms of **47**, **48**, and **51** measured at 77 K. All isotherms were categorized as type I gas sorption profiles by IUPAC classification [140]. It seemed that the isotherms of **47** and **51** exhibited H4-like hysteresis loops, implying the existence of narrow slit-like mesopores. Scanning electron microscope (SEM) observations were carried out, and the SEM images of **47** and **48** are shown in Fig. 19. CMP **47** revealed the presence of aggregates of masses and plates due to the partial two-dimensional structures and slit-like pores created by the step structure of [2.2]paracyclophane, whereas the morphology of **48** mainly exhibited aggregates of spherical particles with submicrometer diameters derived from growth of the network in three dimensions.

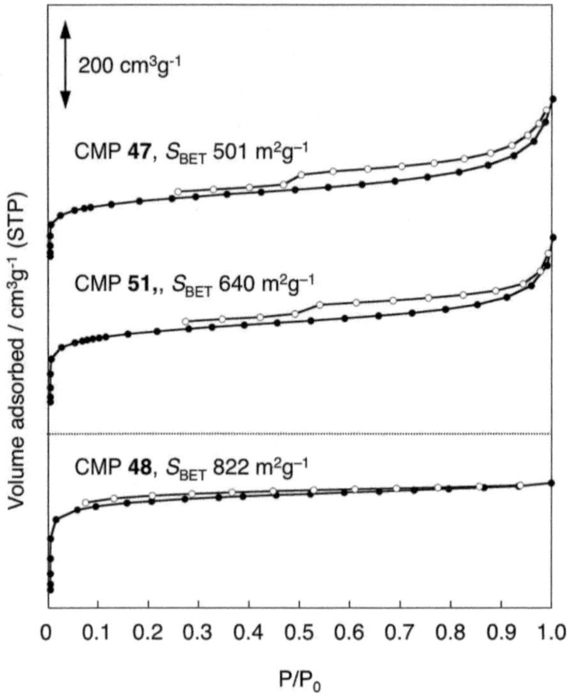

Fig. 18 N_2 adsorption–desorption isotherms of **47**, **48**, and **51** at 77 K; a *filled circle* and an *open circle* indicate adsorption and desorption isotherms, respectively

47 **48**

Fig. 19 SEM images of **47** and **48**

Conclusion

This chapter focused on the synthesis and fundamental properties of π-stacked polymers, which consist of the stacked π-electron systems created by the [*m.n*]cyclophane, in particular [2.2]paracyclophane, in the polymer main chain. As described above, cyclophane forms the π-electron systems and stacks them in proximity; thus, cyclophane acts as a bridging unit, when it is incorporated into the π-conjugated polymer backbone. Pseudo-*para*-, pseudo-*ortho*-, and pseudo-*geminal*-disubstituted [2.2]paracyclophanes allow the construction of various π-stacked conformations such as straight, zigzag, and fully stacked structures. Optically active π-stacked polymers comprising the planar chiral pseudo-*ortho*-disubstituted [2.2]paracyclophanes open a new frontier in chiral polymer chemistry. It should be emphasized again that the properties of the cyclophane-based polymers depend on the properties of the stacked π-electron systems rather than the cyclophanes.

Common π-conjugated polymers have a set of HOMO (valence band) and LUMO (conduction band) energy bandgaps, whereas in the π-stacked polymers each π-electron system has its own HOMO–LUMO energy bandgap. Various aromatic groups can be incorporated into the polymers; therefore, energy and charge transfer through the polymer chain can be controlled by appropriate tuning of the bandgaps and energy levels of the stacked π-electron system. It was shown that π-stacked polymers based on pseudo-*para*-disubstituted [2.2]paracyclophane carried energy effectively due to the proximity of the π-electron systems, the overlap between the absorption and photoluminescence spectra, and alignment of the dipole moments of the stacked π-electron systems. π-Stacked polymers based on pseudo-geminal-disubstituted [2.2]paracyclophane would exhibit high charge mobility because of the fully stacked structure of the π-electron systems. Furthermore, for example, we could make the Z-scheme in a single polymer chain as can be seen in a photosynthetic system and use the transferred electrons for chemical reduction. We hope that this new class of π-stacked polymers can make a fundamental contribution to the field of molecular electronics in the form of single molecular wires.

Acknowledgments Our works introduced herein are mainly supported by Grant-in-Aid for Young Scientists (B) (No. 16750096) and Young Scientists (A) (No. 21685012 and No. 24685018) from the Ministry of Education, Culture, Sports, Science and Technology, Japan.

References

1. Brown CJ, Farthing AC (1949) Nature 164:915
2. Cram DJ, Steinberg H (1951) J Am Chem Soc 73:5691
3. Vogtle F (1993) Cyclophane chemistry: synthesis, structures and reactions. Wiley, Chichester
4. Gleiter R, Hopf H (eds) (2004) Modern cyclophane chemistry. Wiley-VCH, Weinheim
5. Morisaki Y, Chujo Y (2006) Angew Chem Int Ed 45:6430
6. Morisaki Y, Chujo Y (2008) Prog Polym Sci 33:346
7. Hopf H (2008) Angew Chem Int Ed 47:9808
8. Morisaki Y, Chujo Y (2008) Bull Chem Soc Jpn 82:1070
9. Morisaki Y, Chujo Y (2011) Polym Chem 2:1249
10. Morisaki Y, Chujo Y (2012) Chem Lett 41:840
11. Mizogami S, Yoshimura S (1985) J Chem Soc Chem Commun 427
12. Mizogami S, Yoshimura S (1985) J Chem Soc Chem Commun 1736
13. Guyard L, Audebert P (2001) Electrochem Commun 3:164
14. Guyard L, Audebert P, Dolbier WR Jr, Duan JX (2002) J Electroanal Chem 537:189

15. Salhi F, Lee B, Metz C, Bottomley LA, Collard DM (2002) Org Lett 4:3195
16. Salhi F, Collard DM (2003) Adv Mater 15:81
17. Diederich F, Stang PJ (eds) (1998) Metal-catalyzed cross-coupling reactions. VCH, Weinheim
18. Miyaura N (ed) (2002) Cross-coupling reaction: a practical guide. Springer, Berlin
19. Negishi E (ed) (2002) Handbook of organopalladium chemistry for organic synthesis. Wiley, New York
20. de Meijere A, Diedrich F (eds) (2004) Metal-catalyzed cross-coupling reactions. Wiely-VCH, Weinheim
21. Tohda Y, Sonogashira K, Hagihara N (1977) Synthesis 777
22. Sonogashira K (2002) In: Negishi E (ed) Handbook of organopalladium chemistry for organic synthesis. Wiley, New York, pp 493–529
23. Morisaki Y, Chujo Y (2002) Macromolecules 35:587
24. Mizoroki T, Mori K, Ozaki A (1971) Bull Chem Soc Jpn 44:581
25. Heck RF, Nolley JP Jr (1972) J Org Chem 37:2320
26. Morisaki Y, Chujo Y (2004) Macromolecules 37:4099
27. Miyaura N, Yamada K, Suzuki A (1979) Tetrahedron Lett 20:3437
28. Miyaura N, Suzuki A (1995) Chem Rev 95:2457
29. Morisaki Y, Chujo Y (2005) Bull Chem Soc Jpn 78:288
30. Morisaki Y, Ueno S, Saeki A, Asano A, Seki S, Chujo Y (2012) Chem Eur J 18:4216
31. Bubeck C (1998) In: Müllen K, Wegner G (eds) Electronic materials: the oligomer approach, Wiley-VCH, Weinheim pp 449–478
32. Meier H, Stalmach U, Kolshorn H (1997) Acta Polym 48:379
33. Luo Y, Norman P, Ruud K, Agren H (1998) Chem Phys Lett 285:160
34. Dembinski R, Bartik T, Bartik B, Jaeger M, Gladysz JA (2000) J Am Chem Soc 122:810
35. Gibtner T, Hampel F, Gisselbrecht JP, Hirsch A (2002) Chem Eur J 8:408–432
36. Meier H (2006) Carbon-rich compounds. In: Haley MM, Tykwinski RR (eds) Molecules to materials. Wiely-VCH, Weinheim, pp 476–528
37. Nakano T (2010) Polym J 42:103
38. Nakano T, Takewaki K, Yade T, Okamoto Y (2001) J Am Chem Soc 123:9182
39. Nakano T, Yade T (2003) J Am Chem Soc 125:15474
40. Nakano T, Yade T, Yokoyama M, Nagayama N (2004) Chem Lett 33:296
41. Nakano T, Yade T, Fukuda Y, Yamaguchi T, Okumura S (2005) Macromolecules 38:8140
42. Yade T, Nakano T (2006) J Polym Sci A Polym Chem 44:561
43. Nakano T, Yade T (2008) Chem Lett 37:258–259
44. Nakano T, Tanikawa M, Nakagawa O, Yade T, Sakamoto T (2009) J Polym Sci A Polym Chem 47:239
45. García Martínez A, Osío Barcina J, de Fresno CA, Schlüter AD, Frahn J (1999) Adv Mater 11:27
46. Caraballo-Martínez N, Colorado Heras MR, Mba Blázquez M, Osío Barcina J, García Martínez A, Torres Salvador MR (2007) Org Lett 9:2943
47. Osío Barcina J, Colorado Heras MR, Mba M, Gómez Aspe R, Herrero-García N (2009) J Org Chem 74:7148
48. Bartholomew GP, Bazan GC (2001) Acc Chem Res 34:30
49. Oldham WJ Jr, Miao YJ, Lachicotte RJ, Bazan GC (1998) J Am Chem Soc 120:419
50. Bazan GC, Oldham WJ Jr, Lachicotte RJ, Tretiak S, Chernyak V, Mukamel S (1998) J Am Chem Soc 120:9188
51. Wang S, Bazan GC, Tretiak S, Mukamel S (2000) J Am Chem Soc 122:1289
52. Bartholomew GP, Bazan GC (2002) J Am Chem Soc 124:5183
53. Bartholomew GP, Bazan GC (2002) Synthesis 1245
54. Ruseckas A, Namdas EB, Lee JY, Mukamel S, Wang S, Bazan GC, Sundström V (2003) J Phys Chem A 107:8029–8034
55. Bazan GC (2007) J Org Chem 72:8645
56. Förster T (1946) Naturwissenschaften 33:166
57. van der Laan GP, de Haas MP, Hummel A, Frey H, Sheiko S, Möller M (1994) Macromolecules 27:1897

58. Hoofman RJOM, de Haas MP, Siebbeles LDA, Warman JM (1998) Nature 392:54
59. Warman JM, Gelinck GH, de Haas MP (2002) J Phys Condens Matter 14:9935
60. Grozema FC, Siebbeles LDA, Warman JM, Seki S, Tagawa S, Scherf U (2002) Adv Mater 14:228
61. Kocherzhenko AA, Patwardhan S, Grozema FC, Anderson HL, Siebbeles LDA (2009) J Am Chem Soc 131:5522
62. Pingel P, Zen A, Abellón RD, Grozema FC, Siebbeles LDA, Neher D (2010) Adv Funct Mater 20:2286
63. Saeki A, Tsuji M, Seki S (2011) Adv Energy Mater 1:661
64. Saeki A, Seki S, Shimizu Y, Yamao T, Hotta S (2010) J Chem Phys 132:134509–134511
65. Prasanthkumar S, Saeki A, Seki S, Ajayaghosh A (2010) J Am Chem Soc 132:8866
66. Yamamoto Y, Zhang G, Jin W, Fukushima T, Ishii N, Saeki A, Seki S, Tagawa S, Minari T, Tsukagoshi K, Aida T (2009) Proc Natl Acad Sci 106:21051
67. Saeki A, Seki S, Takenobu T, Iwasa Y, Tagawa S (2008) Adv Mater 20:920
68. Saeki A, Fukumatsu T, Seki S (2011) Macromolecules 44:3416
69. Fratiloiu S, Grozema FC, Koizumi Y, Seki S, Saeki A, Tagawa S, Dudek SP, Siebbeles LDA (2006) J Phys Chem B 110:5984
70. Feng X, Marcon V, Pisulal W, Hansen MR, Kirkpatrick J, Grozema F, Andrienko D, Kremer K, Müllen K (2009) Nat Mater 8:421
71. Morisaki Y, Lin L, Chujo Y (2009) Chem Lett 38:734
72. Morisaki Y, Lin L, Chujo Y (2009) J Polym Sci A Polym Chem 47:5979
73. Greenham NC, Moratti SC, Bradley DDC, Friend RH, Holmes AB (1993) Nature 365:628
74. Lin L, Morisaki Y, Chujo Y (2010) Int J Polym Sci 2010:908128
75. Walker SD, Barder TE, Martinelli JR, Buchwald SL (2004) Angew Chem Int Ed 43:1871
76. Reich HJ, Cram DJ (1969) J Am Chem Soc 91:3517
77. Bondarenko L, Dix I, Hinrichs H, Hopf H (2004) Synthesis 16:2751
78. Morisaki Y, Wada N, Arita M, Chujo Y (2009) Polym Bull 62:305
79. Cram DJ, Laainger NL (1955) J Am Chem Soc 77:6289
80. Rozenberg V, Sergeeva E, Hopf H (2004) In: Gleiter R, Hopf H (eds) Modern cyclophane chemistry. Wiley-VCH, Weinheim, pp 435–462
81. Rowlands GJ (2008) Org Biomol Chem 6:1527
82. Gibson SE, Knight JD (2003) Org Biomol Chem 1:1256
83. Aly AA, Brown AB (2009) Tetrahedron 65:8055
84. Paradies J (2011) Synthesis 3749
85. Pye PJ, Rossen K, Reamer RA, Tsou NN, Volante RP, Reider PJ (1997) J Am Chem Soc 119:6207
86. Rossen K, Pye PJ, Maliakal A, Volante RP (1997) J Org Chem 62:6462
87. Zhuravsky R, Starikova Z, Vorontsov E, Rozenberg V (2008) Tetrahedron Asym 19:216
88. Jiang B, Zhao XL (2004) Tetrahedron Asym 15:1141
89. Jones PG, Hillmer J, Hopf H (2003) Acta Crystallogr E59:o24
90. Pamperin D, Hopf H, Syldatk C, Pietzsch M (1997) Tetrahedron Asym 8:319
91. Pamperin D, Ohse B, Hopf H, Pietzsch M (1998) J Mol Cat B Enzym 5:317
92. Braddock DC, MacGilp ID, Perry BG (2002) J Org Chem 67:8679
93. Morisaki Y, Hifumi R, Lin L, Inoshita K, Chujo Y (2012) Chem Lett 41:990
94. Hitchcock PB, Rowlands GJ, Parmar R (2005) Chem Commun 4219
95. Parmar R, Coles MP, Hitchcock PB, Rowlands GJ (2010) Synthesis 4177
96. Lockard JP, Schroeck CW, Johnson CR (1973) Synthesis 485
97. Morisaki Y, Hifumi R, Lin L, Inoshita K, Chujo Y (2012) Polym Chem 3:2727
98. Satrijio A, Meskers SCJ, Swager TM (2006) J Am Chem Soc 128:9030
99. Wilson JN, Steffen W, McKenzie TG, Lieser G, Oda M, Neher D, Bunz UHF (2002) J Am Chem Soc 124:6830
100. Oda M, Nothofer HG, Lieser G, Scherf U, Meskers SCJ, Neher D (2000) Adv Mater 12:362
101. Oda M, Nothofer HG, Scherf U, Šunjic V, Richter D, Regenstein W, Meskers SCJ, Neher D (2002) Macromolecule 35:6792

102. Yu JM, Sakamoto T, Watanabe K, Furumi S, Tamaoki N, Chen Y, Nakano T (2011) Chem Commun 47:3799
103. Watanabe K, Sakamoto T, Taguchi M, Fujiki M, Nakano T (2011) Chem Commun 47:10996
104. Peeters E, Christiaans MPT, Janssen RAJ, Schoo HFM, Dekkers HPJM, Meijer EW (1997) J Am Chem Soc 119:9909
105. Langeveld-Voss BMW, Janssen RA, Christiaans MPT, Meskers SCJ, Dekkers HPJM, Meijer EW (1996) J Am Chem Soc 118:4908
106. Fukao S, Fujiki M (2009) Macromolecules 42:8062
107. Goto H, Akagi K (2005) Angew Chem Int Ed 44:4322
108. Hayasaka H, Miyashita T, Tamura K, Akagi K (2010) Adv Funct Mater 20:1243
109. Nakano Y, Liu Y, Fujiki M (2010) Polym Chem 1:460
110. Kawagoe Y, Fujiki M, Nakano Y (2010) New J Chem 34:637
111. Jagtap SP, Collard DM (2010) J Am Chem Soc 132:12208
112. Jagtap SP, Collard DM (2012) Polym Chem 3:463
113. Li H, Eddaoudi M, O'Keeffe M, Yaghi OM (1999) Nature 402:276
114. Cheetham AK, Férey G, Loiseau T (1999) Angew Chem Int Ed 38:3268
115. Eddaoudi M, Kim J, Rosi N, Vodak D, O'Keeffe M, Yaghi OM (2002) Science 295:469
116. Yaghi OM, O'Keeffe M, Ockwig NW, Chae HK, Eddaoudi M, Kim J (2003) Nature 423:705
117. Matzger A, O'Keeffe M, Yaghi OM (2004) Nature 427:523
118. Kitagawa S, Kitaura R, Noro S (2004) Angew Chem Int Ed 43:2334
119. Kitagawa S, Noro S, Nakamura T (2006) Chem Commun 701
120. Kitagawa S, Matsuda R (2007) Coord Chem Rev 251:2490
121. Maji TK, Kitagawa S (2007) Pure Appl Chem 79:2155
122. Férey G (2008) Chem Soc Rev 37:191
123. Uemura T, Yanai N, Kitagawa S (2009) Chem Soc Rev 38:1228
124. Jiang JX, Cooper AI (2010) Top Curr Chem 293:1
125. Thomas A (2010) Angew Chem Int Ed 49:8328
126. Thomas A, Kuhn P, Weber J, Titirici MM, Antonietti M (2009) Macromol Rapid Commun 30:221
127. McKeown NB, Budd PM (2006) Chem Soc Rev 35:675
128. Cooper AI (2009) Adv Mater 21:1291
129. Jiang JX, Su F, Trewin A, Wood CD, Campbell NL, Niu H, Dikckinson C, Ganin AY, Rosseinsky MJ, Khimyak YJ, Cooper AI (2007) Angew Chem Int Ed 46:8574
130. Jiang JX, Su F, Trewin A, Wood CD, Niu H, Jones TA, Khimyak YZ, Cooper AI (2008) J Am Chem Soc 130:7710
131. Jiang JX, Su F, Niu H, Wood CD, Campbell NL, Khimyak YZ, Cooper AI (2008) Chem Commun 486
132. Dawson R, Su F, Niu H, Wood CD, Jones JTA, Khimyak YZ, Cooper AI (2008) Macromolecules 41:1591
133. Palkovits R, Antonietti M, Kuhn P, Thomas A, Schüth F (2009) Angew Chem Int Ed 48:6909
134. Chen L, Honsho Y, Seki S, Jiang DL (2010) J Am Chem Soc 132:6742
135. Jiang JX, Laybourn A, Clowes R, Khimyak YZ, Bacsa J, Higgins SJ, Adams DJ, Cooper AI (2010) Macromolecules 43:7577
136. Dawson R, Laybourn A, Khimyak YZ, Adams DJ, Cooper AI (2010) Macromolecules 43:8524
137. Jiang JX, Wang C, Laybourn A, Hassell T, Clowes R, Khimyak YZ, Xiao J, Higgins SJ, Adams DJ, Cooper AI (2011) Angew Chem Int Ed 50:1072
138. Morisaki Y, Gon M, Tsuji Y, Kajiwara Y, Chujo Y (2011) Tetrahedron Lett 52:5504
139. Morisaki Y, Gon M, Tsuji Y, Kajiwara Y, Chujo Y (2012) Macromol Chem Phys 213:572
140. Sing KSW, Everett DH, Haul RAW, Moscou L, Pierotti RA, Rouquérol J, Siemieniewska T (1985) Pure Appl Chem 57:603

π-Stacked Oligomers as Models for Semiconducting Conjugated Organic Materials

David M. Collard

Introduction

Background

Semiconducting π-conjugated oligomers and polymers show great promise in applications ranging from field-effect transistors to light-emitting diodes and photovoltaic cells [1–3]. Across this broad class of materials the electronic and optical properties of thin films are highly dependent on intermolecular interactions between the π systems of each chain. Accordingly, optimization of these properties may be achieved by the design of new materials and by exerting control over the packing of the conjugated chains in the solid state.

The packing of conjugated chains in the crystalline state may vary from a perpendicular herringbone arrangement to a face-to-face assembly. Benzene, linear fused polynuclear aromatic hydrocarbons (e.g., naphthalene, anthracene, tetracene), and paraphenylenes (e.g., biphenyl, *p*-terphenyl, *p*,*p*-quaterphenyl) all crystallize in a herringbone arrangement. The edge-to-face arrangement of chains in these crystalline materials is favored by interactions between the hydrogen atoms of one molecule and the π system of another [4]. The angle between the planar molecules decreases as they become longer, which leads to a greater amount of intermolecular π–π interaction. The presence of side chains on such linear structures interferes with the edge-to-face interaction of the conjugated units, thereby leading to a face-to-face arrangement. This is illustrated with reference to two well-characterized molecular semiconductors: pentacene and sexithiophene. Pentacene molecules pack in a tilted herringbone motif with a significant amount of intermolecular π–π

D.M. Collard (✉)
School of Chemistry and Biochemistry, Georgia Institute of Technology,
Atlanta, GA 30332-0400, USA
e-mail: david.collard@chemistry.gatech.edu

T. Nakano (ed.), *π-Stacked Polymers and Molecules: Theory, Synthesis, and Properties*,
DOI 10.1007/978-4-431-54129-5_4,

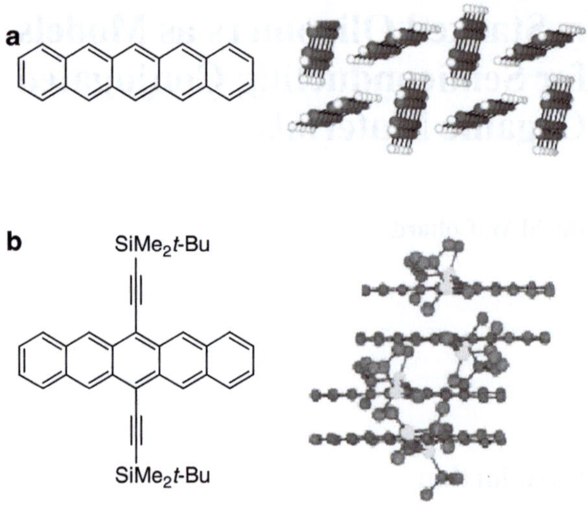

Fig. 1 Molecular packing of pentacenes. (**a**) Tilted herringbone packing of pentacene; (**b**) face-to-face brickwork packing of 6,13-bis((*tert-butyldimethylsilyl*)ethynyl) pentacene. Reprinted with permission from Anthony et al. [5]. Copyright 2002 American Chemical Society

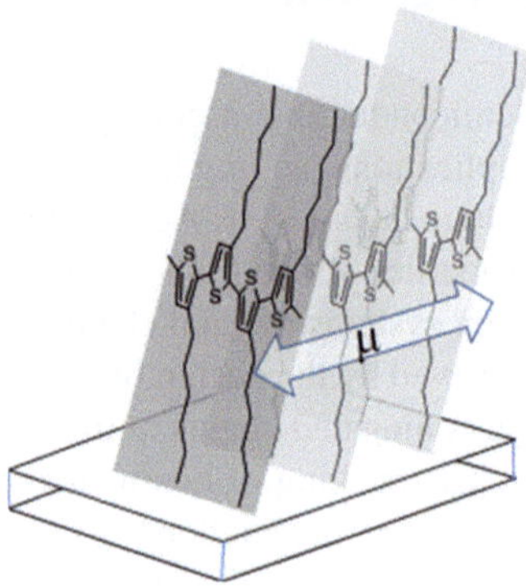

Fig. 2 Face-to-face packing of thiophene oligomer and polymer chains

interaction that accounts for its semiconducting properties. Installation of substituents on the long edges of pentacene (most commonly at the 6 and 13 positions) switches the packing to a face-to-face arrangement (Fig. 1) [5–7]. Similarly, sexithiophene and α,ϖ-dialkylsexithiophenes crystallize in a herringbone arrangement [8, 9], and poly(3-alkylthiophene)s crystallize in a lamellar fashion with stacks of conjugated backbones that are separated by the side chains (Fig. 2) [10, 11].

The importance of interactions between conjugated chains is illustrated by the effect of orientation of poly(3-alkylthiophene)s on the mobility of charge carriers in field-effect transistors [12, 13]. Thin films in which the polythiophene backbone of a poly(3-alkylthiophene) is oriented perpendicular to the surface of the gate electrode (see Fig. 2) yield a higher measured hole mobility than the case in which the conjugated chains are primarily oriented parallel to the surface. This is interpreted in terms of the orientation of π stacks that define a pathway for facile charge

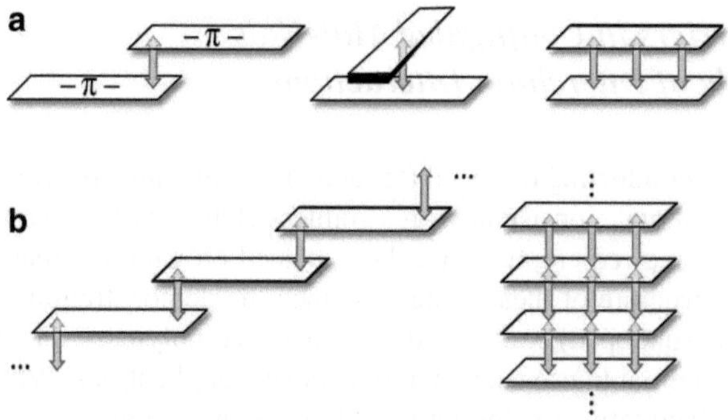

Fig. 3 Organization of conjugated segments in stacked conjugated oligomers and polymers. (**a**) Stacked "dimers," *left* to *right*: end-on-end, cross, and fully stacked; (**b**) polymers with end-on-end "staircase" arrangement of conjugated segments (*left*) and fully stacked segments (*right*)

transport. When the conjugated polymer backbones stand perpendicular to the gate electrode the π stacking axis lies along the source-drain direction, thereby creating a pathway for charge transport through the stacks.

Significant efforts have been devoted to exploring the influence of interactions between oligomeric π systems on the electronic structure of molecules in which pairs of conjugated segments are held in close proximity by a rigid scaffold. The arrangement of conjugated segments in such molecules may mimic the stacked organization of conjugated oligomers and polymer chains in semiconducting thin films. Accordingly, these molecules may serve as models to provide a greater understanding of the effect of intermolecular interactions on the nature of excitons and charge carriers in conjugated materials.

Studies of such molecules are reviewed in this chapter with a regard for how judicious choice of the scaffold provides control over the distance, orientation, and extent of overlap between the conjugated segments (Fig. 3a). The discussion is focused on well-defined structures in which conjugated segments that resemble common linear conjugated polymers (e.g., polythiophene [14], poly(phenylene vinylene) [15], and poly(phenylene ethynylene) [16]) are held in a stacked organization. An emphasis is placed on molecules in which [2.2]paracyclophane, bicyclo[4.4.1]undecane and extended [*n*,*n*]heterocyclophanes serve as a scaffold to hold oligomeric conjugated segments in a face-to-face arrangement. Comparisons of the spectroscopic and electrochemical properties of these stacked compounds to those of analogous unstacked linear oligomers illustrate the effects of interactions between π systems. In addition to the oligomeric structures described in this chapter, there is significant interest in the preparation and characterization of analogous π-stacked polymers—macromolecules consisting of relatively short conjugated segments that are held in a stacked arrangement, as depicted in Fig. 3b. These materials are described by Chujo and Morisaki of Kyoto University in "Cyclophane-based π-Stacked Polymers" in this volume.

Charge Carriers in Conjugated Materials and the Role of Interchain Interactions

Early studies of conducting oxidized ("p-doped") conjugated polymers revealed the lack of an ESR signal, consistent with a spinless charge carrier that is analogous to a bipolaron. A renewed emphasis has been placed on gaining greater insights into the electronic structure of these materials since the demonstration of semiconductivity of neutral (undoped) conjugated oligomers and polymers, and with the recognition of the potential to use such materials in new applications. While conjugation is important in facilitating the formation of low-energy resonance-stabilized charged species, intermolecular interactions are clearly important in defining the nature of these species and to the process of charge migration in these materials.

A number of species may account for the formation of spinless positive charge carriers in conjugated materials. These may result from (1) intrachain pairing of resonance-stabilized radical cations to form a bipolaron-like delocalized dicationic species [17–21], (2) reversible formation of sigma bonds between radical cations to form dicationic "sigma dimers" [22–27], or (3) reversible non-covalent dimerization of radical cations to form spinless dicationic "π dimers." In particular, the latter species has received a significant amount of interest in explaining charge transport in conjugated materials

Oligothiophenes that bear substituents at the terminal α-positions undergo oxidation to afford resonance-stabilized radical cations that are sufficiently stable to allow for spectroscopic characterization. Detailed analyses of oligothiophenes and related compounds by Miller [28–33], Baüerle [34–36], Tour [37, 38], Garnier [39], Meijer [40], Zotti [41, 42], Hotta [43, 44], and Roncali [45–47] indicate that the radical cations of these compounds associate in solution to form π dimers. The dimerization process overcomes the Coulombic repulsion between the delocalized monocations. The π dimers give rise to new absorption bands in the near-IR region of the electromagnetic spectrum and are ESR silent. This behavior is illustrated by the changes in the UV–vis-near-IR spectrum of the monocation of 3″,4″,5,5″″-tetramethyl-2,2′:5′,2″:5″,2‴:5‴,2″″-quinquethiophene that are illustrated in Fig. 4 [48]. In methylene chloride, the neutral oligomer gives rise to a π–π* transition at 400 nm. Oxidation of the oligothiophene by addition of $FeCl_3$ to the solution provides the resonance-delocalized radical cation that results in the appearance of two new transitions (731 and 1,322 nm, Fig. 4, dashed line). These peaks disappear upon cooling the solution and two new peaks appear: a new π–π* transition (620 nm) and a charge transfer band in the near-IR region of the spectrum (1,085 nm). This spectral change is reversed upon warming the solution back to room temperature.

The equilibrium between the radical cation and π dimer is also driven by the choice of solvent. For example, oxidation of 3″,4″,5,5″″-tetramethyl-2,2′:5′,2″:5″,2‴:5‴,2″″-quinquethiophene to the monocationic state in acetonitrile provides a solution that displays characteristics of the radical cation π dimer even at room temperature. Whereas the radical cations of oligothiophenes and related conjugated oligomers often undergo reversible dimerization in solution, the neutral and

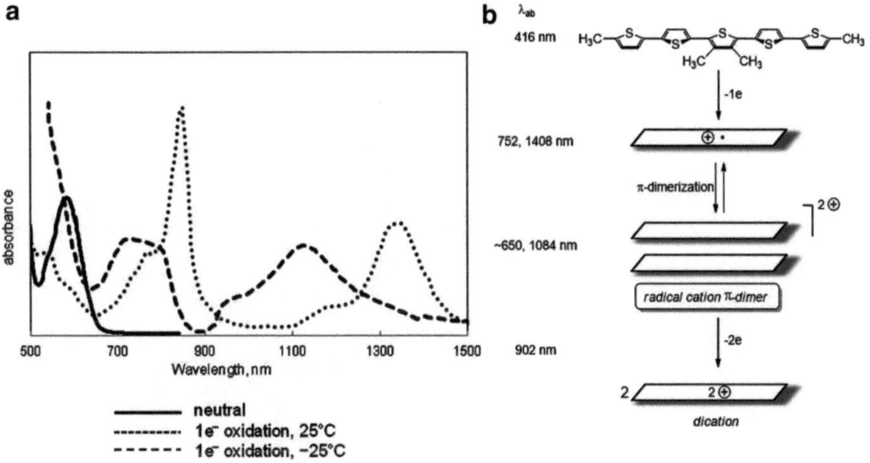

Fig. 4 Dimerization of delocalized radical cations. (**a**) Spectrum of radical cation (*dashed line*) and radical cation π dimer (*solid line*) of 3,3″,4″,5″″-tetramethyl-2,2′:5′,2″:5″,2‴:5‴,2″″-quinquethiophene formed upon cooling to −25 °C; (**b**) schematic representation of redox chemistry and formation of a radical cation π dimer

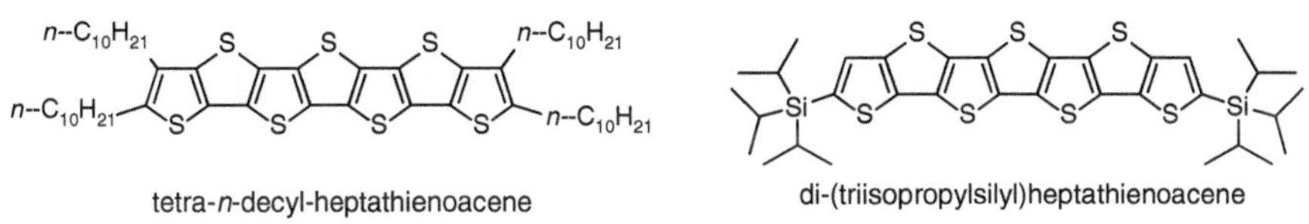

Fig. 5 Structures of substituted heptathienoacenes

dicationic states of these oligomers generally do not display this behavior (although evidence has been presented for the dimerization of the dicationic state of a duodecithiophene [19]).

Even large phenyldimethylsilyl substituents [31] and bicyclo[2.2.2]octane groups [50] at the terminal positions of oligothiophenes do not interfere with the dimerization process. Miller suggests that such large substituents are accommodated by a twisting of the oligomers with respect to one another within the π dimer [28]. However, an analysis of fused heptathienoacenes suggests that the association of radical cations is impeded by particularly bulky substituents at the termini of the oligomer. Whereas the radical cation of tetra(*n*-decyl)heptathienoacene (Fig. 5) shows spectral characteristics that are consistent with formation of a spinless π dimer, the bis(triisopropylsilyl) compound does not [51].

Spectroscopic studies of the equilibrium between oligomeric resonance-delocalized radical cations and spinless dicationic π dimers do not provide direct evidence for the structure of the latter species or for the role of radical cation π dimers in charge transport in solid materials. Evidence for the potential role of π

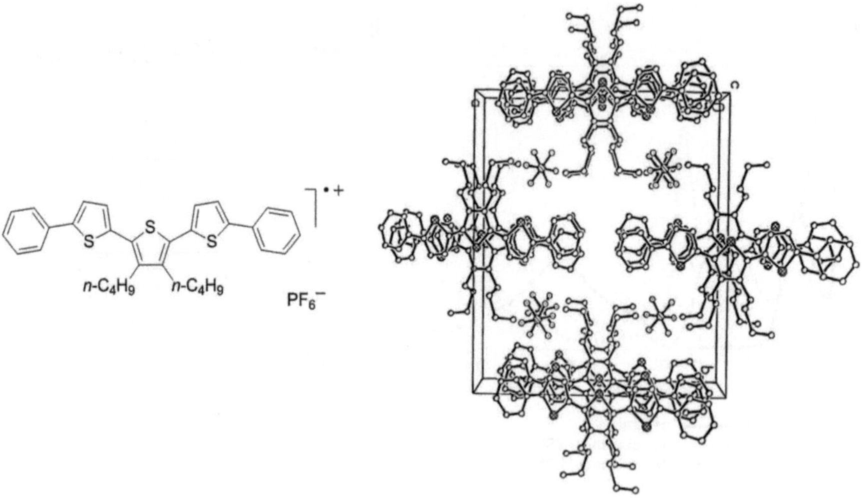

Fig. 6 Crystal structure of the hexafluorophosphate salt of the radical cation of 3′,4′-dibutyl-2,5″-diphenyl-5,2′:5′,2″-terthiophene. Reprinted with permission from Graf et al. [33]. Copyright 1996 American Chemical Society

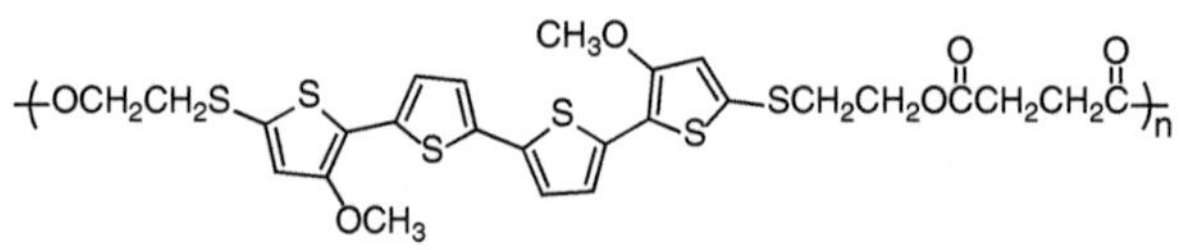

Fig. 7 Structure of quaterthiophene-containing polyester

dimers in the solid state comes from single crystal X-ray diffraction of electrocrystallized oligothiophene radical cation salts [33]. The chemical oxidation of 3′,4′-dibutyl-2,5″-diphenyl-5,2′:5′,2″-terthiophene with $NOPF_6$, or by electrochemical oxidation in the presence of n-Bu_4PF_6, affords diffraction-quality crystals of the radical cation hexafluorophosphate salt. The crystals possess the spectral characteristics of radical cation π dimers and exhibit a conductivity of 10^{-2} S cm^{-1}. The X-ray crystal structure shows that the planar radical cations assemble into tilted stacks (Fig. 6).

The spectral characteristics of a radical cation π dimer are also noted for a polyester that contains oxidized quaterthiophene units (Fig. 7) [52]. The isolated quaterthiophene units of this polymer are too short for the formation of a stable dicationic oxidation state. However, oxidation of the polymer affords a material that has a strong optical absorption spectrum that resembles that of a radical cation π dimer, a weak ESR signal, and a conductivity of 0.8 S cm^{-1}. Thus, it is posited that spinless charge carriers arise from the association of the mono(radial cation)s derived from oxidation of the quaterthiophene units within the solid polymeric material.

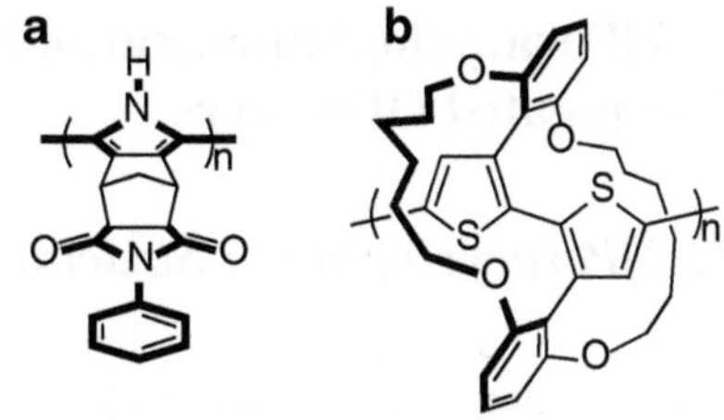

Fig. 8 Noninteracting molecular wires. (**a**) Canopied polypyrrole; (**b**) a self-threading polythiophene

The aggregation of conjugated chains, both in solution and in the solid state, is hindered by incorporation of particularly rigid and bulky substituents along the edges of the oligomers. Minimizing the interactions between conjugated chains in this manner results in distinct changes in the electronic and optical properties of the materials. For example, Swager has prepared polypyrroles in which each heterocyclic repeat unit is canopied with a rigid bicyclic unit (Fig. 8a) [53]. Upon oxidation, the canopied polypyrrole exhibits similar conductivity and changes in the UV–visible spectrum as the unsubstituted polypyrrole. However, the conductivity reaches a maximum at an intermediate level of oxidation, at which the intensity of the ESR signal is at a maximum. This is in contrast to the behavior of polypyrrole which reaches a maximum value of conductivity at a higher level of oxidation, at which the material exhibits a small ESR signal. Accordingly, the conductivity of polypyrrole may be ascribed to a bipolaron-like charge carrier for which interchain interactions play a significant role. For the canopied analog, resonance-stabilized radical cations that are isolated on individual noninteracting conjugated chains appear to account for conductivity.

A second example of a conjugated polymer in which side chains impede chain–chain contacts is the self-threading polythiophene shown in Fig. 8b [54]. The polymer, in which the conjugated backbone is sheathed by its cyclic side chains, exhibits a UV–visible spectrum, both in solution and in the solid state, that is consistent with noninteracting polymer chains. In the absence of interchain interactions the high conductivity of the material is ascribed to intrachain charge transport. Sheathing of conjugated polymer backbones to prepare such insulated molecular wires [55] may also be achieved by rotaxination with cyclodextrins [56] and crown ethers [57, 58].

Extensive studies of the aggregation of conjugated radical cations to form spinless dicationic π dimers in solution and extended π stacks in the solid state provide some insights into the roles of intermolecular interactions in conjugated materials. However, the equilibrium nature of the dimerization process and the limited amount of structural information about the packing of conjugated units in π dimers present significant challenges. This has led a number of research groups to prepare new compounds in which pairs of conjugated segments are held in well-defined stacked arrangements by attachment to a molecular scaffold. These mimic the face-to-face packing of chains that is present in conjugated materials and may provide new opportunities to explore aspects of the charge transport and photophysics of organic semiconductors.

[2.2]Paracyclophane-Scaffolded π-Stacked Conjugated Oligomers

[2.2]Paracyclophane: Structure

A significant number of molecules in which two conjugated segments are held in a stacked arrangement are based on the use of [2.2]paracyclophane as a scaffold. [2.2] Paracyclophane consists of two benzene rings that are held in a stacked arrangement by ethano bridges which are in para positions to one another on each ring. X-ray crystallography reveals that the two benzene rings are slightly bent out of planarity (Fig. 9a) and twisted by ca. 6° relative to one another around an axis that passes through their centers (Fig. 9b) [59]. This is accompanied by a small twist in the ethano bridges and a relatively long benzyl–benzyl σ bond. As a consequence, pairs of carbon atoms in the stacked benzene rings are held at a distance of 2.8 and 3.1 Å. Larger paracyclophanes (e.g., [3,3] [60, 61] and [4.4] [62]), along with simple metacyclophanes and orthocyclophanes, are conformationally flexible such that the aromatic subunits are often not held in such a stacked arrangement (Fig. 9b).

The placement of conjugated arms on a [2.2]paracyclophane scaffold has provided a large number of conjugated polymers and oligomers in which the stacked benzene rings of the cyclophane are part of the conjugated segments (Fig. 10). In the

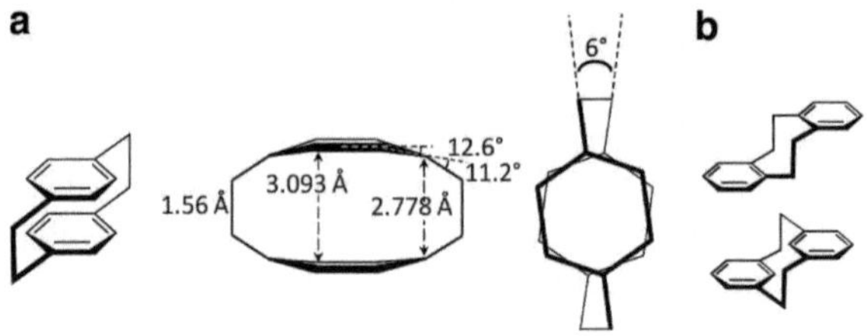

Fig. 9 Structural features of [2.2]cyclophanes. (**a**) [2.2]Paracyclophane structural parameters; (**b**) low-energy unstacked *anti* conformations of [2.2]orthocyclophane and [2.2]metacyclophane

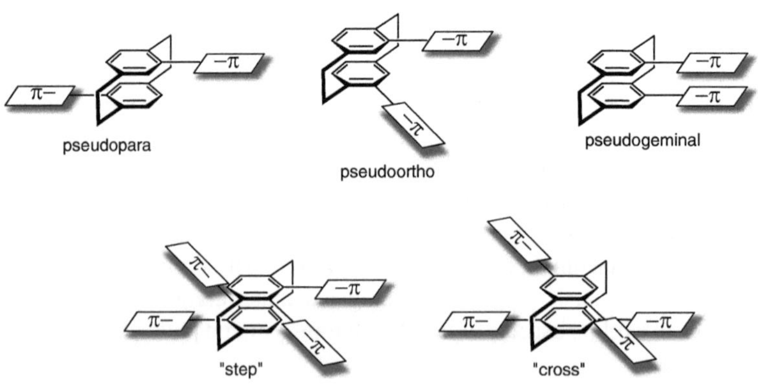

Fig. 10 General structures of [2.2]paracyclophane-scaffolded conjugated oligomers

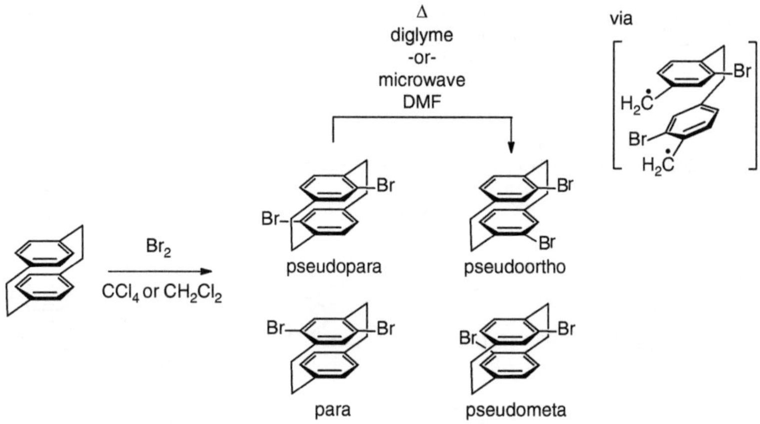

Fig. 11 Synthesis of pseudopara, pseudometa, pseudoortho, and para dibromo[2.2] paracyclophanes

pseudopara and pseudoortho disubstituted compounds,[1] only the terminal benzene rings of the two conjugated segments are held in a stacked arrangement (i.e., those within the cyclophane core itself). In the "cross" and "step" [63] tetrasubstituted compounds the point of interaction between π systems is at the center of each conjugated segment, and it is still restricted to the single pair of benzene rings within the cyclophane scaffold. In pseudogeminal analogs there is a potential for interaction between the conjugated segments over extended lengths. As such, this molecular geometry might be a better mimic of the interaction of conjugated chains in organic semiconductors.

General Synthetic Routes to Functional [2.2]Paracyclophane Cores

The preparation of substituted [2.2]paracyclophanes has a long and rich history [64–68]. Electrophilic aromatic bromination of [2.2]paracyclophane proceeds rapidly in the presence of Lewis acids to afford a mixture of pseudopara, pseudoortho, pseudometa, and para isomers of dibromo[2.2]paracyclophane (Fig. 11). Repeated recrystallization of the mixture from chloroform provides the pseudopara isomer, albeit usually in modest yield (~25 %) [69, 70]. Cram showed that the pseudopara isomer is equilibrated with the pseudoortho compound upon heating to 200 °C in triglyme [71]. The reaction proceeds by heterolytic cleavage of a two-carbon bridge to form a diradical and subsequent ring closure to afford a mixture of the two

[1] Identification of the position of substituents using the pseudo prefix, along with the "cross" and "step" descriptors, avoids the use of conflicting and potentially ambiguous nomenclatures [63].

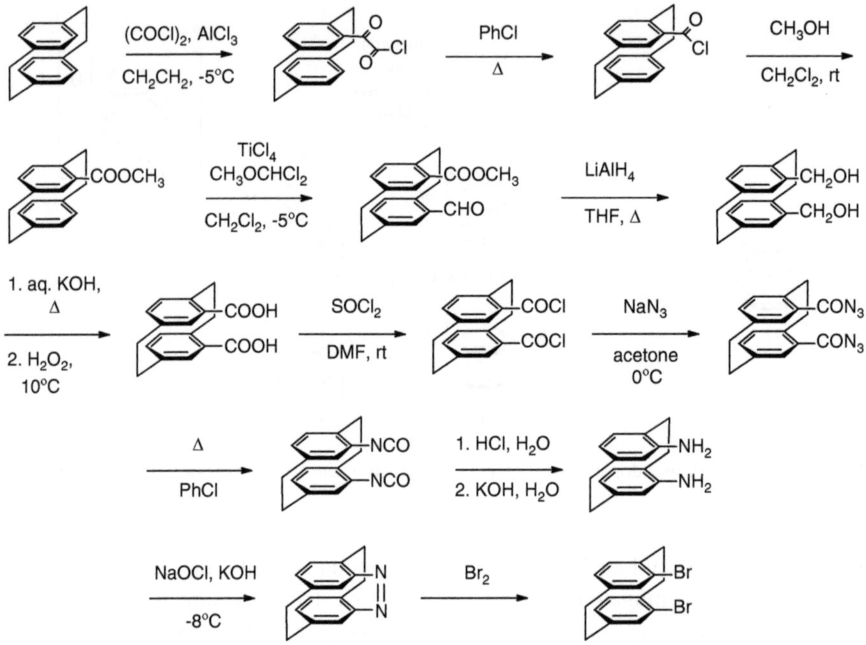

Fig. 12 Synthesis of pseudogeminal dibromo[2.2]paracyclophane

isomers, as shown in Fig. 11. Pye optimized the conditions for this reaction (230 °C, 3 h, triglyme) to afford a 69 % isolated yield of the pseudoortho isomer on a multigram scale [72]. A further improvement was realized by Braddock through the use of microwave irradiation (180 °C, 6 min, DMF), with simple separation of the two isomers based on precipitation of the pseudopara isomer from the reaction mixture upon cooling to room temperature and removal of the solvent to afford the pseudoortho isomer [73]. Preparation of the pseudometa isomer is achieved by chromatographic separation of the mixture of isomers obtained from the dibromination reaction [74].

Preparation of the pseudogeminal dibromo isomer requires a multistep synthesis that was first described by Hopf [75]. The entire synthetic route from [2.2]paracyclophane is outlined in Fig. 12. Methyl [2.2]paracyclophane-4-carboxylate is prepared by Friedel–Crafts acylation of [2.2]paracyclophane with oxalyl chloride [76], which is followed by thermal decarbonylation and quenching with excess of methanol to give the methyl [2.2]paracyclophane carboxylate [77]. The pseudogeminal substitution pattern is attained by Rieche formylation upon treatment of the methyl ester with α,α′-dichloromethyl methyl ether in the presence of $TiCl_4$. The formylation is regioselective with the reaction primarily taking place at the carbon atom below the ester-substituted position. Presumably, this regioselectivity arises from chelation of the ester with the titanium which delivers the electrophilic reagent to the proximate position. Having established the pseudogeminal substitution

Fig. 13 Preparation of pseudoortho dibromide, a tribromo compound, and "step" and "cross" tetrabromo[2.2] paracyclophanes

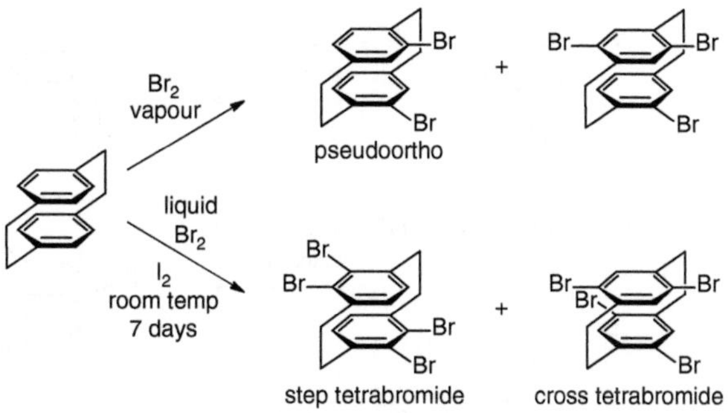

pattern, the remaining steps in the synthesis involve a series of functional group interconversions. Reduction of both the ester and the aldehyde in a single step ($LiAlH_4$ in THF) provides the symmetrically substituted bis(hydroxymethyl) analog. Oxidation of the diol upon treatment with H_2O_2 and KOH affords the corresponding diacid. Conversion of the diacid to the acyl azide followed by a Curtius rearrangement and hydrolysis of the intermediate isocyanate gives pseudogeminal diamino[2.2]paracyclophane [77]. Diazotization followed by reaction with bromine affords the pseudogeminal dibromo[2.2]paracyclophane [75]. A number of more highly ring-brominated [2.2]paracyclophanes have also been prepared. These include tribromo compounds [78, 79] and the "cross" and "step" tetrabromo compounds (Fig. 13) [79].

The pseudopara dichloro- and dibromo- [2.2]paracyclophanes are also available in modest yields (~20 %), and free from other regioisomers, from Winberg 1,6-elimination-dimerization [80, 81], of (2-halo-4-methylbenzyl)trimethylammonium hydroxides (Fig. 14) [82]. Similarly, reaction of (2,5-dihalo-4-methylbenzyl)trimethylammonium hydroxides affords the "cross" tetrahalo[2.2]paracyclophane substitution pattern previously reported by de Meijere [79]. Additional value of this route might also lie in the ability to prepare new highly halogenated [2.2]paracyclophanes, including 4,8,12,16-tetrabromo[2.2]paracyclophane and 4,5,7,8,12,13,15,16-octabromo[2.2] paracyclophane.

In addition to the brominated compounds, the vinyl-, ethynyl-, and formyl-substituted [2.2]paracyclophanes have proven useful for the preparation of π-stacked oligomers. The pseudopara and pseudoortho divinyl compounds are available from Heck coupling reactions of the corresponding dibromo[2.2]cyclophanes with ethylene or by Stille coupling [83]. The diethynyl[2.2]paracyclophanes are available by Sonogashira coupling of the dibromo compounds with trimethylsilylacetylene, followed by fluoride-promoted desilylation (Fig. 15a) [84, 85]. More highly ethynylated analogs are available from ethynylation of the corresponding tetrabromides (Fig. 15b) [86]. Pseudogeminal dibromo[2.2]paracyclophane resists Heck and Sonogashira coupling reactions [87].

The pseudopara and pseudoortho diformyl[2.2]paracyclophanes are available by formylation of the corresponding dibromo compounds (Fig. 16a). The

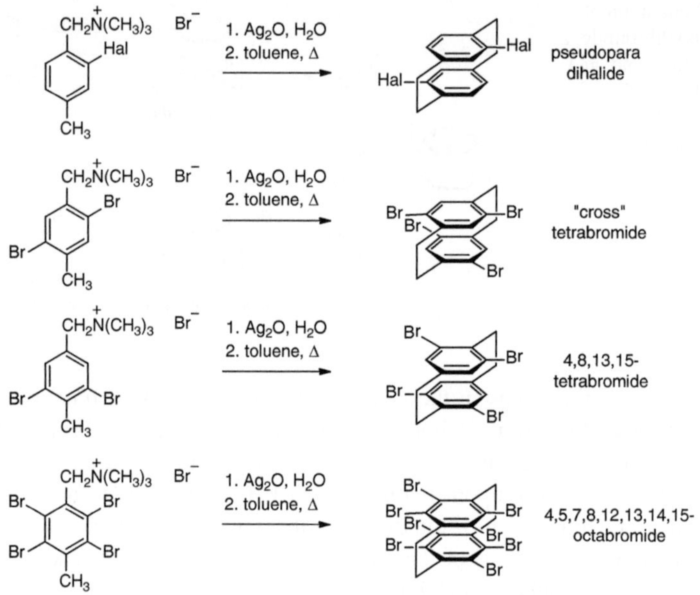

Fig. 14 Winberg 1,6-elimination-dimerization to prepare halo-substituted [2.2]paracyclophanes [82]

pseudogeminal dialdehyde is conveniently prepared by oxidation of the bis(hydroxymethyl) compound (a synthetic intermediate in Fig. 12) with dichloro-dicyanoquinone (DDQ) or pyridinium chlorochromate (PCC) (Fig. 16b). Another approach to prepare the diformyl compounds is by the reaction of 1,2,4,5-hexatriene and 2-propynal (propiolic aldehyde) [88, 89]. Recrystallization and chromatographic separation of the crude mixture of products provide access to multigram quantities of the pseudopara, pseudometa, pseudoortho, and pseudogeminal dicarboxaldehydes (Fig. 16d) [90]. A variety of "step" tetrasubstituted [2.2]paracyclophanes bearing electron-withdrawing substituents are available from the analogous reaction with other disubstituted alkynes (Fig. 16e) [91].

A Wittig reaction of the diformyl[2.2]paracyclophane with methyltriphenylphosphonium bromide affords the corresponding divinyl[2.2]paracyclophane (Fig. 16c) [92, 93]. Treatment of the dialdehyde with the Bestmann–Ohira reagent ($CH_3COCN_2PO(OCH_3)_2$) provides the corresponding diethynyl[2.2]paracyclophane [94–96].

Coupling of [2.2]Paracyclophane Cores and Conjugated Arms

The bromo-, formyl-, vinyl-, and ethynyl-substituted [2.2]paracyclophanes serve as versatile precursors for the preparation of a variety of π-stacked oligomers. Arylene vinylene arms may be installed onto the [2.2]paracyclophane core by a Wittig

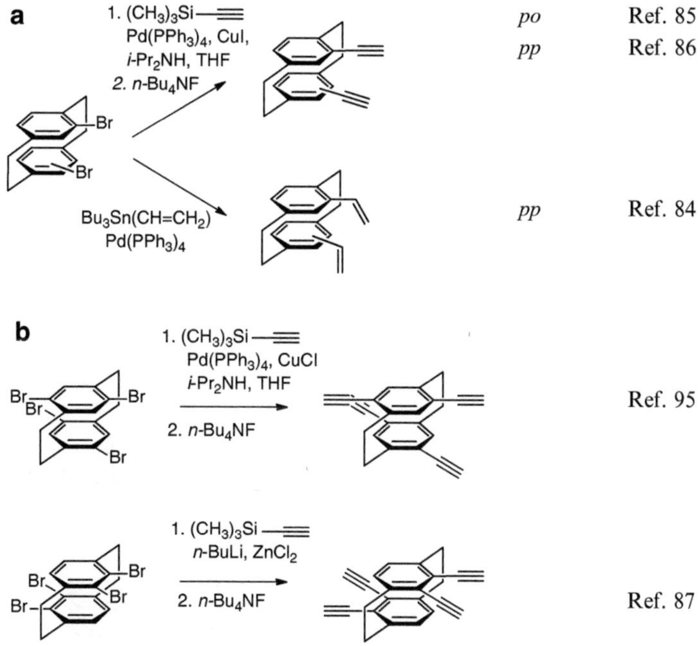

Fig. 15 Preparation of ethynyl and vinyl [2.2]paracyclophanes: (**a**) Divinyl and diethynyl compounds from dibromo[2.2]paracyclophanes; (**b**) "cross" and "step" tetra(ethyne)s prepared from the corresponding tetrabromo compounds

reaction of the diformyl[2.2]paracyclophane with a benzyl phosphonium ylide or by a Heck reaction of the vinyl-substituted cyclophanes with an aryl iodide (Fig. 17). The pseudopara and pseudoortho dibromo[2.2]paracyclophanes, and the "cross" and "step" tetrabromo[2.2]cyclophanes (see "General Synthetic Routes to Functional [2.2]paracyclophane Cores" section), undergo Heck coupling reactions with styrenes to allow for the installation of arylene vinylene arms on each ring of the cyclophane core. Arylene ethynylene arm may be installed by Sonogashira coupling of the dibromo- or diethynyl- [2.2]paracyclophane.

π-Stacked Conjugated Oligomers Based on Pseudopara and Pseudoortho [2.2]Paracyclophane Cores

Following a seminal report from the group of Armin de Meijere at the University of Göttingen [79], the group of Guillermo Bazan of the University of California, Santa Barbara, has performed extensive studies of stacked conjugated systems (e.g., phenylene vinylenes and phenylene ethynylenes) that are scaffolded by a [2.2]paracyclophane core [97, 98]. The original motivation for these studies was to address the role

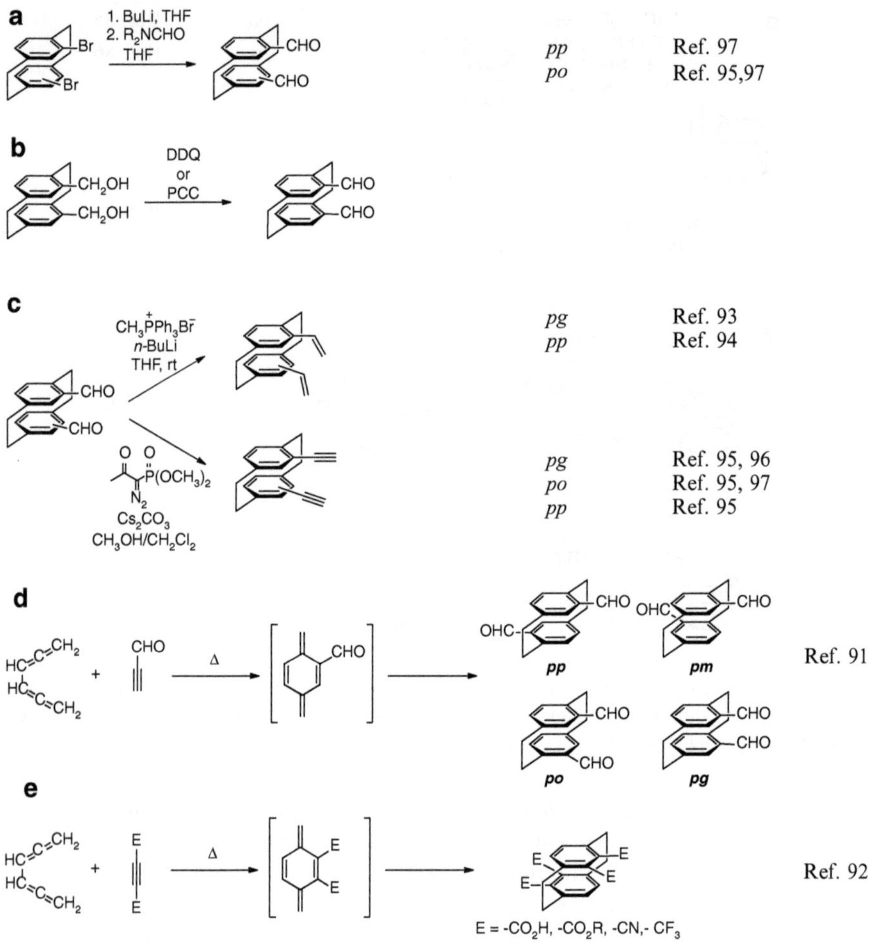

Fig. 16 Preparation and reactions of diformyl[2.2]paracyclophanes. (**a**) formylation of dibromo [2.2]paracyclophanes; (**b**) preparation of pseudogeminal diformyl[2.2]paracyclophane; (**c**) Wittig and Bestmann–Ohira reactions of aldehydes to afford divinyl and diethynyl compounds, respectively; (**d**) preparation of 4,5,12,13-substituted [2.2]paracyclophanes by Diels–Alder reaction between 1,2,4,5-hexatetrene and substituted alkynes; (**e**) preparation of "step" tetrasubstituted compounds from 1,2-disubstituted ethynes

of aggregates of polymer chains on the photophysics of poly(phenylene vinylene) thin films in organic light-emitting diodes [99, 100]. The substituted [2.2]paracyclophanes bring together two conjugated segments in a stacked arrangement, thereby making it possible to explore the effect of interchain interactions in chemically well-defined systems. Spectroscopic features of the stacked phenylene vinylene analogs based on the pseudopara and pseudoortho disubstituted [2.2]paracyclophane scaffolds are provided in Table 1 [101, 102].

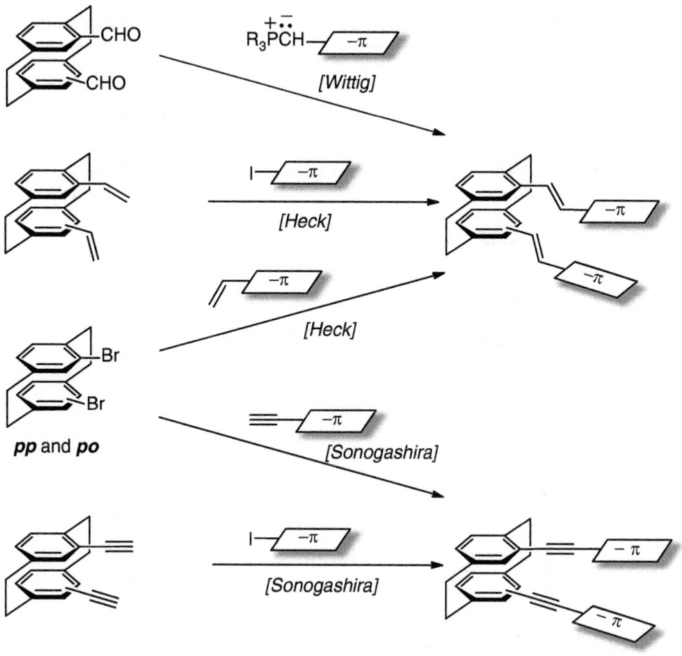

Fig. 17 General scheme for the installation of phenylene vinylene and phenylene ethynylene arms on the [2.2]paracyclophane core

Comparison of the absorption maximum of stilbene (λ_{abs} = 244 nm) and vinyl[2.2] paracyclophane (λ_{abs} = 281 nm) provides the first indication of the effect of stacking phenylene units in this series of compounds [102]. The red-shifted absorption of the cyclophane-stacked compound compared to the unstacked analog is a common feature for this series of compounds. Substitution of both rings of the cyclophane core with vinyl groups in either a pseudoortho (***po***) or pseudopara (***pp***) arrangement provides stacked arrangements of two styrene segments (denoted here as phenylene vinylene, PV) that serve as models for a pair of interacting repeat units of poly(phenylene vinylene). The pseudopara and pseudoortho analogs of **CP[PV]$_2$** exhibit absorption maxima at 254 and 288 nm, respectively, that are both red shifted compared to styrene.

Extending the substituents on the [2.2]paracyclophane core from vinyl substituents to styryl units affords a model for stacked stilbene units, "PV$_2$"—here the subscript identifies the number of phenyl groups in the oligo(phenylene vinylene) segment. As noted for the vinyl compounds, the absorption maximum of the monostyryl-substituted [2.2]paracyclophane (λ_{abs} = 318 nm) is shifted to lower energy than the unstacked dimethylstilbene, **PV$_2$** (λ_{abs} = 294 nm). Holding two stilbene units (PV$_2$) in a stacked arrangement based on the cyclophane in ***pp*-CP[PV$_2$]$_2$** and ***po*-CP[PV$_2$]$_2$** results in more complex red-shifted absorption peaks, with contributions in the spectra of both compounds at ~310 nm and ~330 nm (Fig. 18a). For the pseudopara analog the

Table 1 [2.2]Paracyclophane phenylene vinylenes and unstacked model compounds[a]

Structure	Compound	Spectra	Ref.
Styrene (PV) analogs			[103]
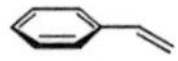	styrene	$\lambda_{ab} = 244$ nm	
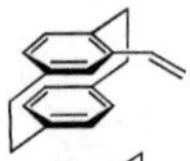	vinyl[2.2]paracyclophane	$\lambda_{ab} = 281$ nm $\lambda_{em} = 374$ nm[br]	
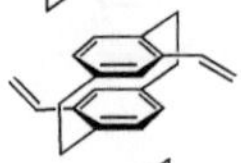	***pp*-CP[PV]₂**	$\lambda_{ab} = 254$ nm $\lambda_{em} = 394$ nm[br]	
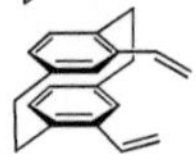	***po*-CP[PV]₂**	$\lambda_{ab} = 288$ nm $\lambda_{em} = 386$ nm[br]	
Stilbene (PV_2) analogs			[103]
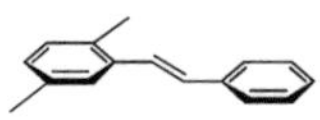	**PV_2**	$\lambda_{ab} = 294$ nm $\lambda_{em} = 338, 355$ nm[vib]	
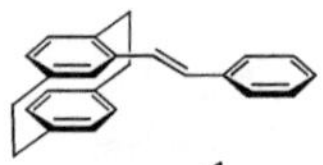	styryl[2.2]paracyclophane	$\lambda_{ab} = 318$ nm $\lambda_{em} = 386$ nm	
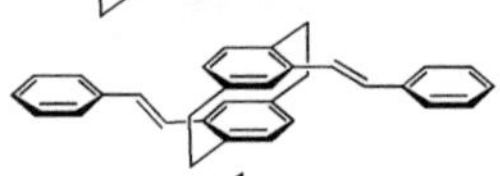	***pp*-CP[PV_2]$_2$**	$\lambda_{ab} = 307$, ~330[sh] nm $\lambda_{em} = 412$ nm[br]	
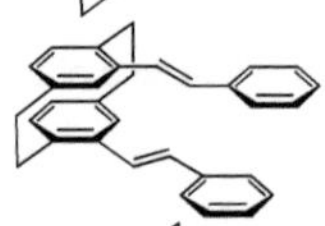	***po*-CP[PV_2]$_2$**	$\lambda_{ab} = 325$ nm $\lambda_{em} = 401$ nm[br]	
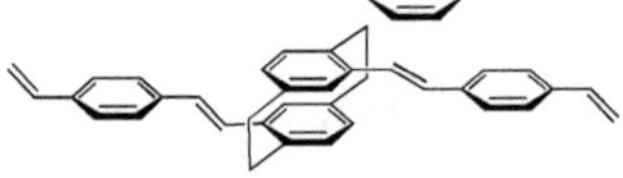	***pp*-CP[(PV_2)V]$_2$**	$\lambda_{ab} = 338$ nm $\lambda_{em} = 390$[sh], 409 nm[b]	
p-Distyrylbenzene (PV_3) analogs			[103]
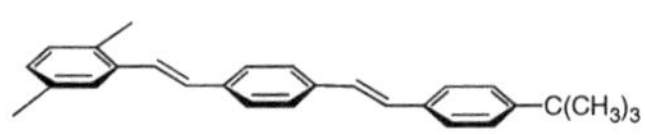	**PV_3**	$\lambda_{ab} = 349$ nm $\lambda_{em} = 391$ nm[vib]	
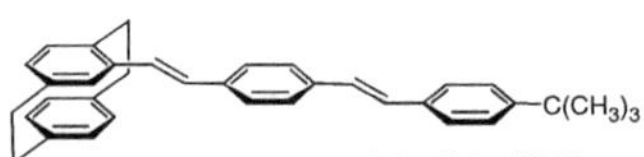	4-styryl(styryl)[2.2]paracyclophane	$\lambda_{ab} = 366$ nm $\lambda_{em} = 403, 427$ nm[vib]	
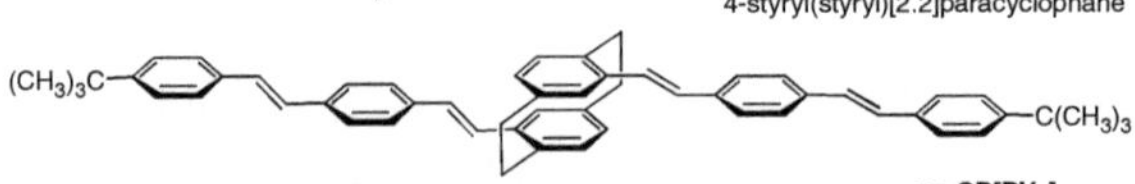	***pp*-CP[PV_3]$_2$**	$\lambda_{ab} = 369$ nm $\lambda_{em} = 407, 430$ nm[vib]	
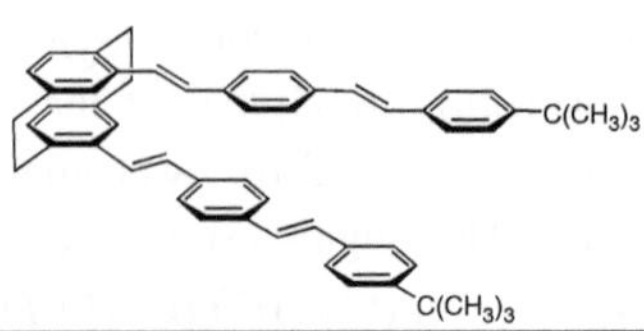	***po*-CP[PV_3]$_2$**	$\lambda_{ab} = 355$ nm $\lambda_{em} = 408, 431$ nm[vib]	

[a]Spectra obtained in hexane
[b]Broad at room temperature, vibronic at 77 k
[br]Broad, [sh]Shoulder, [vib]Vibronic

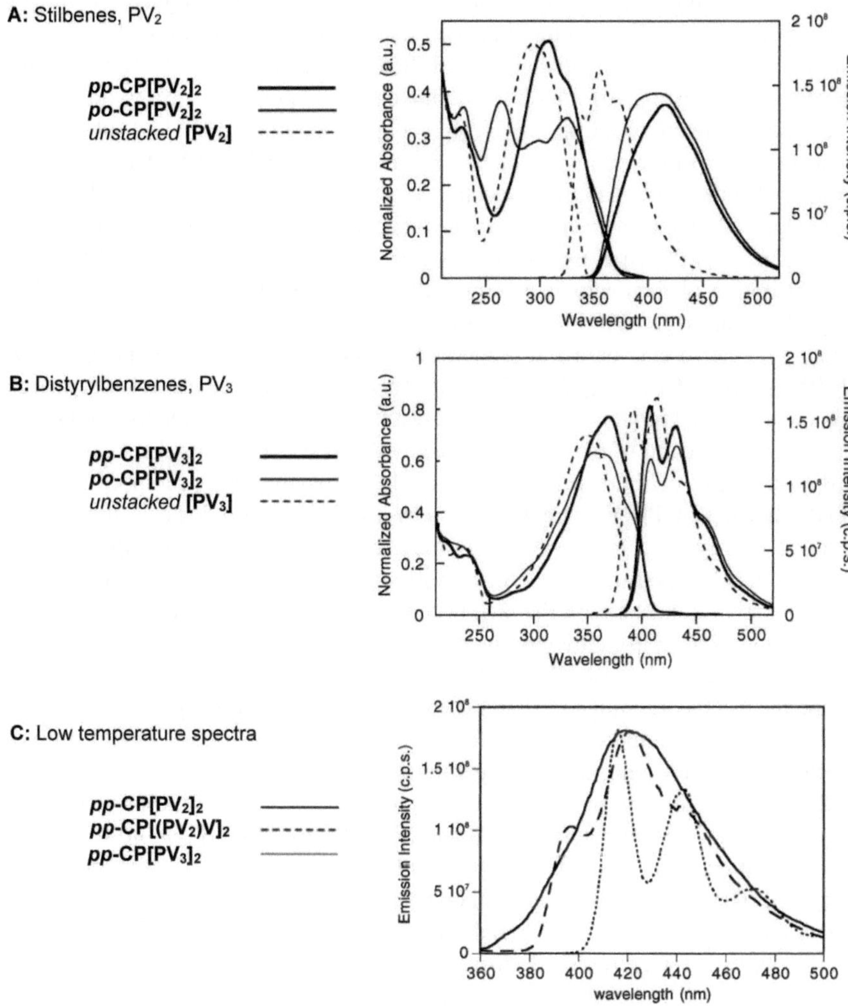

Fig. 18 UV–vis and fluorescence spectra of pseudopara and pseudoortho phenylene vinylenes and model unstacked oligomer: (**a**) stilbene (PV_2) analogs; (**b**) distyrylbenzene (PV_3) analogs; (**c**) low-temperature emission spectra (77 K in 2-methyltetrahydrofuran glass). Modified with permission from Bazan et al. [102]. Copyright 1998 American Chemical Society

contribution at shorter wavelength predominates (λ_{abs} = 307 nm) and the longer wavelength contribution appears as a shoulder. For the pseudoortho isomer the situation is reversed, with the longer wavelength contribution (λ_{abs} = 325 nm) being stronger. This is in contrast to the unstacked analog, stilbene, which has an absorption maximum at 294 nm and does not have the longer wavelength contribution in the absorption spectrum. Thus, the lower-energy contribution appears to be a result of stacking the stilbenoid segments. A more dramatic difference is observed in the fluorescence spectra of the stilbene analogs $\mathbf{PV_2}$, ***pp*-CP[PV$_2$]$_2$**, and ***po*-CP[PV$_2$]$_2$**. The fluorescence spectrum of the unstacked model compound displays a vibronic progression with the 0–0 peak at 338 nm. The spectra of the two stacked analogs exhibit broad featureless emission peaks that are significantly red shifted, with a maximum at 412 nm for the pseudopara analog and 401 nm for the pseudoortho compound.

Further extension of the ***pp*-CP[PV$_2$]$_2$** analog by incorporating vinyl substituents at the 4 position of the peripheral phenyl rings to afford ***pp*-CP[(PV$_2$)V]$_2$** results in a further red shift in the absorption spectra, as expected upon increasing the length of the conjugated segments. In contrast to the broad emission peaks of ***pp*-CP[PV$_2$]$_2$**, the emission peak of the vinyl-extended analog has a significant shoulder at ~390 nm.

Substitution of the cyclophane core with 4-styrl(styryl) arms to afford stacked distyrylbenzene segments (PV$_3$) results in absorptions at lower energies. In common with the shorter pseudopara and pseudoortho analogs, the absorption peaks of ***pp*-CP[PV$_3$]$_2$** and ***po*-CP[PV$_3$]$_2$** (λ_{abs} = 369 and 355, respectively) appear at lower energy than the unstacked model (349 nm). Both display low-energy shoulders that are not present in the spectra of the shorter stacked PV and PV$_2$ compounds. Also in contrast to the case of the shorter stacked analogs, the fluorescence spectra of ***pp*-CP[PV$_3$]$_2$** and ***po*-CP[PV$_3$]$_2$** are dominated by a series of vibronic peaks that appear at lower energy (λ_{em}(0–0) = ~408 nm) than those of the model compound **PV$_3$** (λ_{em}(0–0) = 391 nm) (Fig. 18b). The vibronic structure of the emission peak of the stacked distyrylbenzene ***pp*-CP[PV$_3$]$_2$** becomes even more pronounced at low temperature (77 K in a 2-methyltetrahydrofuran glass), whereas the emission of the shorter stilbene analog ***pp*-CP[PV$_2$]$_2$** remains broad (Fig. 18c). The pseudopara compound that has a conjugation length between these two analogs, ***pp*-CP[(PV)$_2$ V]$_2$**, which has a broad emission at room temperature with a low-energy shoulder, gives a distinctly vibronic peak at low temperature.

Thus, all of the stacked PV and PV$_2$ analogs (vinyl[2.2]paracyclophane and styryl[2.2]paracyclophane, ***pp*-CP[PV]$_2$**, ***po*-CP[PV]$_2$**, ***pp*-CP[PV$_2$]$_2$**, and ***po*-CP[PV$_2$]$_2$**) give a broad emission peak that is significantly red shifted relative to the absorption maximum. The emissions of these compounds are also significantly red shifted relative to their unstacked analogs (i.e., **PV** and **PV$_2$**). Similar shifts in the emission maximum are also observed for a series of [2.2]paracyclophane-terminated oligo(arylene ethynylene)s [103, 104]. This behavior is in contrast to that of the longer stacked PV$_3$ analogs 4- styrylstyryl[2.2]paracyclophane, ***pp*-CP[PV$_3$]$_2$** and ***po*-CP[PV$_3$]$_2$**. The latter all display vibronic structure in their fluorescence spectra, have smaller Stokes shifts, and are less red shifted from the emission of the unstacked analog **PV$_3$**.

Accordingly, the spectra of the cyclophanes with shorter stacked segments (PV, PV$_2$) resemble those of excimers and are distinct from those of the corresponding unstacked analogs, whereas the spectra of compounds with the longer stacked PV$_3$ segments closely resemble those of the unstacked model. ***pp*-CP[(PV$_2$)V]$_2$** appears to represent a transition between these two cases, with a broad excimer-like emission spectrum at room temperature and a vibronic emission at low temperature.

The spectroscopic behavior of the pseudopara and pseudoortho phenylene vinylene analogs presented in Table 1 can be explained by the energy diagram shown in Fig. 19. The similarity of the absorption spectra of the stacked compounds and those of the corresponding unstacked linear oligomers suggests that the electronic structure

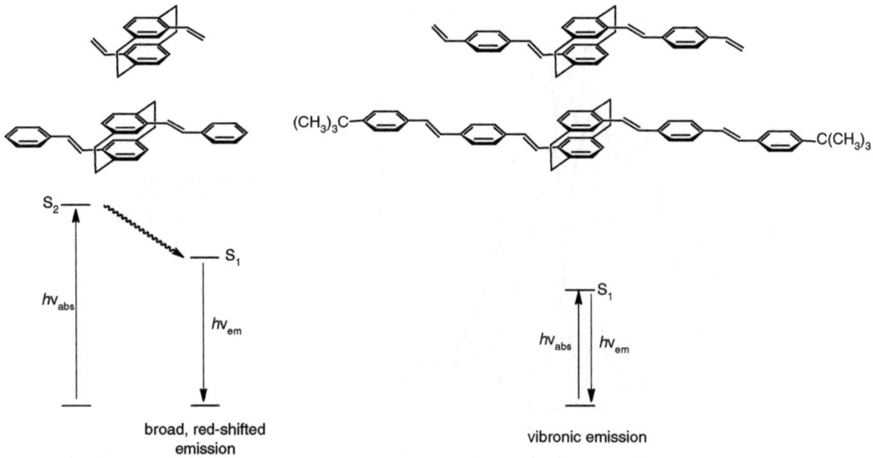

Fig. 19 Qualitative energy level diagram accounting for the electronic spectral properties of stacked PV analogs [102]

of the absorbing chromophore in the stacked compounds resembles that of a single conjugated segment. Stacking of two segments within the [2.2]paracyclophane scaffold results in only minor perturbations that give rise to the small red shift in the absorption maximum and emergence of a low-energy tail in the spectrum. For the stacked compounds that consist of relatively short conjugated segments (PV and PV_2), the appearance of broad peaks in the fluorescence spectra that are significantly red shifted from the absorption maximum and from the emission of the corresponding unstacked analogs suggests the presence of a second, lower-energy excited state. Thus, it is proposed that absorption affords an excited state that is primarily localized on a single segment and this is followed by rapid internal conversion to the lower-energy excited state that is delocalized over both tiers of the cyclophane core. Bazan named this a "phane" state. This latter state is responsible for the broad, featureless low-energy emission. For stacked compounds consisting of longer conjugated segments, the segment-localized excited state is lower in energy than the phane state. Accordingly, there is no driving force for internal conversion, and the single-segment chromophore is responsible for both absorption and emission. Thus, the absorption and fluorescence spectra of the longer stacked compounds resemble those of the unstacked analogs in terms of energy and the presence of vibronic splitting.

The effect of changing the site of interaction between two stacked PV_3 segments is illustrated by the difference between the spectra of ***pp*-CP[PV$_3$]$_2$**, in which the terminal rings of the conjugated segments are held in a stacked arrangement by the cyclophane core, and the "crossed" tetrasubstituted compound ***X*-CP[PV$_3$]$_2$**, in which the central rings of each PV_3 segment are stacked with one another (see Fig. 20 and Table 2) [105]. The absorption spectra of compounds with the latter architecture display two peaks with wavelengths greater than 300 nm. A high-energy peak in this region (R = *tert*-butyl, λ_{abs} = ~350 nm) resembles the peak in the absorption spectrum of terminally stacked ***pp*-CP[PV$_3$]$_2$** and the unstacked analogs

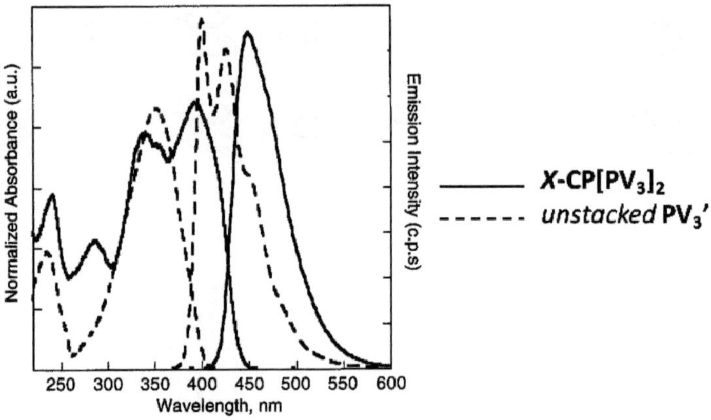

Fig. 20 UV–vis and fluorescence spectra of ***X*-CP[PV$_3$]$_2$**. Modified with permission from Bartholemew and Bazan [97]. Copyright 2001 American Chemical Society

PV$_3$ (Table 1) and **PV$_3$′** (Table 2), which all have absorption maxima in the range of 349–369 nm. However, there is an additional low-energy peak at ~410 nm in the absorption spectrum of *X*-**CP[PV$_3$]$_2$**, which is in the same region of the spectrum as the low-energy shoulder to the absorption peak of ***pp*-CP[PV$_3$]$_2$**. The emission maximum of *X*-**CP[PV$_3$]$_2$** (450 nm) is also red shifted relative to that of the terminally stacked and unstacked PV$_3$ compounds (which all appear in the range of 390–410 nm). In contrast to the PV$_3$ analogs in which the terminal ring of each segment is held in a stacked arrangement within cyclophane core (i.e., ***pp***- and ***po***-CP[**PV$_3$**]$_2$), the emission band of *X*-**CP[PV$_3$]$_2$** lacks vibronic structure (Fig. 20). Thus, changing the location of the interaction between the conjugated segments from the terminal ring of the PV$_3$ segments in ***pp*** and ***po*-CP[PV$_3$]$_2$** to the central rings in ***X*-CP[PV$_3$]$_2$** has a strong influence on the electronic structure in this series of compounds. In particular, stacking of the central rings of the PV$_3$ conjugated segments in ***X*-CP[PV$_3$]$_2$** results in a low-energy emissive state, as previously reported for the terminally stacked PV and PV$_2$ analogs, but which is not observed for the longer terminally stacked PV$_3$ compounds. Extension of the "cross" phenylene vinylene to the "pentamer" ***X*-CP[PV$_5$]$_2$** results in a red-shifted emission relative to the unstacked model **PV$_5$′**, as in the case of the PV$_3$ analog. However, in the longer analog the vibronic nature of the emission peak is restored, although it is red shifted by ~40 nm from that of the unstacked analog (Table 2).

The effect of the location of the point of contact between segments on the spectra of the stacked PV$_3$ compounds may be explained by the distribution of the highest occupied molecular orbital (HOMO) and lowest unoccupied molecular orbital (LUMO) of an isolated PV$_3$ segment [97]. The strong contributions of the central ring to the frontier orbitals are consistent with the impact of intersegment contact at this position in ***X*-CP[PV$_3$]$_2$** compared to the smaller effect of stacking PV$_3$ segments by their terminal rings in the pseudopara or pseudoortho arrangements.

Accordingly, the model presented in Fig. 19 must be modified to account for the location of contact between the conjugated segments. A new model, presented in

Table 2 Four-armed [2.2]paracyclophane phenylene vinylenes and unstacked model compounds

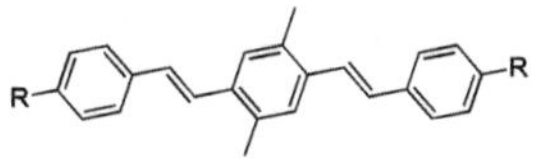 **PV$_3$'**	R=–H λ_{ab}=353 nm λ_{em}=404 nm^{vib}	[79]
	R=–C(CH$_3$)$_3$ λ_{ab}=~350 nm λ_{em}=~400, ~430 nm^{vib}	[105]
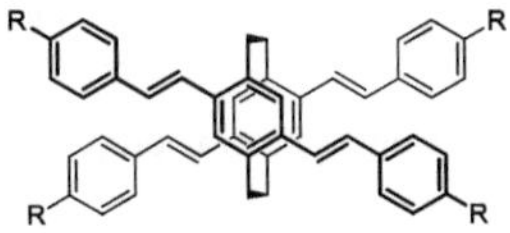 ***x*-CP[PV$_3$]$_2$**	R=–H λ_{ab}=328, 394 nm λ_{em}=465 nm^{br}	[79]
	R=–C(CH$_3$)$_3$ λ_{ab}=~340, ~390 nm λ_{em}=~450 nm^{br}	[105]
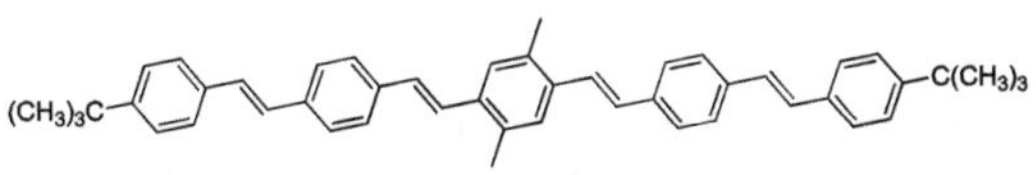 **PV$_5$'**	λ_{ab}=~370 nm λ_{em}=~450, ~480 nm^{vib}	[105]
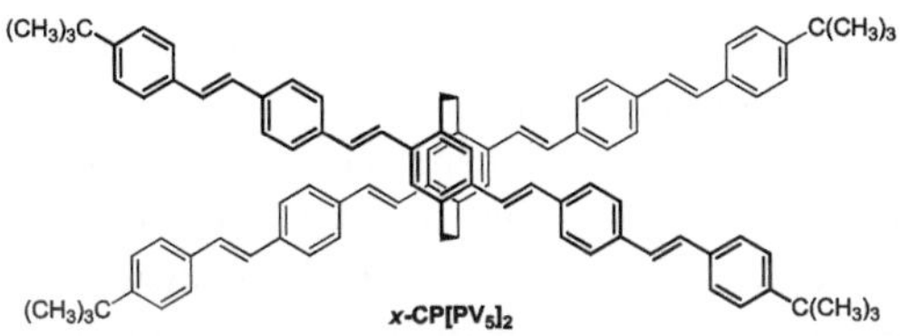 ***x*-CP[PV$_5$]$_2$**	λ_{ab}=429 nm λ_{em}=~490, ~510 nm^{vib}	[105]

^{br}Broad, ^{vi}Vibronic

Fig. 21, recognizes that the S_1 and S_2 states each have contributions from (1) an electronic state that is dominated by the single segment (named TB, for "through bond," by Bazan) and (2) a second state that is associated with the [2.2]paracyclophane scaffold (TS, for "through space") [98]. For cyclophane-scaffolded compounds that consist of short conjugated segments that are stacked by a single terminal ring (e.g., ***po***- and ***pp***- **CP[PV]$_2$** and **CP[PV$_2$]$_2$** analogs) the energy of the TB state is higher than that of the TS state, as shown in Fig. 21a. Internal conversion from the TB-dominated (i.e., localized) S_2 state populates the delocalized TS-dominated S_1 state, from which emission takes place at a relatively low energy and without vibronic coupling. For longer conjugated segments that are stacked in this manner (e.g., ***po***- and ***pp***-**CP[PV$_3$]$_2$**) the energy of the TB state is lower than the TS state (Fig. 21b). Thus, the S_1 state more closely resembles the localized TB state such that absorption and emission involve this state and the spectra are similar to those of the corresponding unstacked model compounds. For the "cross" PV$_3$ analog ***X***-**CP[PV$_3$]$_3$** a match in the energy of the TB and TS states means that the excited state is delocalized over the entire molecule, resulting in a broad, low-energy emission.

The redox properties of cyclophane-stacked phenylene vinylene compounds have received less attention. ***X***-**CP[PV$_3$]$_2$** undergoes a sequence of four reduction steps at $E_{½}$ values of −2.05, −2.27, −2.52, and −2.99 V (versus SCE), whereas the

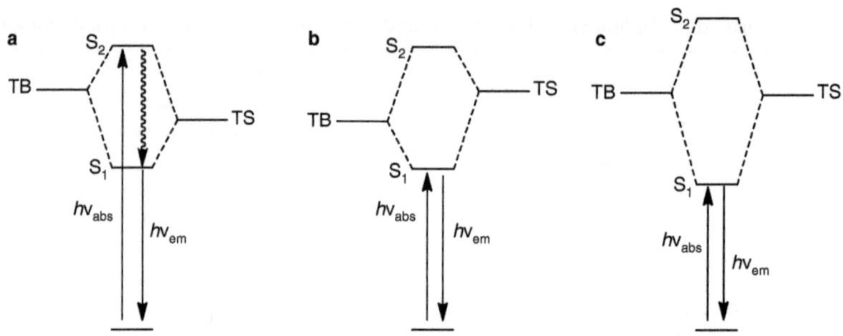

Fig. 21 Refined model for the description of the electronic spectral properties of stacked PV analogs [98]

unstacked analog undergoes two one-electron reductions at −2.25 and −2.54 V [106]. Changes in the chemical shifts of signals in the ^{13}C NMR spectrum that take place upon reduction with alkali metals indicate a delocalization of the negative charge towards the peripheral rings of the molecule and away from the point of interaction between the two charges conjugated segments.

The effects of conjugation length on "cross" cyclophane chromophores are also apparent in the series of tetraamino-substituted analogs of ***X*-CP[PV$_n$]$_2$** (Table 3) [107]. The absorption maxima of the amino-substituted "cross" [2.2]paracyclophane-scaffolded analogs are all red shifted by ~25 nm relative to the respective unstacked linear models. In each case, the emission spectrum consists of a vibronic progression. While the stacked and unstacked "trimers" (i.e., the shortest member in this series) display the largest difference in emission energy, the amino-substituted stacked compound displays vibronic emission. This is in contrast to the emission of the analogous hydrocarbon "trimer." The longer stacked PV_7 analog exhibits absorption and emission maxima that are similar to those of the PV_5 analog, suggesting that the minimum energy for the excited state is reached at five rings in this particular molecular geometry. In addition, a number of analogs of ***X*-CP[PV$_n$]$_2$** bearing electron-donating amino groups and electron-accepting nitro groups [108, 109] and water-soluble analogs [110–112] have been prepared (Table 3).

A further extension of the spectroscopic characterization of [2.2]paracyclophane-stacked oligomers reveals that interchain interactions between conjugated segments have distinctly different effects on one- and two-photon absorption [107, 112, 113]. This finding may provide impetus for the design of new two-photon absorbing chromophores and for further exploration of the role of aggregation of these dyes.

Fewer [2.2]paracyclophane-scaffolded phenylene ethynylene (PE) analogs have been prepared and characterized (Table 4). The pseudopara cyclophane ***pp*-CP[PE$_2$]$_2$** displays a broad red-shifted emission peak (λ_{em} = 405 nm) relative to the maximum of the unstacked dixylyl acetylene analog, **PE$_2$** (λ_{em} = 340 nm), consistent once again with internal conversion to a lower-lying phane state [97]. As in the case of the

Table 3 Amino-substituted and water-soluble "cross" [2.2]paracyclophane-scaffolded phenylene vinylenes

Tetraamino ***X*-CP[PV$_n$]$_2$** 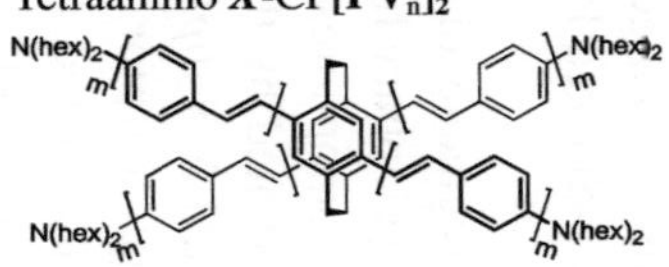	$m=1$ PV$_3$	$\lambda_{ab}=439$ nm $\lambda_{em}=490$ nm^{vib}	[107][a]
	$m=2$ PV$_5$	$\lambda_{ab}=457$ nm $\lambda_{em}=510$ nm^{vib}	
	$m=3$ PV$_7$	$\lambda_{ab}=454$ nm $\lambda_{em}=506$ nm^{vib}	
Water-soluble ***X*-CP[PV$_n$]$_2$** analogs 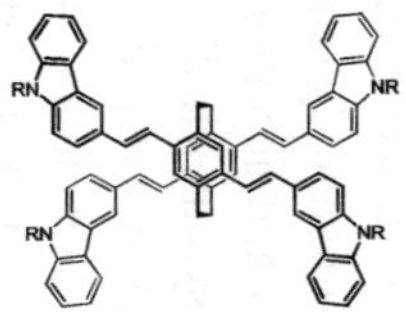 X = $-(CH_2)_6\overset{\oplus}{N}(CH_3)_2\overset{\ominus}{Br}$ [112]	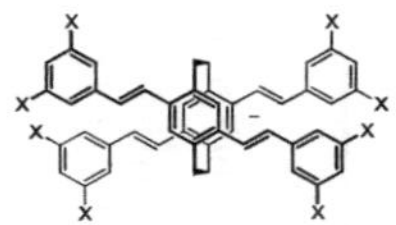 X = $-O(CH_2)_4SO_3^{\ominus}(n-Bu)_4N^{\oplus}$ [110, 111] X = $-(CH_2)_6\overset{\oplus}{N}(CH_3)_3\overset{\ominus}{Br}$ [111]	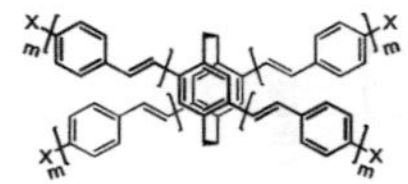 $-N[(CH_2)_6\overset{\oplus}{N}(CH_3)_3\overset{\ominus}{I}]_2$ [112] $\overset{\oplus}{}\ \overset{\ominus}{}$ $-N[(4\text{-}C_4H_6\text{-}(CH_2)_6N(CH_3)_3I]_2$ [111, 112]	
Donor–acceptor 4-amino/4-nitro analogs of ***X*-CP[PV$_2$]$_2$**	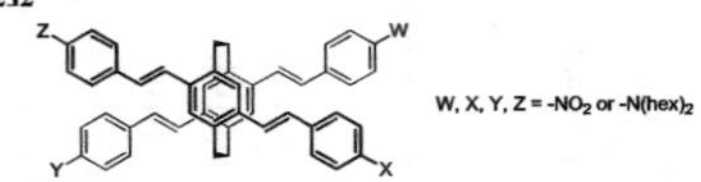		[107]

[a]Spectra recorded in toluene
[vib]0–0 peak of vibronic progression

Table 4 [2.2]Paracyclophane phenylene ethynylenes and unstacked models

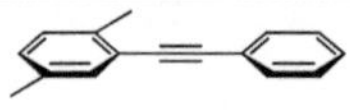	**PE$_2$**	λ_{ab} = ~280, 300 nm λ_{em} = ~340 nm	[97]
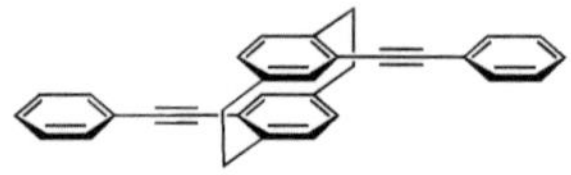	***pp*-CP[PE$_2$]$_2$**	λ_{ab} = ~295 nm λ_{em} = ~405 nm^{br}	
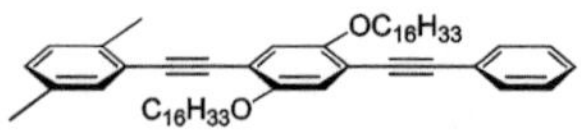	**PE$_3$**	λ_{ab} = ~305 nm λ_{em} = ~400, ~420 nm^{vib}	
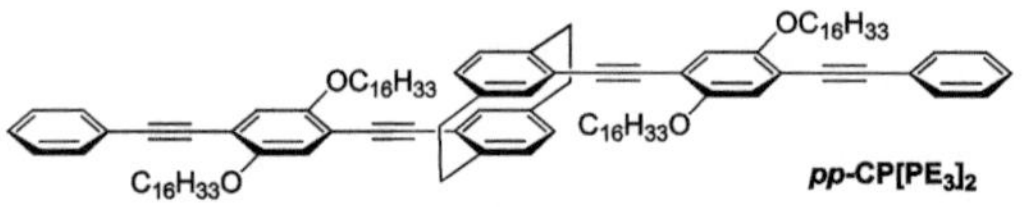	***pp*-CP[PE$_3$]$_2$**	λ_{ab} = ~380 nm λ_{em} = ~405, 420 nm^{vib}	
"Cross" analog			
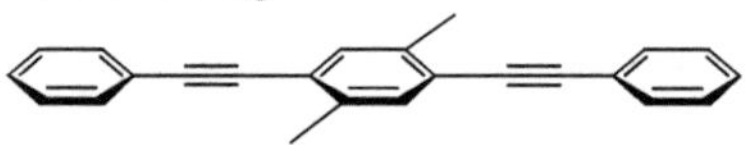		λ_{ab} = 328 nm λ_{em} = 353 nm	[79]
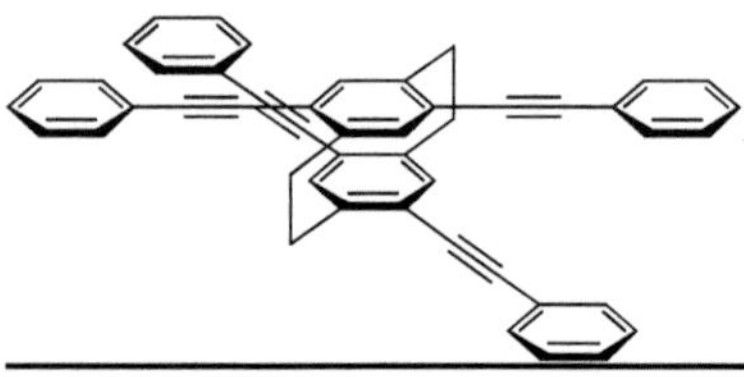		λ_{ab} = 346 nm λ_{em} = 415 nm^{br}	

phenylene vinylene oligomers, the absorption and emission spectra of the longer phenylene ethynylene analog, ***pp*-CP[PE$_3$]$_2$**, are similar to those of the unstacked analog, **PE$_3$**. Thus, lowering of the localized TB state relative to the TS state leads to emission from the segment-localized excited state. Also in common with the phenylene vinylene series, moving the point of contact from the terminal ring of ***pp*-CP[PE$_3$]$_2$** to the central ring of corresponding "cross" stacked compound ***X*-CP[PE$_3$]$_2$** leads to excimer-like emission. The peak maximum of the broad emission from the "cross" PE$_3$ compound (λ_{em} = 415 nm) is strongly red shifted from its absorption maximum (λ_{abs} = 346 nm) and relative to the unstacked analog (λ_{em} = 353 nm) [79]. Chujo and Morisaki have reported additional pseudopara PE$_3$ analogs as models for analogous stacked conjugated polymer (e.g., general structures shown in Fig. 1b as described in "Cyclophane-based π-Stacked Polymers" in this volume) [114, 115].

In addition to its use as a scaffold for linear conjugated segments that may serve as models for semiconducting organic materials, [2.2]paracyclophane has also been used to explore the electronic structure and dynamics of chromophores that consist of combinations of electron donors and acceptors, as illustrated by the examples shown in Fig. 22. These include symmetrically substituted *N,N-diphenylamino-substituted* ***pp*-CP[PE$_2$]$_2$** analogs [116–118], donor–bridge–acceptor triads [119, 120], and ***pp*-CP** analogs that are substituted with metal-centered redox active groups [121, 122]. Finally, "wiring" of the two benzene tiers of a [2.2]paracyclophane core with conjugated bridges provides macrocycles in which the stacked π–π

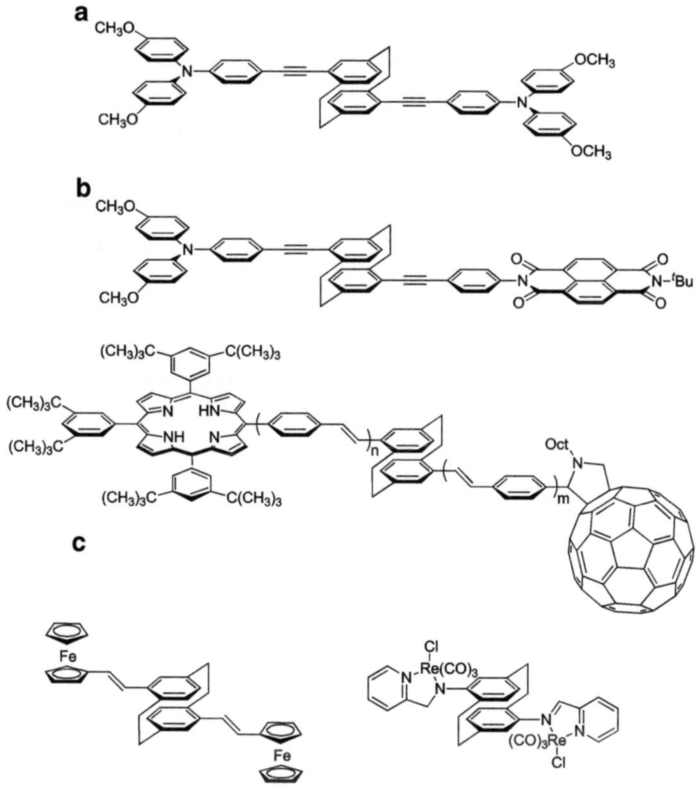

Fig. 22 Examples of [2.2]paracyclophane-bridged donor-acceptor compounds: (**a**) symmetrical *N*,*N*-di(4-methoxyphenyl)amino-substituted ***pp*-[PE$_2$]$_2$**; (**b**) donor–bridge–acceptor ***pp*-[PE$_2$]$_2$** and ***pp*-[PV$_2$]$_2$** triads; (**c**) metal-centered redox active substituted [2.2]paracyclophane

contact brings together the middle or two ends of a conjugated chain in a series of intriguing molecules (Fig. 23) [123, 124].

Increasing the Extent of π-Stacking: Pseudogeminal [2.2] Paracyclophane-Scaffolded Phenylene Vinylenes

While the pseudopara and pseudoortho disubstituted and "cross" tetrasubstituted [2.2]paracyclophanes hold two conjugated segments in close proximity, only the benzene rings within the paracyclophane core of these compounds are arranged in a stacked arrangement. This results in minimal interaction between the two conjugated segments. This is in contrast to the situation in thin films of semiconducting

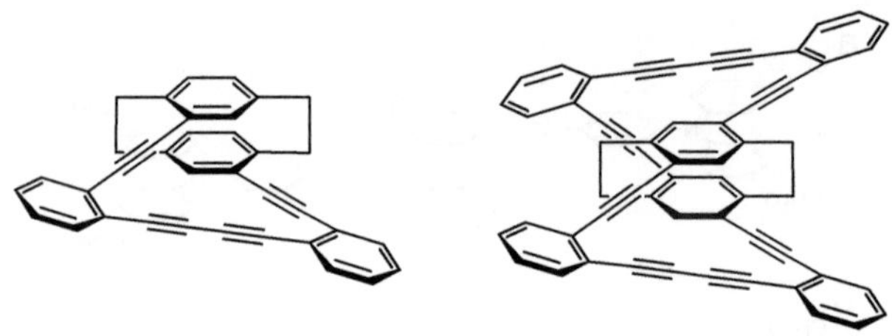

Fig. 23 "Wiring" of the two benzene tiers of a [2.2]paracyclophane core with conjugated bridges

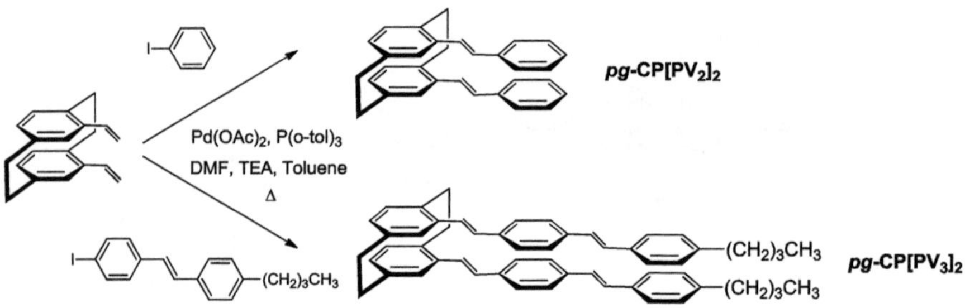

Fig. 24 Heck coupling of PV arms onto pseudogeminal divinyl[2.2]paracyclophane [125]

conjugated materials in which there is stacking of the π systems over extended lengths. To increase the length over which pairs of conjugated segments are held in a stacked arrangement, Subodh Jagtap at the Georgia Institute of Technology prepared new stacked phenylene vinylene analogs that are based on a pseudogeminal (*pg*) disubstituted [2.2]paracyclophane scaffold. The U-turn provided by the pseudogeminal substitution pattern of the cyclophane core holds conjugated segments in a stacked arrangement over an extended length (although crystal structures are not yet available to determine the precise arrangement of the segments or to discern any distortion of the [2.2]paracyclophane core arising from steric interactions between the pseudogeminal substituents).

A Heck reaction between pseudogeminal divinyl[2,2]paracyclophane and 1-iodobenzene in the presence of $Pd(OAc)_2$ and $P(o\text{-}tol)_3$ affords ***pg*-CP[PV$_2$]$_2$** (Fig. 24). The longer stacked analog, ***pg*-CP[PV$_3$]$_2$**, was similarly prepared by Heck coupling with 1-iodo-4-styrylbenzene [125]. The longer PV_3 analog was prepared with *n*-butyl groups at the para position of the terminal phenyl rings to impart solubility without introducing sterically bulky substituents that might interfere with stacking. In contrast to the pseudopara and pseudoortho analogs, the pseudogeminal compounds are sensitive to light. This observation is in accord with the finding of Hopf that similar cyclophanes in which conjugated units are held in pseudogeminal positions undergo intramolecular photochemical cyclizations, as illustrated by the example shown in Fig. 25 [126].

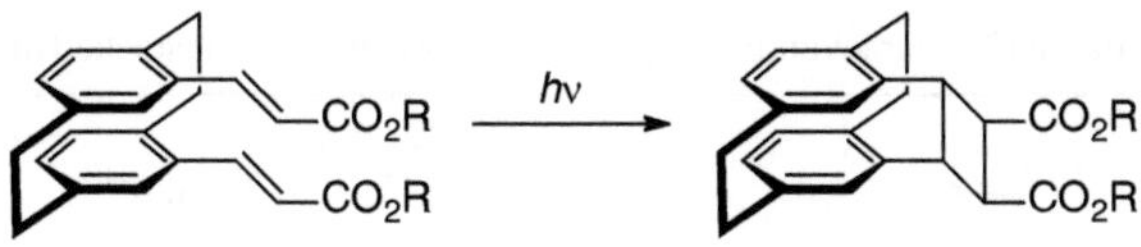

Fig. 25 Photochemical intramolecular [2+2] cycloaddition of cinnamate units of a pseudogeminal substituted [2.2]paracyclophane [126]

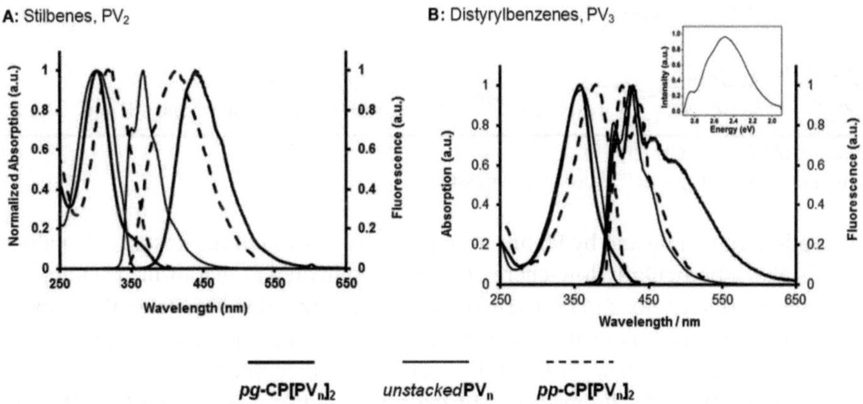

Fig. 26 UV–vis and fluorescence spectra of pseudogeminal and pseudopara phenylene vinylenes and model unstacked oligomer: (**a**) stilbene (PV_2) analogs; (**b**) distyrylbenzene (PV_3) analogs (*inset*: long wavelength peak from deconvolution of the emission spectrum of ***pg*-CP[PV$_3$]$_2$**) [125]

The stacked pseudogeminal phenylene vinylene compounds and their unstacked analogs were characterized by UV-vis and fluorescence spectroscopy to explore the effect of interaction between the stacked segments over extended lengths (Fig. 26 and Table 5). As with the pseudopara and pseudoortho series, the pseudogeminal stacked compounds display absorption spectra that are similar to those of the corresponding unstacked linear analogs. The absorption maximum of the pseudogeminal stacked analog ***pg*-CP[PV$_2$]$_2$** and the unstacked chromophore **MePV$_2$** are similar (297 and 296 nm, respectively), as are the maxima of the stacked and unstacked PV_3 analogs (352 and 350 nm, respectively) [125]. However, in both cases there are significant low-energy tails to the absorption peaks of the stacked compounds that are not present in the spectra of the corresponding unstacked oligomers. These absorption tails are also much more prominent than in the spectra of the corresponding pseudopara and pseudoortho analogs. Deconvolution of the absorption peaks reveals contributions for the pseudogeminal stacked PV_2 and PV_3 analogs at 358 and 404 nm, respectively (Fig. 26).

The fluorescence spectrum of the unstacked model **PV$_2$** consists of a series of vibronic peaks at 345, 359, and 381 nm, with the 0–0 peak shifted by 48 nm from the absorption maximum. In contrast, the spectrum of ***pg*-CP[PV$_2$]$_2$** consists of a broad peak with a maximum at 435 nm. This is red shifted by 139 nm from the

Table 5 Pseudogeminal [2.2]paracyclophane phenylene vinylenes and unstacked model compounds[a]

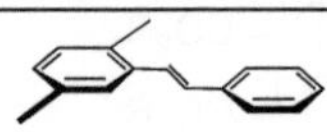	**PV$_2$**	λ_{ab} = 297 nm λ_{em} = 345, 359, 381 nm[vib]	[125]
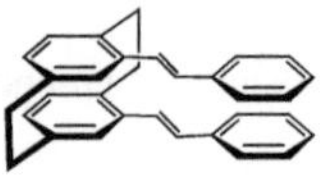	***pg*-CP[PV$_2$]$_2$**	λ_{ab} = 296, 358[tl] nm λ_{em} = 435 nm[br]	
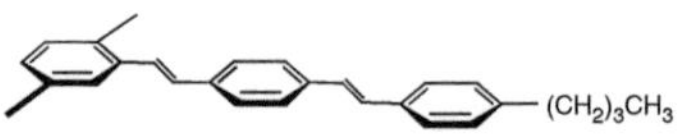	**PV$_3$"**	λ_{ab} = 352 nm λ_{em} = 396, 416, 440 nm[vib]	
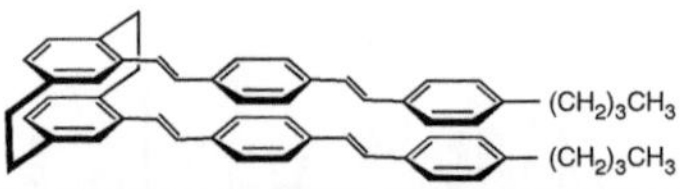	***pg*-CP[PV$_3$]$_2$**	λ_{ab} = 350, 404[sh] nm[tl] λ_{em} = 404, 427, 446 nm[vib] and 481 nm[br]	

[a]Spectra obtained in CH_2Cl_2
[br]Broad, [tl]Low-energy tail, [vib]Vibronic progression

absorption maximum and by 90 nm from the 0–0 emission of the unstacked analog. This is also a particularly low-energy emission compared to that of the pseudopara and pseudoortho analogs (412 and 401 nm), which are red shifted from the respective absorption maxima of these compounds by a smaller amount (74 and 63 nm, respectively, Table 1) than for the fully stacked pseudogeminal analog.

The intensity of the fluorescence of the longer stacked analog ***pg*-CP[PV$_3$]$_2$** is greatly reduced compared to that of the unstacked model compound **Me$_2$PV$_3$**. However, the peak maxima of a series of vibronic peaks (404, 427, 446 nm) and the Stokes shift (54 nm) are all similar for the stacked and unstacked analogs, as well as for the pseudoortho and pseudopara compounds. However, in addition to these peaks, there is a significant broad contribution in the spectrum of the pseudogeminal compound with a maximum at 481 nm that is absent from the spectrum of the other analogs. This longer wavelength peak resembles the emission peak for the "cross" isomer ***X*-CP[PV$_3$]$_2$** (λ_{em} ~ 460 nm) and can be ascribed to emission from a low-energy excimer-like state. This is in contrast to the segment-localized emission of the pseudopara and pseudoortho stacked PV_3 compounds.

Time-dependent DFT calculations conducted by Sukrit Mukhopadhyay and Veaceslav Coropceanu in the group of Jean-Luc Brédas support the notion that for ***pp*-CP[PV$_2$]$_2$** the absorption populates a segment-localized S_2 state. Rapid internal conversion to the excimer-like S_1 state takes place, and the fluorescence spectrum is dominated by the low-energy emission from this state [125]. These calculations indicate that the fully stacked pseudogeminal isomer undergoes a significant change in geometry upon excitation, with a decrease in the distance between the conjugated segments. Natural transition orbitals corresponding to the $S_0 \rightarrow S_1$ (i.e., absorption) and relaxed $S_1 \rightarrow S_0$ transitions (i.e., fluorescence) of the PV_2 and PV_3 analogs of the pseudopara and pseudogeminal isomers are shown in Fig. 27. This demonstrates that the relaxed S_1 state of ***pg*-CP[PV$_2$]$_2$**, which corresponds to an excimer-like state, is delocalized over the entire molecule (Fig. 27b). This is in contrast to the phane-like S_1 state of ***pp*-CP[PV$_2$]$_2$** where the delocalization is primarily limited to the cyclophane core (Fig. 27b). For the longer stacked PV_3 analog, a single distyrylbenzene segment-localized excited state dominates both the absorption and emission for the

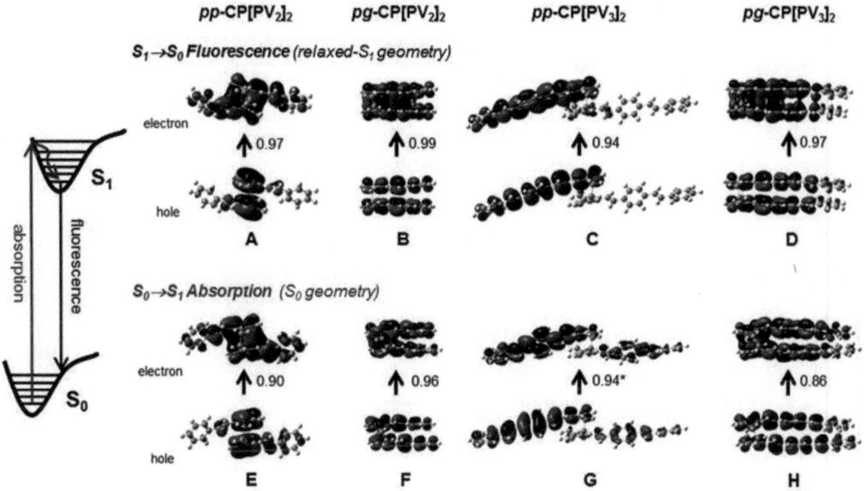

Fig. 27 Natural transition orbitals (NTOs) of the relaxed $S_1 \rightarrow S_0$ transitions (**a–d**) and $S_0 \rightarrow S_1$ transitions (**e–h**) for *pp* and *po* arrangements of PV_2 and PV_3 segments (from [125])

pseudopara analog (Fig. 27c, g). For the pseudogeminal PV_3 isomer, the new broad low-energy emission arises from the relaxed S_1 excited state that is once again extensively delocalized over the entire length of both tiers of the molecule (Fig. 27d, h).

Further extension of the concept of stacking conjugated oligomers over their extended lengths led Jagtap to prepare polymeric structures based on the pseudogeminal disubstituted [2.2]paracyclophanes [87]. Most other [2.2]paracyclophane-containing π-stacked polymers are based on a pseudopara disubstituted core (along with a few pseudoortho analogs) that provides only a limited amount of overlap between each pair of conjugated segments. The pseudopara and pseudoortho polymers (**poly-*pp*-CP[PE$_3$]** and **poly-*po*-CP[PE$_3$]**) exhibit vibronic emission, with the 0–0 peak shifted by ~30 nm from the absorption maximum [127] (Table 6). In contrast, the fully stacked and multilayered pseudogeminal polymer (**poly-*pg*-CP[PV$_3$]**) displays a broad emission peak that appears at a particularly low energy (λ_{em} = 530 nm), which is shifted by 171 nm from the absorption maximum. The PE_3 segment has a sufficiently large conjugation length that there is no driving force for internal conversion from the segment-localized S_2 state to the excimer-like S_1 state in the unstacked and terminally stacked arrangements (i.e., pseudopara and pseudoortho), thereby resulting in vibronic emission.

Table 6 Comparison of spectroscopic behavior of PE_3 oligomers in various arrangements

	λ_{abs}	λ_{em}	Δ_{peak}	References
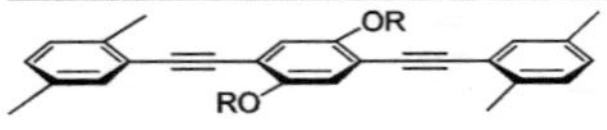 **Me_4PE_3**	321, 374 nm	402, 413 nm^{vib}	28 nm	[87]
pp-**CP[PE_3]$_2$** (Table 1)	~380 nm	~400, ~420 nm^{vib}	~40 nm	[96]
X-**CP[PE_3]$_2$** (Table 3)	346 nm	415 nm^{br}	75 nm	[79]
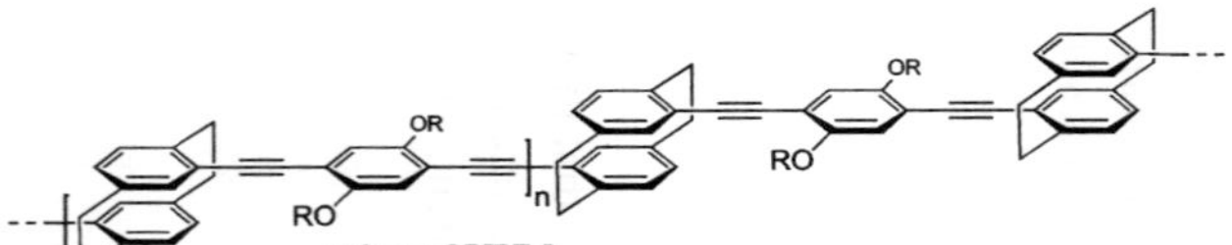 **poly-*pp*-CP[PE_3]**	319, 386 nm	414, 438 nm^{vib}	28 nm	[127]
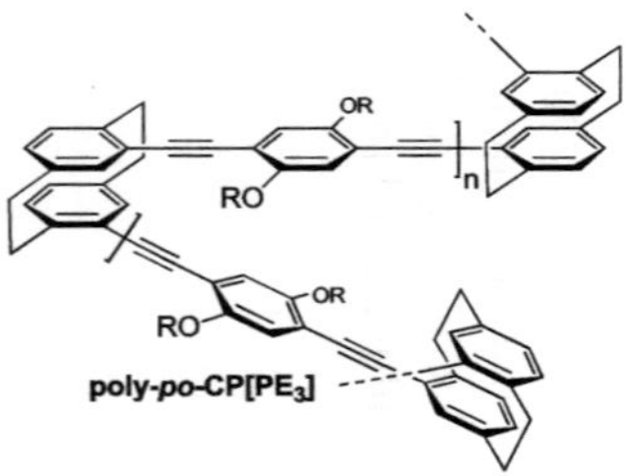 **poly-*po*-CP[PE_3]**	319, 377 nm	411, 434 nm^{vib}	34 nm	[127]
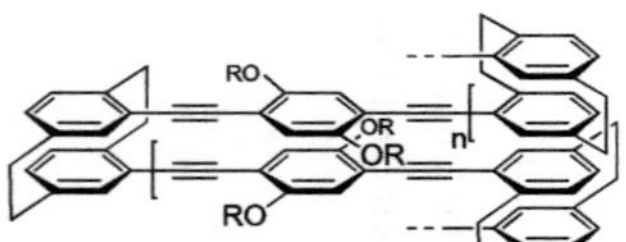 **poly-*pg*-CP[PE_3]**	320, 359 nm^{sh}	530 nm^{br}	171 nm	[87]

^{br}Broad, ^{vib}Vibronic, ^{sh}Strong shoulder to 450 nm

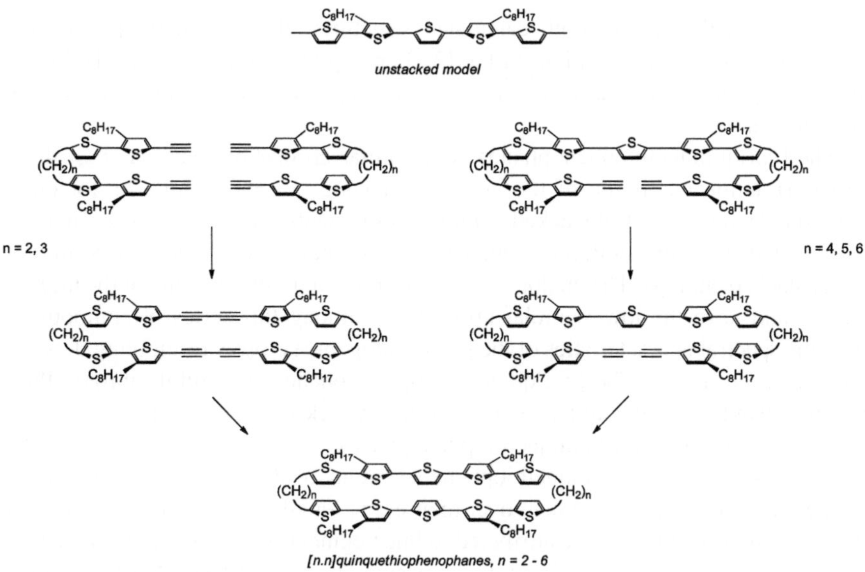

Fig. 28 Otsubo's synthesis of [*n.n*]quinquethiophenophanes [129]

Extended Heterocyclophanes as Models for Spinless Charge Carriers

[n.n]Quinquethiophenophanes

Tetsuo Otsubo et al. of Hiroshima University have explored the electronic structures of intramolecular radical cation π dimers in molecules consisting of stacked quinquethiophene segments that are incorporated into a macrocyclic cyclophanes (Fig. 28). In these structures the conjugated segments are held in a stacked arrangement along their entire length, as might be imagined for the pseudogeminal [2.2] paracyclophane-scaffolded oligomer (see previous section).

Otsubo's synthetic approach to the [*n.n*]quinquethiophenophanes makes use of the Ellington oxidative homocoupling of ethynylbithiophenes to form diacetylenes that are subsequently subjected to cyclization upon treatment with sodium sulfide to form the central thiophene ring of each quinquethiophene unit (Fig. 28) [128, 129].

While single crystal structures of the [*n.n*]quinquethiophenophanes are not available, small upfield shifts in the ^{1}H NMR spectra for the β-thienyl hydrogen atoms of the cyclic compounds relative to the uncyclized model compound are consistent with the expected effects of ring currents in the stacked structure. The stacked arrangement of the oligothiophenes in the [2.2] and [3.3]quinquethiophenophanes is also apparent in the energy-minimized structures obtained from quantum chemical calculations. In both calculated structures, the quinquethiophene tiers adopt an all-trans conformation. In the [2.2] analog the two tiers are stacked in an anti arrangement, whereas in the [3.3] analog they are stacked in a syn arrangement.

The [*n.n*]quinquethiophenophanes have similar absorption maxima ($\lambda_{abs}=401$–413 nm) as the maximum of the unstacked analog ($\lambda_{abs}=416$ nm). As for other compounds that consist of stacked conjugated segments, the extinction coefficients for the stacked quinquethiophenophanes are consistently lower than twice that of the unstacked analogs. The analogs with shorter alkylene spacers emit at the highest energy and have the lowest molar absorptivity among this series of compounds, with the spectral characteristics of the [6.6] analog most closely resembling those of the unstacked model. The [*n.n*]quinquethiophenophanes all exhibit vibronic fluorescence bands that are similar to those of the unstacked model compound (model, $\lambda_{em}=503$ and 533; [*n.n*]quinquethiophenophanes, $\lambda_{em}=508$–513 nm and 536–540 nm). Although the quinquethiophene segments of the [2.2] and [3.3] analogs are held in a stacked arrangement over their entire length, only small differences in their emission spectra may be attributed to intersegment interactions. This behavior is in contrast to the broad excimer-like emission observed for ***pg*-CP[PE$_3$]$_2$**, which has a similar geometry. This may arise from differences in the electronic structure of the π systems of the thiophene and phenylene ethynylene segments and suggests the need for further studies to gain a more complete understanding of the effect of the structure and packing of conjugated segments on the properties of stacked π systems. While there are only small differences in the absorption and emission spectra of the neutral [*n.n*]quinquethiophenophanes and model quinquethiophene, the effect of interchain interactions is clearly observed in the spectra of the corresponding radical cations.

The unstacked oligomer exhibits two one-electron oxidation waves, corresponding to sequential oxidation from the neutral molecule to the radical cation ($E_{1/2}=+0.31$ V versus Ag/AgCl) and to the dicationic state ($E_{1/2}=+0.53$ V). For each of the [*n.n*]quinquethiophenophanes there is a two-electron oxidation wave that appears at a lower potential ($E_{1/2}=+0.22$–0.28 V) than for the first oxidation of the unstacked analog (Fig. 29). The lower potential required to oxidize the stacked compound to the dicationic state suggests that this state is stabilized relative to the mono(radical cation) of the unstacked model oligomer. Stabilization of the dicationic state of the stacked compounds must overcome the Coulombic destabilization that might be expected in the more highly charged state.

In all cases, the stacked compounds undergo further oxidation to the tetracationic state at higher potentials than for that of the second oxidation of the unstacked model. These higher potentials are consistent with Coulombic repulsion between highly charged segments and destabilization of the more highly charges species, even in the presence of long alkylene spacers. The [3.3] analog of this series displays the lowest first oxidation potential, the highest second oxidation potential, and a distinct splitting of the second redox wave into two one-electron peaks.

The UV–visible-near-IR spectra of the [*n,n*]quinquethiophenophane dication for the compounds with $n=3$–6 (Fig. 30a) are similar to those of the π dimer of the

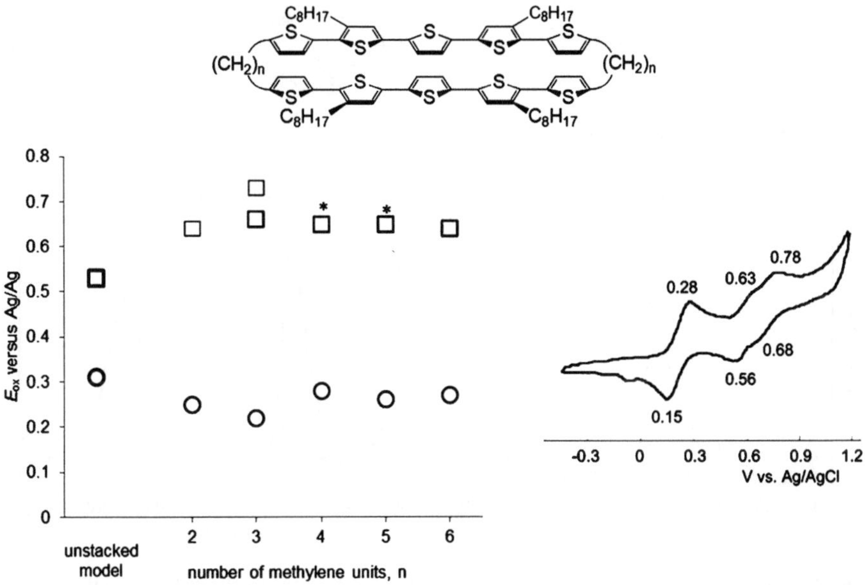

Fig. 29 Electrochemistry of [*n.n*]quinquethiophenophanes. *Left*, plot of $E_{1/2}$ for reversible first (*open circles*) and second oxidations (*open squares*) compared to unstacked model compound. *Right*, Cyclic voltammogram of [3.3]quinquethiophenophane. *Asterisk* denotes some splitting of the peak. Used with permission from Sakai et al. [129]. Copyright 2005 American Chemical Society

radical cations of the model linear quinquethiophene in solution (Fig. 30b). Removal of one electron per molecule from the unstacked quinquethiophene affords a spectrum at room temperature that has peaks at 752 and 1,408 nm, corresponding to an isolated radical cation. Upon cooling to 248 K these peaks diminish but do not entirely disappear, and new peaks at ca. 650 and 1,084 nm grow in size. These latter peaks can be ascribed to the formation of the radical cation π dimer (Fig. 31a). The equilibrium between the radical cation and the π dimer is supported by the presence of peaks corresponding to both of these species in the UV-vis-near-IR spectrum and the reversibility of these changes.

Two-electron oxidation of the [3.3]quinquethiophenophane provides a spectrum at room temperature that consists of two strong peaks (at ca. 670 and 1,230 nm, Fig. 30b) that resemble those of the radical cation π dimer of the linear oligomer (650 and 1,084 nm, Fig. 30a). Differences in the values of the absorption maximum of the radical cation π dimer of the linear oligomer and of the stacked dication might be attributed to differences in the arrangement of the conjugated segments in the non-covalently linked dimer and in the cyclophane structure (Fig. 31b). The presence of a radical cation π dimer electronic state is further supported by the lack of an ESR signal for the dicationic state of the stacked compound. When [3.3]quinquethiophenophane is treated with one equivalent of oxidant, the resulting spectrum indicates the presence of the neutral, monocationic, and dicationic states, as shown in Fig. 30b

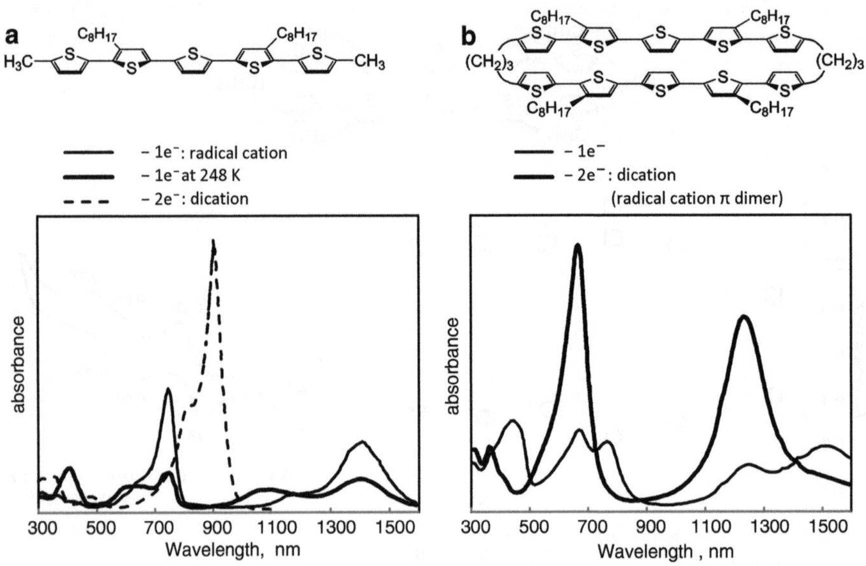

Fig. 30 Ultraviolet–visible-near-IR spectra (CH_2Cl_2 solutions). (**a**) Unstacked quinquethiophene; (**b**) [3.3]quinquethiophenophane (*arrows* indicate changes in peak intensity upon lowering the temperature. Modified with permission from Sakai et al. [129]. Copyright 2005 American Chemical Society

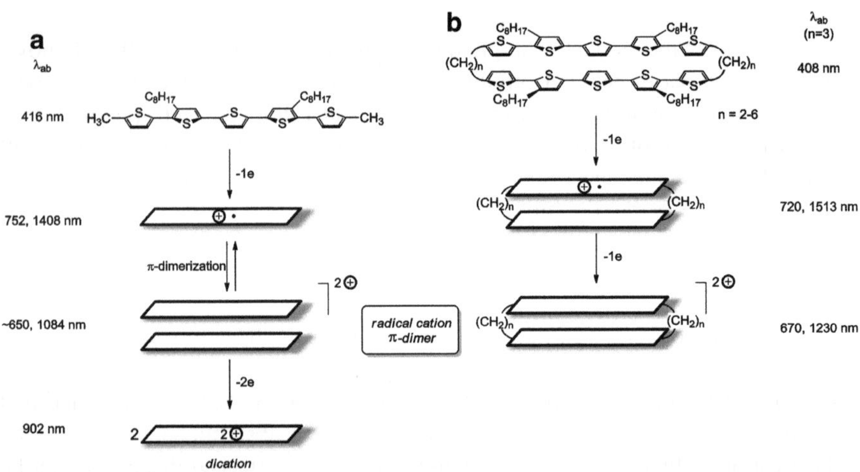

Fig. 31 Sequential oxidation of quinquethiophenes. (**a**) Unstacked linear quinquethiophene; (**b**) [*n.n*]quinquethiophenophane, with absorption maxima for [3.3]quinquethiophenophane

In a second approach to explore the interactions between oxidized conjugated segments, Otsubo explored the spectra of various oxidation states of α,ϖ-di(quinquethiophenyl)alkanes (Fig. 32) [130]. The absorption and emission spectra of the neutral compounds (n=2–6) are similar to those of the unstacked quinque-

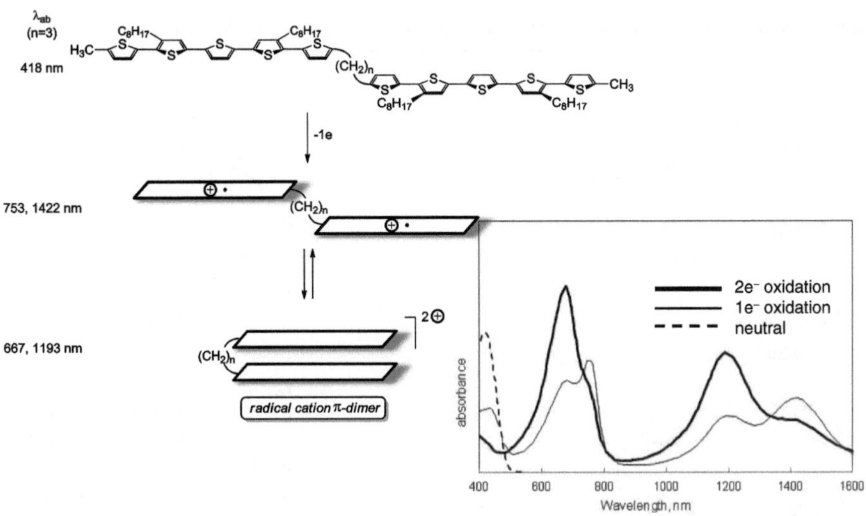

Fig. 32 Oxidation of 1,3-di(quinquethiophenyl)propane. Modified with permission from Satou et al. [130]. Copyright 2004 American Chemical Society

thiophene model. As expected, one-electron oxidation of the 1,2-di(quinquethiophenyl)ethane (i.e., $n=2$) with $FeCl_3$ provides a UV-vis-near-IR spectrum at room temperature that has peaks which are consistent with equal amounts of neutral quinquethiophenyl unit (418 nm; compared to 417 nm for the neutral linear oligomer) and radical cation (peaks at 752 and 1,419 nm; compared to 751 and 1,412 nm for the radical cation of the linear oligomer). Two-electron oxidation provides a spectrum in which the peak assigned to the neutral quinquethiophenyl segment has disappeared, and those of the quinquethiophenyl radical cation have doubled in intensity.

The spectrum obtained after two-electron oxidation of the analog with a 1,3-propanediyl spacer (i.e., $n=3$, shown in the inset of Fig. 32) has a prominent pair of peaks at 676 and 1,193 nm. These correspond to the formation of a radical cation π dimer: the spectrum of the radical cation π dimer of the linear quinquethiophene oligomer obtained at low temperature has peaks at ~650 and 1,084 nm. One-electron oxidation affords a spectrum that has a peak which corresponds to the presence of the neutral quinquethiophene unit ($\lambda_{abs}=418$ nm), a pair of peaks for the radical cation of the quinquethiophene unit (753 and 1,422 nm), and the pair of peaks for the radical cation π dimer. Accordingly, it appears that all three species shown in Fig. 32 are present in equilibrium. Similar sets of peaks are observed upon oxidation of the analogs with longer spacers ($n=4$–6). Thus, linking two quinquethiophenyl units with an alkylene chain ($n=3$–6) facilitates formation of the radical cation π dimer. π dimerization is hindered in the case of the 1,2-ethanediyl spacer by torsional strain within the short alkylene bridge.

A number of pseudopara thiophene-substituted [2.2]paracyclophane-stacked dimers and polymers have also been reported [131–134].

Table 7 Spectroscopic characterization of diphenylfuran cyclophanes, cyclophenes, and linear model

Linear model

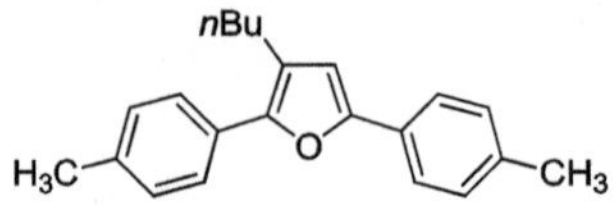

$\lambda_{ab} = 324$ nm[a]
$\lambda_{em} = 362, 379, 398$ nm[a, vib]

[2.2]cyclophane

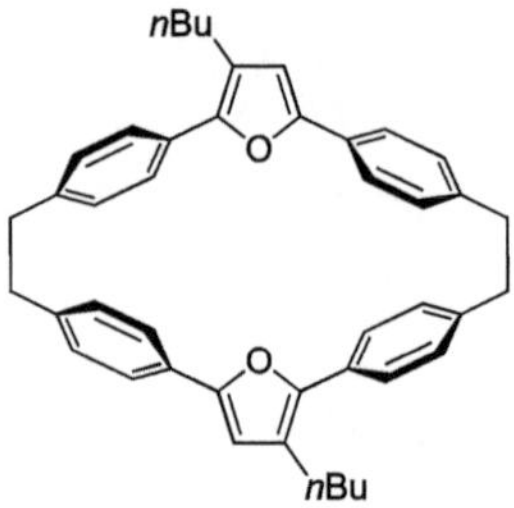

$\lambda_{ab} = 314$ nm[a]
$\lambda_{em} = 370, 389, 410$ nm [a, vib]

Cyclophenes

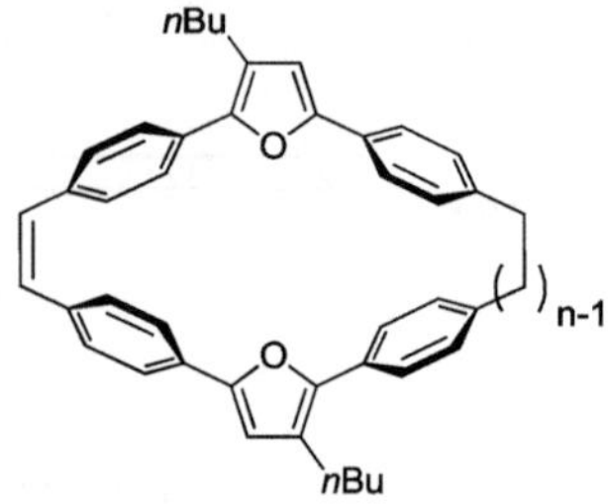

	λ_{ab}	λ_{em}
$n = 2$[a]	321 nm	499 nm[br]
	372, 390, 416 nm[vib]	
$n = 2$[b]	325 nm[c]	501 nm[br]
$n = 3$[b]	327 nm[c]	506 nm[br]
$n = 4$[b]	330 nm[c]	496 nm[br]
$n = 5$[b]	330 nm[c]	498 nm[br]
$n = 6$[b]	331 nm[c]	498 nm[br]
$n = 12$[b]	337 nm[c]	494 nm[br]

Other [2.2]cyclophanes with extended π systems

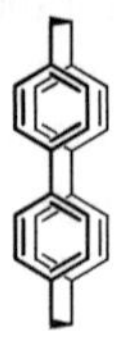
[2.2] (4,4') diphenylophane

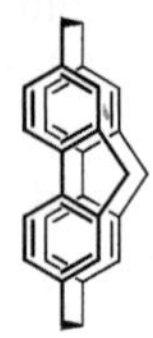
syn-[2.2] (2,7) fluorenophane

anti-[2.2] (2,7) fluorenophane

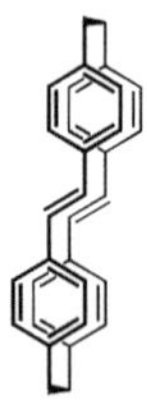
stilbenophane

[2.2] (2,7) phenanthrenophane

[2,2] (2,7) pyrenophane

[a]In EtOAc [135]
[b]In $CHCl_3$ [136]
[c]With increasing shoulder at ~380 nm as *n* increases
[br]Broad, [vib]Vibronic

[n,n]Diphenylfuranophanes

Tien-Yau Luh of National Taiwan University has reported a series of related cyclophanes that hold 2,5-diphenylfuran chromophores in close proximity [135–138] (Table 7). The absorption maximum of the [2.2]cyclophane ($\lambda_{ab} = 314$ nm) and the linear model oligomer (λ_{ab} = 321 nm) are similar, as is the vibronic nature of the fluorescence peaks at ~360–410 nm for the two compounds, as shown in Table 7 (left). The corresponding stacked compound in which one of the two-carbon bridges is unsaturated gives a similar absorption maximum (λ_{abs} = 321 nm). However, the peak exhibits a significant tail out to ca. 430 nm. The fluorescence spectrum of this compound is dominated by a strong broad red-shifted emission with a peak maximum at 499 nm and a very weak set of vibronic peaks in the region of the emission of the [2.2]cyclophane and the linear model compound (370–420 nm) [135]. For a series of analogs of the alkene-bridged cyclophanes (Table 7, right), the absorption maximum increases slightly as the length of the saturated bridge increases, suggesting that a relief of ring strain allows for greater conjugation. There is also an increase in the strength of the tail of the absorption peak. These analogs all display a broad red-shifted emission peak (494–506 nm), with the quantum yield increasing monotonically with an increase in bridge length [136]. While calculations suggest that these compounds do not adopt a stacked conformation, the spectral properties of the cyclophene analogs suggest significant contributions from two excited states. The similarity in the absorption maxima for this series of compounds may be interpreted as the formation of a single segment-localized excited state. Internal conversion to a lower-energy excited state, resembling a phane state or an extended excimer-like state, is responsible for the broad red-shifted emission [137].

Other [2.2]cyclophanes in which extended π-systems are held in a stacked arrangement by bridges at their termini include [2.2] (4,4′) diphenylophane, *syn*-[2.2] (2,7) fluorenophane, *anti*-[2.2] (2,7) fluorenophane [139], stilbenophane [140, 141], [2.2] (2,7) phenanthrenophane [142], and [2,2] (2,7) pyrenophane [143–145].

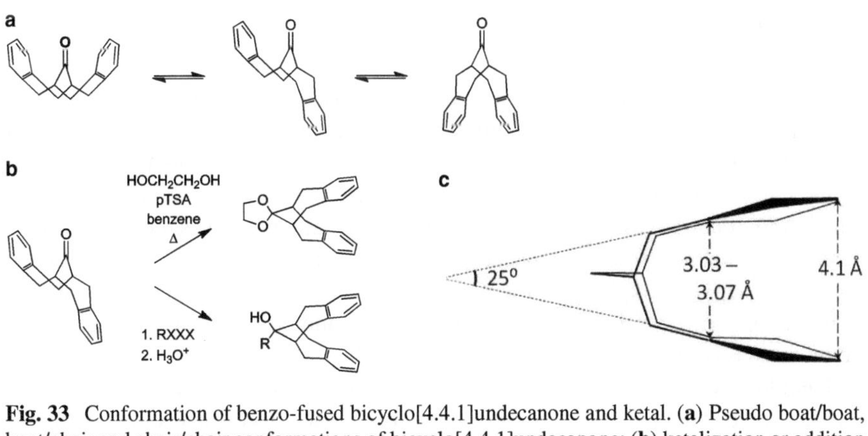

Fig. 33 Conformation of benzo-fused bicyclo[4.4.1]undecanone and ketal. (**a**) Pseudo boat/boat, boat/chair and chair/chair conformations of bicyclo[4.4.1]undecanone; (**b**) ketalization or addition of Grignard reagent locks bicyclic framework into pseudochair/chair conformation in which the fused arenes are stacked atop one another; (**c**) structural parameters for pseudo chair–chair conformation of benzo-fused bicyclo[4.4.1]undecanone core

π-Stacked Bicyclo[4.4.1]Undecanone-Fused Conjugated Oligomers

Bisarene-Fused Bicyclo[4.4.1]Undecanes: Structure

In another approach to prepare well-defined conjugated oligomers in which planar conjugated segments are stacked over their entire length, Kurt Knoblock [48] at Georgia Tech was drawn to the idea of investigating π–π interactions in ketals of arene-fused bicyclo[4.4.1]undecanes (Fig. 33). Orthocyclophanes such as these are often conformationally flexible and the arene units typically do not interact (as illustrated for [2.2]orthocyclophane shown in Fig. 9b). Shuntaro Mataka of Kyushu University showed that dibenzo[*c*,*h*]bicyclo[4.4.1]undedec-3,8-dien-11-one exists as a mixture of conformations in which the two seven-member rings of the bicycloundecane core exist in any of the three possible combinations of pseudochair and pseudoboat conformations (Fig. 33a). However, conversion of the ketone into a ketal with ethylene glycol introduces sufficient steric bulk to the one-carbon bridge so as to confine the bicyclic structure in a pseudochair/pseudochair conformation in which the arene rings are held in a stacked arrangement (Fig. 33b) [146]. The ^{1}H NMR spectrum of the ketone consists of a series of very broad peaks that are consistent with the conformational flexibility of the bicycloundecanone framework. In contrast, the spectrum of the ketal exhibits a series of sharp peaks, including separate signals for the two sets of diastereotopic benzylic hydrogen atoms in non-exchanging pseudoequatorial and pseudoaxial positions. X-ray crystallography of the benzo-fused ethylene ketal reveals that the individual benzene rings are held in a stacked arrangement at an angle of ~25° and with an intercentoid distance of ~3.4 Å (Fig. 33c) [147].

Synthesis of Bicyclo[4.4.1]Undecane-Scaffolded Conjugated Oligomers

Mataka's synthetic approach to stacked arene-fused bicyclo[4.4.1]undecanes is illustrated for the benzo-fused compound in Fig. 34a [148]. Briefly, dimethyl 1,3-acetonedicarboxylate is alkylated with two equivalents of a 1,2-bis(bromomethyl) arene (1,2-bis(bromomethyl)benzene in this case) to form the bicycloundecane core. Saponification followed by thermal decarboxylation removes the ester groups. Finally, ketalization affords the stacked compound. Variation of the bis(bromomethyl) arene, and stepwise alkylation with two different bis(bromomethyl)arenes, as shown in Fig. 34b, allows for the synthesis of a series of symmetric and unsymmetric compounds, as illustrated by the examples shown in Fig. 34c.

Knoblock's study began with the preparation of the bisthieno-fused bicyclo[4.4.1]undecanone by alkylation of dimethyl 1,3-acetonedicarboxylate with 3,4-bis(bromomethyl)thiophene followed by saponification, decarboxylation, and

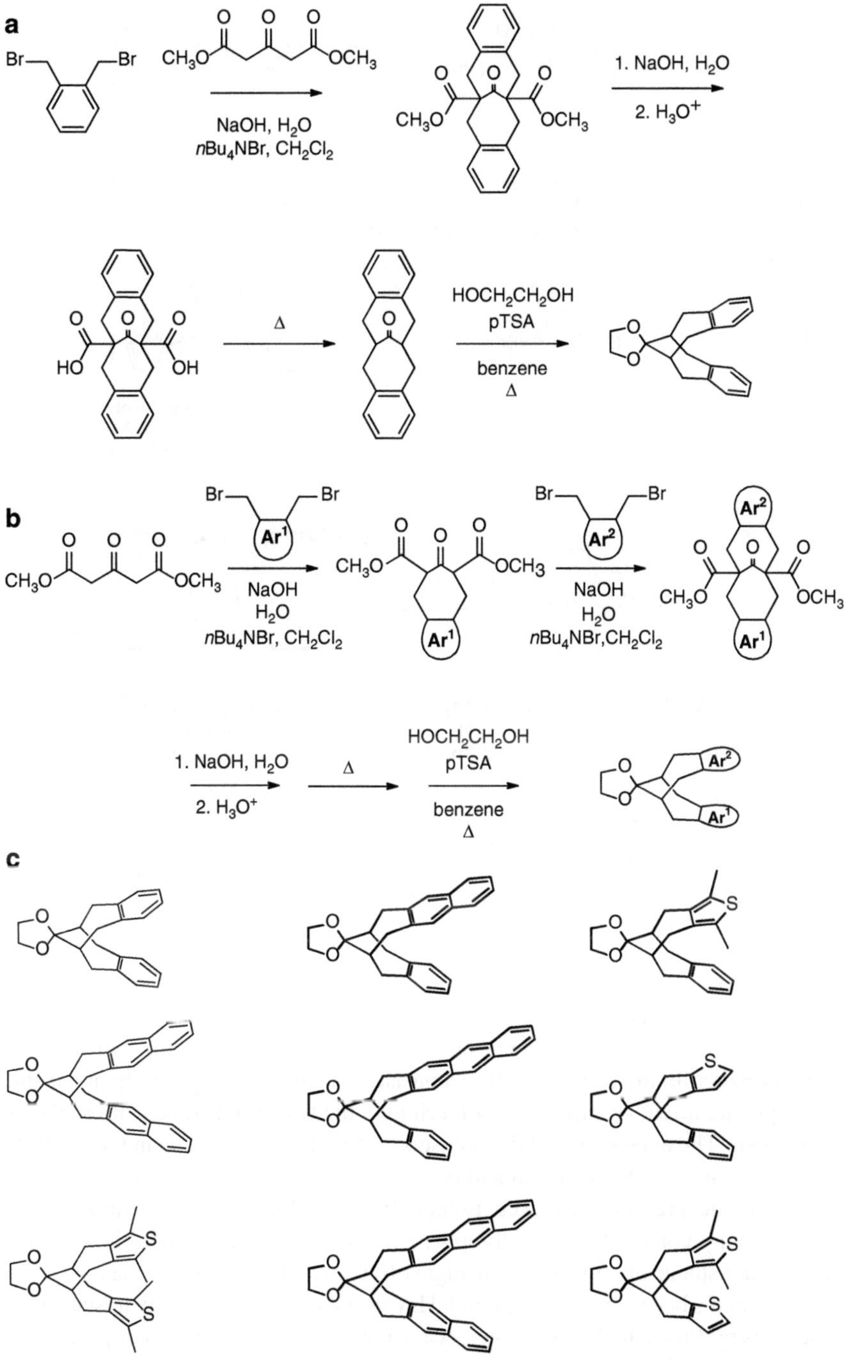

Fig. 34 Mataka's stacked arene-fused [4.4.1]bicycloundecanes: (**a**) synthesis of benzo-fused bicycloundecane; (**b**) synthesis of unsymmetrical stacked arene-fused bicycloundecanes; (**c**) representative symmetrical (*left*) and unsymmetrical (*right*) bicycloundecane-scaffolded stacked arenes

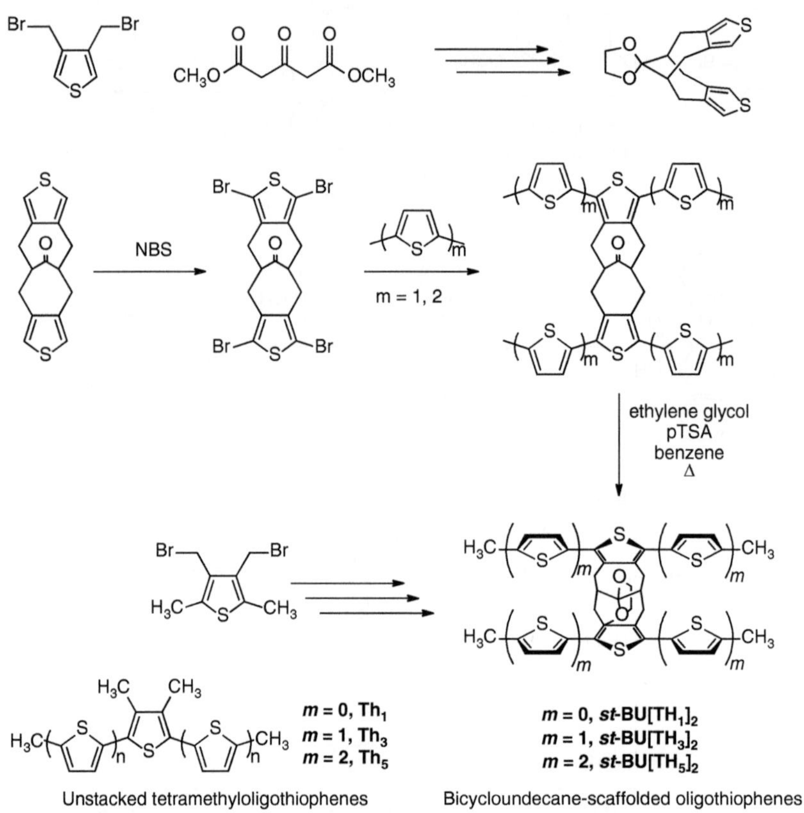

Fig. 35 Thieno-fused bicyclo[4.4.1]undecanes. (**a**) Preparation of bisthieno-fused bicyclo[4.4.1]undecan-11-one and stacked ketal; (**b**) installation of thienyl arms to prepare ***st*-BU[Th$_n$]$_2$**, and model unstacked tetramethyloligothiophenes

ketalization (Fig. 35a) [149]. Ketalization with ethylene glycol locks the bicycloalkane into a pseudochair/pseudochair conformation.

An upfield shift for the α-thienyl protons in the ^{1}H NMR spectrum of the ketal (δ6.5) compared to the ketone (δ7.0) is consistent with the change in chemical shift anisotropy that might be expected upon bringing the two fused arenes into a stacked arrangement. The appearance of distinct sets of doublets of doublets in the ^{1}H NMR spectrum for the pseudoequatorial and pseudoaxial hydrogen atoms provides further support for the presence a locked pseudochair/pseudochair conformation. X-ray crystallography confirms that the thiophene rings of the ketal are held in a stacked fashion at a distance of ~3.6 Å and an angle of 28°, which is similar to the conformation of Mataka's benzo-fused compound. Having established the ability to stack thiophene units by fusion to the bicycloundecane core, we set about the incorporation of conjugated arms. The thieno-fused ketone undergoes bromination to provide the tetrabromide (Fig. 35b), which was subjected to Stille coupling with 2-trimethylstannyl-5-methylthiophene followed by ketalization to give the stacked terthiophene

Table 8 Bicyclo[4.4.1]undecane-scaffolded stacked oligothiophenes

	F First and second oxidation potentials E_1/V (no. e^-)	 E_2/V (no. e^-)	
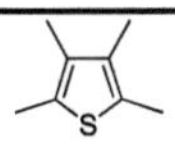 **Th₁**	+1.36 (1)	–	$\lambda_{abs}=242$ nm $\lambda_{em}=307$ nm
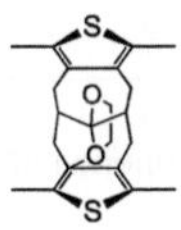 ***st*-BU[Th₁]₂**	+1.09 (1)[a] +1.59 (1)[a]	–	$\lambda_{abs}=241$ nm $\lambda_{em}=308$ nm
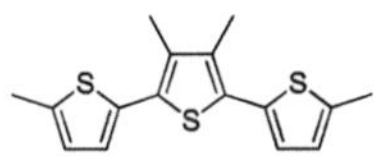 **Th₃**	+0.96 (1)	+1.23 (1)	$\lambda_{abs}=345$ nm $\lambda_{em}=423$, 442 nm[vib]
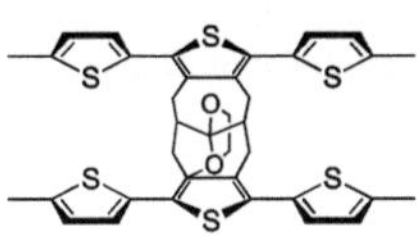 ***st*-BU[Th₃]₂**	+0.70 (1)[a] +0.90 (1)[a]	+1.49 (2)	$\lambda_{abs}=335$ nm $\lambda_{em}=444$ nm[br]
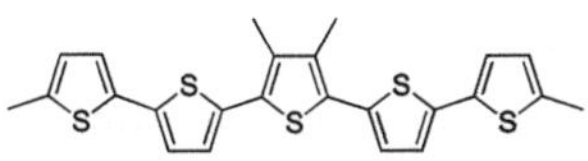 **Th₅**	+0.79 (1)	+1.02 (1)	$\lambda_{abs}=400$ nm $\lambda_{em}=492$, ~500 nm[vib]
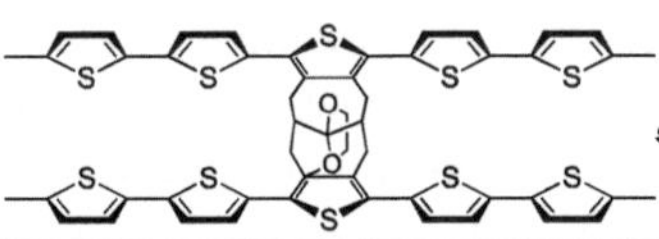 ***st*-BU[Th₅]₂**	+0.67 (1)[a] +0.73 (1)[a]	+1.22 (2)	$\lambda_{abs}=386$ nm $\lambda_{em}=526$ nm[br]

[a]Split wave
[br]Broad excitonic with tail to >600 nm, [vib]vibronic

st-**BU[Th$_3$]$_2$**. Coupling with 5-trimethylstannyl-5′-methyl-2,2′-bithiophene in the same sequence afforded the corresponding stacked quinquethiophene, ***st*-BU[Th$_5$]$_2$**.

The Electronic Structure of Stacked Bicyclo[4.4.1]Undecane-Fused Conjugated Oligothiophenes: Stabilization of the +2 Oxidation State

The unstacked models **Th$_3$** and **Th$_5$** exhibit vibronic emission peaks, whereas the stacked bicycloundecane-scaffolded oligothiophenes display broad emission peaks with tails extending beyond 600 nm (Table 8). Thus, the fluorescence spectra are consistent with excitation to give a segment-localized excited state that undergoes internal conversion to give a lower-energy excimer-like excited state which is

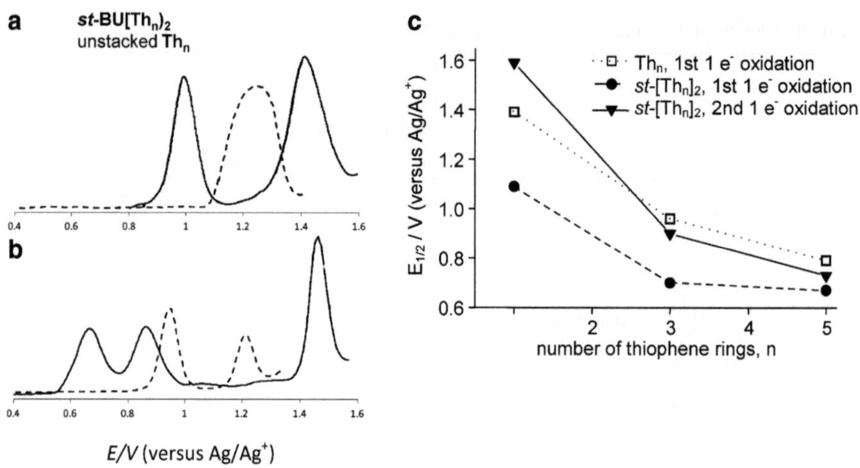

Fig. 36 Differential pulse voltammograms (DPV). (**a**), ***st*-BU[Th$_1$]$_2$** and unstacked model **Th$_1$**; (**b**) ***st*-BU[Th$_3$]$_2$** and unstacked "trimer" **Th$_3$**; (**c**) plot of oxidation potentials of linear (**Th$_n$**) and stacked analogs (***st*-BU[Th$_n$]$_2$**) (from [149])

responsible for a low-energy emission. This behavior is in contrast to the vibronic emission noted by Otsubo for [3.3]quinquethiophenophanes.

The effects of π stacking in this series of compounds are also apparent from their electrochemical behavior. The differential pulse voltammogram of thiophene-stacked compound ***st*-BU[Th$_1$]$_2$** indicates separate one-electron oxidations at +1.09 and +1.59 V (versus Ag/Ag$^+$) for conversion to the +1 and +2 states, respectively (Table 8, Fig. 36a). In contrast, the unstacked analog tetramethylthiophene **Th$_1$** exhibits a single irreversible oxidation wave at +1.39 V. Thus, removal of the first electron from the stacked compound to form the monocation is facilitated by the stacked arrangement of the arene rings. However, the second oxidation of the stacked compound takes place at a higher potential than the oxidation of the unstacked model, suggesting that Coulombic repulsion between the charged aromatic segments raises the energy of the stacked dication.

When comparing the redox behavior of the terthiophene analogs **Th$_3$** and ***st*-BU[Th$_3$]$_2$**, we see that the stacked compound undergoes two one-electron oxidations, which both take place at a lower potential (+0.70 and +0.90 V) than the first oxidation of the unstacked model (+0.96 V) (Fig. 36b). A similar pair of one-electron oxidations at relatively low potentials is also observed for the stacked quinquethiophene ***st*-BU[Th$_5$]$_2$** at + 0.67 V and +0.73 V, compared to the first oxidation of the unstacked model 3,3″,4″,5″″-tetramethyl-2,2′:5′,2″:5″,2‴:5‴,2″″-quinquethiophene, **Th$_5$**) at +0.79 V. The observation that removal of the second electron from ***st*-BU[Th$_3$]$_2$** and ***st*-BU[Th$_5$]$_2$** to give the dication takes place at lower potentials than the first oxidation of the corresponding unstacked analogs suggests that both the cationic and the dicationic state are stabilized by the interaction between stacked segments. The stabilizing effect outweighs the Coulombic repulsion between the charged segments.

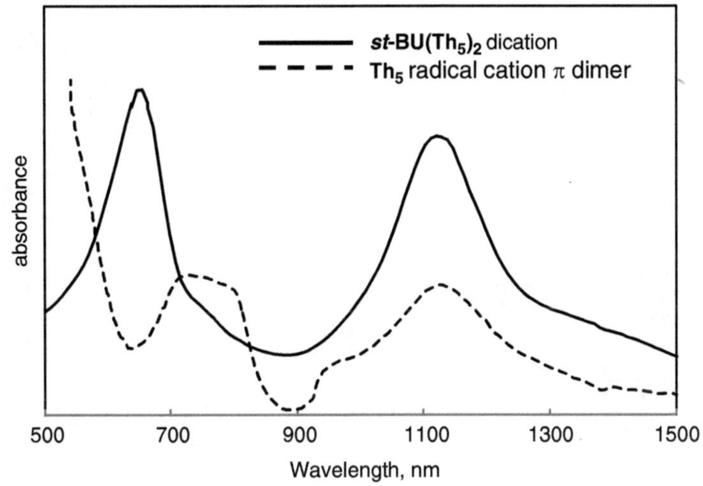

Fig. 37 UV–vis-NIR spectra: dication of ***st*-BU[Th$_5$]$_2$** compared to the π dimer of the model unstacked quinquethiophene **Th$_5$**

Addition of $FeCl_3$ to a solution of ***st*-BU[Th$_5$]$_2$** to afford the dicationic state results in a UV-vis-near-IR spectrum that consists of two strong absorptions, at 654 and 1,123 nm (Fig. 37). The appearance of the spectrum closely resembles that of the radical cation π dimer of the unstacked analog **Th$_5$** that is formed at low temperature and has peaks at 620 and 1,085 nm (Fig. 37). The stacked "trimer" *st*-**BU[Th$_3$]$_2$** displays similar behavior. Thus, the lower potentials for oxidation of the stacked oligothiophenes to form the dications are in accord with the formation of an intramolecular radical cation π dimer that is stabilized relative to unstacked radical cation by interaction between the conjugated segments. The organization of the conjugated segments in the stacked compounds provides direct access to the π dimer electronic state at room temperature. In contrast to the thermally reversible π dimerization of radical cations of linear conjugated oligomers, the attachment of the oligomers to a scaffold to provide a stacked arrangement allows for direct access to, and investigation of, the dicationic species by electrochemical methods. The enhanced stability of the intramolecular radical cation π dimer, as determined by electrochemistry, is consistent with the observation of intermolecular π dimerization of delocalized radical cations in solution.

Photophysical and Redox Properties of Stacked Bicyclo[4.4.1] Undecane-Scaffolded Oligo(Phenylene Ethynylene)s

Subodh Jagtap and Glen Brizius at Georgia Tech further extended the work of Knoblock to prepare bicycloundecane-scaffolded oligo(phenylene ethynylene)s,

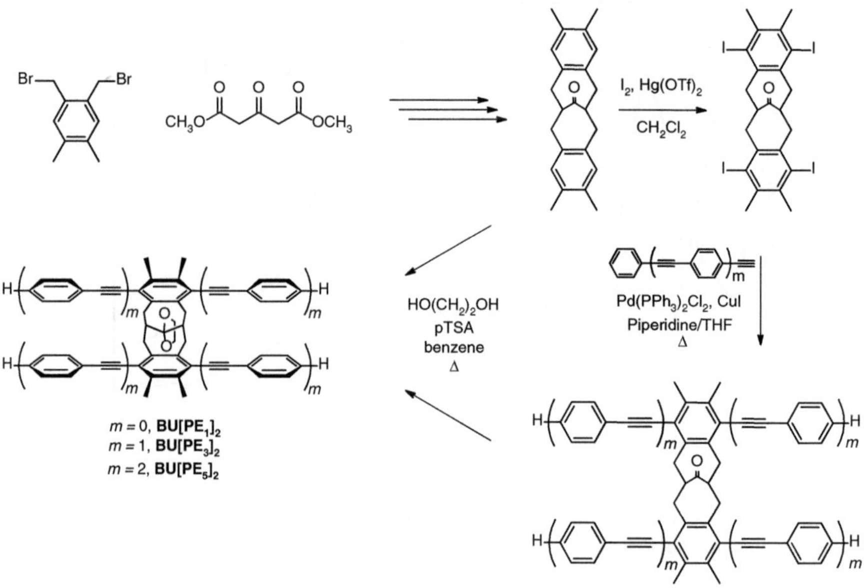

Fig. 38 Synthesis of bicyclo[4.4.1]undecane-scaffolded oligo(phenylene vinylene)s ***st*-BU[PE$_n$]$_2$**, $n = 1,3,5$ [150]

***st*-BU[PE$_n$]$_2$**, as shown in Fig. 38 [150]. The molecular design of the ***st*-BU[PE$_n$]$_2$** series includes methyl groups in the 4 and 5 positions of the bicycloundecane-fused benzene rings. These allow us to direct the functionalization of the fused arene rings to the positions that are ortho to the points of fusion to the bicyclic scaffold and thereby stack two linear phenylene ethynylene segments atop one another. We prepared the dimethylbenzo-fused bicyclo[4.4.1]undecanone following Mataka's general synthetic route, iodinated the aromatic rings, displaced the iodides with the phenylene ethynylene arms using Sonogashira coupling, and ketalized to form ***st*-BU[PE$_n$]$_2$**, $n = 1$, 3, and 5 (Fig. 38).

The methyl groups on the benzo-fused bicycloundecane core of this series of compounds do not interfere with stacking of the arene units, as shown by X-ray diffraction and conformational analysis by ^{1}H NMR spectroscopy. The center-to-center distance between the stacked benzene rings of ***st*-BU[PE$_1$]$_2$** is 3.42 Å; the benzene rings are tilted at an angle of 16.8° (Fig. 39a). The phenylene ethynylene arms also do not interfere with stacking: the central rings of the two conjugated PE$_3$ segments of ***st*-BU[PE$_3$]$_2$** are held 3.42 Å apart (Fig. 39b) at a tilt angle of 17.9°. Thus, the stacked pseudochair/pseudochair conformations of the phenylene ethynylene compounds are similar to those of the parent compound that was originally reported by Mataka. The two segments are held in a stacked arrangement along their entire length (Fig. 39c).

For this series of compounds, as with the other systems described above, only small differences are observed between the absorption spectra of the stacked and unstacked analogs. The absorption maximum of the stacked "trimer" ***st*-BU[PE$_3$]$_2$** ($\lambda_{ab} = 325$ nm) and pentamer ***st*-BU[PE$_5$]$_2$** ($\lambda_{ab} = 355$ nm) appears at slightly higher

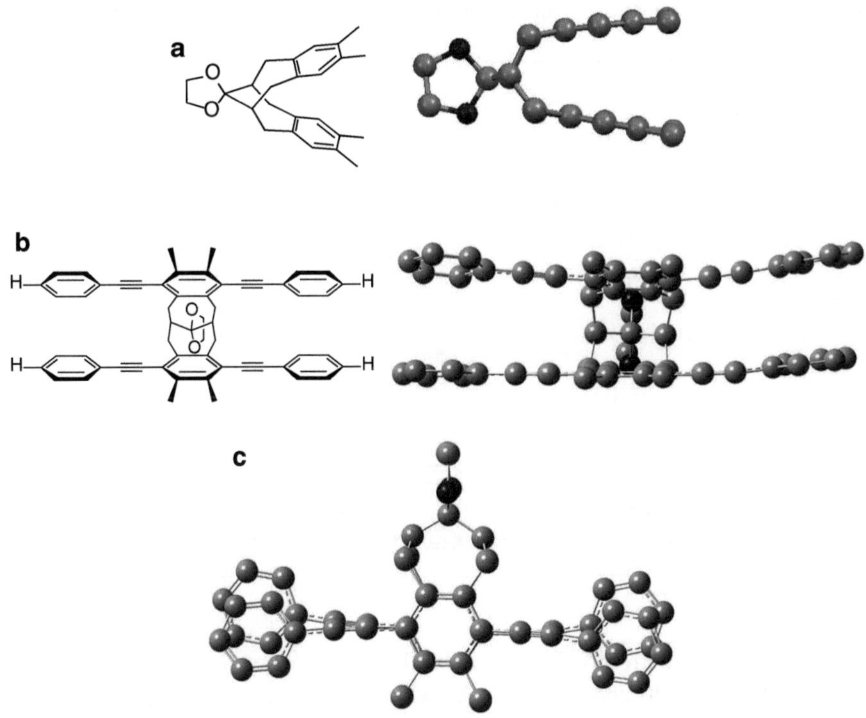

Fig. 39 X-ray crystal structures. (**a**) Ketal ***st*-BU[PE$_1$]$_2$**; (**b**) side view of ***st*-BU[PE$_3$]$_2$**; (**c**) top view of ***st*-BU[PE$_3$]$_2$** (from [150])

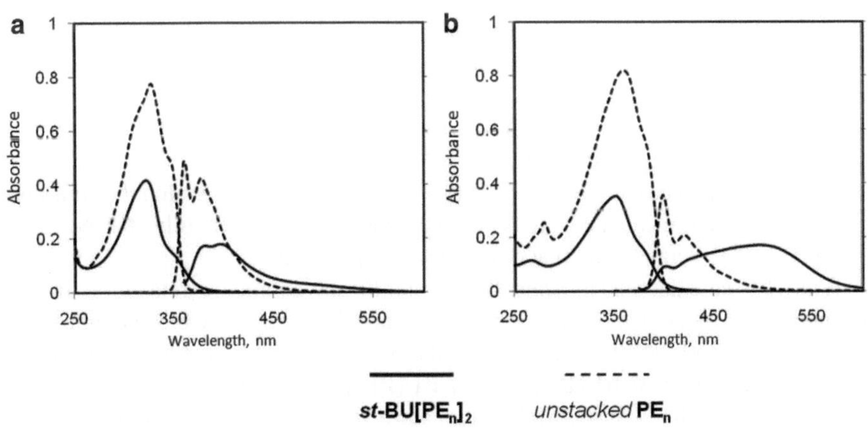

Fig. 40 UV–vis and fluorescence spectra of bicycloundecane-scaffolded oligo(phenylene ethynylene)s, ***st*-BU[PE$_n$]$_2$** and unstacked analogs. (**a**) ***st*-BU[PE$_3$]$_2$** and **PE$_3$**; (**b**) ***st*-BU[PE$_5$]$_2$** and **PE$_5$** (from [150])

Table 9 Bicyclo[4.4.1]undecane-scaffolded oligo(phenylene vinylene)s ***st*-BU[PE$_n$]$_2$** and unstacked models

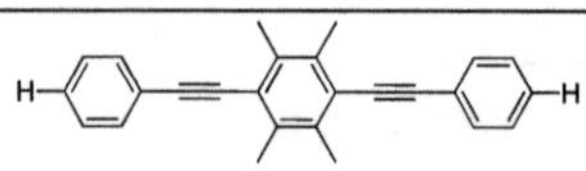	**PE$_3$**	$\lambda_{ab} = 330$ nm $\lambda_{em} = 360, 375$ nm^{vib}	[150]
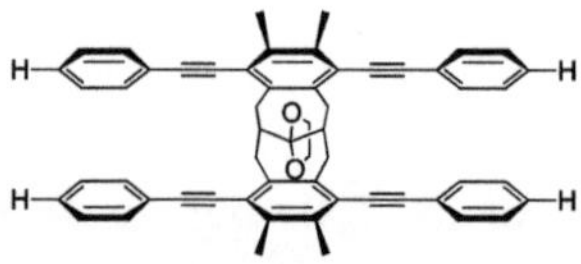	***st*-BU[PE$_3$]$_2$**	$\lambda_{ab} = 325$ nm $\lambda_{em} = 382, 397^{vib}$ 490 nm$^{wk\ br}$	
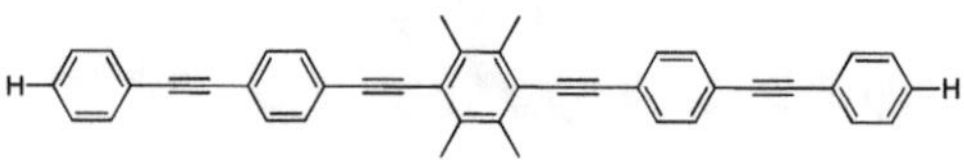	**PE$_5$**	$\lambda_{ab} = 365$ nm $\lambda_{em} = 399, 418$ nm^{vib}	
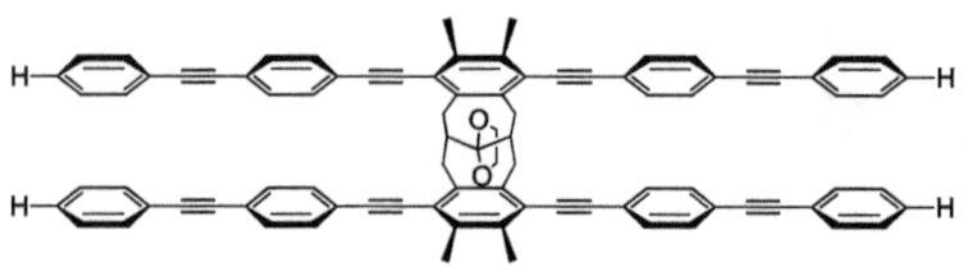	***st*-BU[PE$_5$]$_2$**	$\lambda_{ab} = 355$ nm $\lambda_{em} = 398$ nm$^{wk\ vib}$ 495 nm^{br}	

^{br}Broad, ^{vib}Vibronic, ^{wk}Weak

energy than for the corresponding unstacked linear oligomers ($\lambda_{ab} = 330$ and 365 nm, respectively) (Fig. 40 and Table 9). As with the [2.2]paracyclophane-scaffolded oligomers, the similarity of the UV–visible spectra of the stacked and unstacked analogs suggests that absorption affords a segment-localized excited state. Larger differences are seen in the fluorescence spectra. The emission spectrum of the shorter oligomer, ***st*-BU[PE$_3$]$_2$**, is dominated by a vibronic progression that is red shifted relative to the emission of the unstacked model. On close inspection, there is a long low-energy tail to the emission. Deconvolution of the spectrum reveals that the origin of the tail is a broad and weak emission with a peak at 490 nm. In contrast, the emission spectrum of the longer analog, ***st*-BU[PE$_5$]$_2$**, is dominated by a broad and significantly red-shifted excimer-like emission. In addition, there is a weak vibronic peak in the spectrum that matches the energy 0–0 peak of the unstacked model, **PE$_5$**. Thus, both stacked analogs exhibit emission from both a segment-localized state and an excimer-like excited state. For the shorter analog the emission from the segment-localized state is stronger, and for the longer fully stacked compound it is the excimer state that dominates the spectrum.

TD-DFT calculations indicate that a significant geometric relaxation in the excited state leads to the formation of an excimer-like state in the bicycloundecane-scaffolded oligothiophenes. These calculations indicate that the energies of the excimer-like state of ***st*-BU[PE$_3$]$_2$** and ***st*-BU[PE$_5$]$_2$** are similar, which is consistent with the observation that the peak maxima are similar in both cases (ca. 490 nm). This is explained by the delocalization of the frontier molecular orbitals in the two cases (Fig. 41). In ***st*-BU[PE$_3$]$_2$** these are distributed over the entire length of the molecule, whereas in the stacked pentamer, ***st*-BU[PE$_5$]$_2$**, they are primarily restricted to the central three-ring section of the PE$_5$ segments.

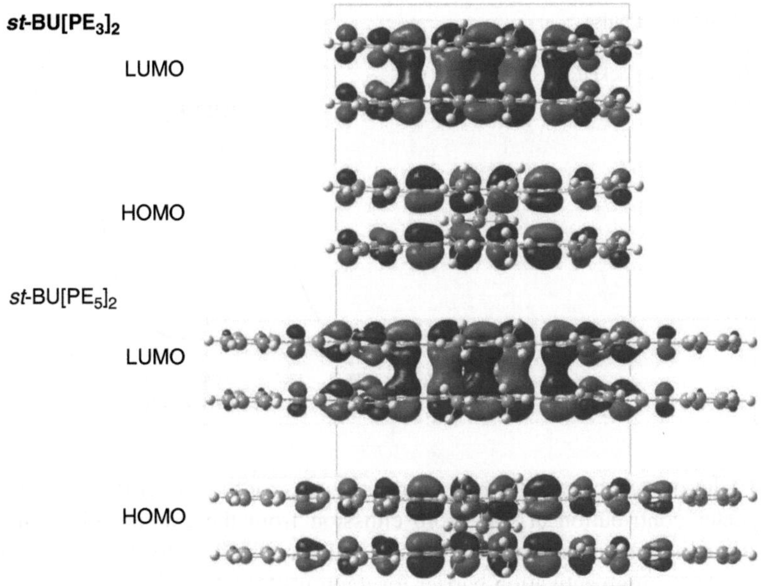

Fig. 41 Frontier molecular orbitals of ***st*-BU[PE$_3$]$_2$** and ***st*-BU[PE$_5$]$_2$** (from [150])

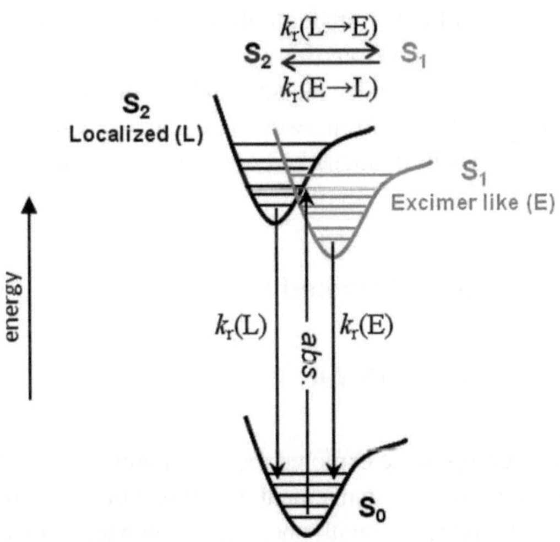

Fig. 42 Simple three-state model depicting emission from the segment-localized (L) and excimer-like (E) excited states, with exchange between them via the forward (k_r(L → E)) and backward (k_r(E → L)) processes (from [150])

The relative strength of the contributions in the fluorescence spectra from the segment-localized state (vibronic, high-energy emission) and excimer-like state (broad, featureless long wavelength emission) can be explained by the simple three-state model depicted in Fig. 42. For the shorter analog, ***st*-BU[PE$_3$]$_2$**, the localized

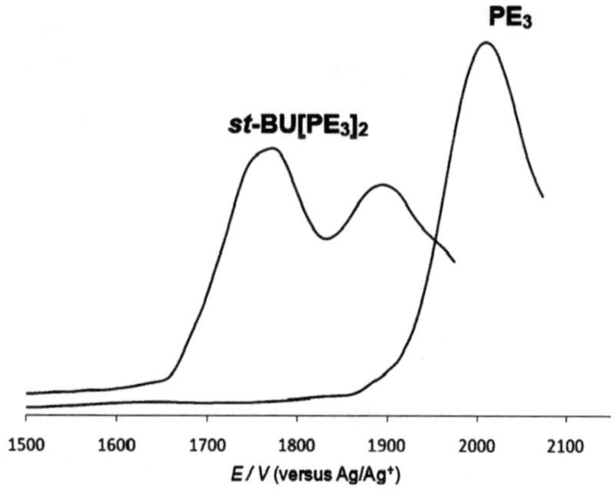

Fig. 43 Differential pulse voltammograms (DPV) of ***st*-BU[PE$_3$]$_2$** and unstacked analog **PE$_3$**

state (L) dominates the emission spectrum, whereas for the longer PE$_3$ analog, there is a greater contribution arising from emission from the excimer-like state (E). A semi-qualitative analysis of the various pathways leading to emission suggests that a relatively high activation barrier for the transition from the localized excited state to excimer-like state of ***st*-BU[PE$_3$]$_2$** means that emission from the higher-energy localized state predominates. In contrast, for ***st*-BU[PE$_5$]$_2$**, with its longer phenylene ethynylene segments, the activation barrier for this transition appears to be lower and there is a significant transfer to the excimer-like state, thereby resulting in a strong low-energy emission from this state.

As with the bicyclo[4.4.1]undecane-scaffolded oligothiophenes, the stacked oligo(phenylene ethynylene)s exhibit two one-electron oxidations, both at lower potential than the corresponding unstacked oligomer, as shown in Fig. 43.

Other Intramolecular π-Stacked Oligomers

Aromatic Edges as Scaffolds for Stacked Segments

A number of research groups have explored the use of arenes as scaffolds to hold conjugated side chains in a stacked arrangement. Crisp demonstrated that substitution of the 1 and 8 positions of anthracene with phenylene ethynylene units does not result in a fixed stacked arrangement of the side arm segments—there is sufficient conformational flexibility to allow for rotation of the arms (Fig. 44a). However, side chains and additional tethers might be used to restrict this rotation such that the side chains are held in a face-to-face manner and largely orthogonal to the anthracene scaffold [151–153].

To reduce the space between the conjugated side chains, Bradley Carson at Georgia Tech explored the synthesis and properties of a series of naphthalenes that bear oligo(phenylene ethynylene)s side arms at the 1 and 8 positions (Fig. 44b). This was based on a study by House in which he showed that phenyl substituents in these positions of a naphthalene are held in a stacked arrangement [154]. However, substituents

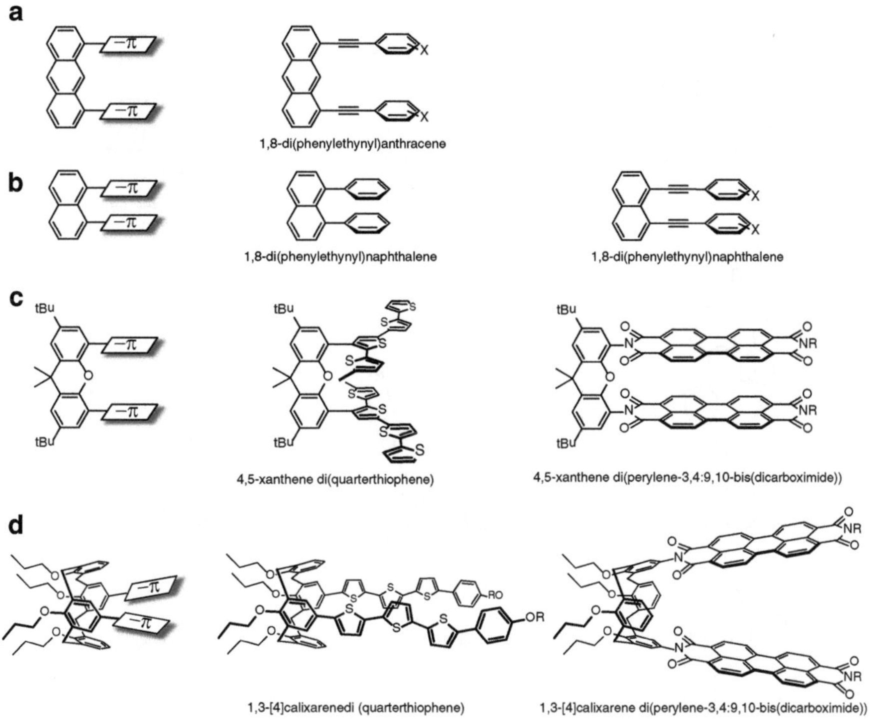

Fig. 44 Arene edge scaffolds: (**a**) 1,8-disubstituted anthracenes; (**b**) 1,8-disubstituted naphthalenes; (**c**) 4,5-disubstituted xanthenes; and (**d**) 1,3-disubstituted [4]calixarenes [151–157, 165, 166]

on the phenylene ethynylene arms have a strong influence on the conformation of the molecules, with significant in-plane and out-of-plane distortions of the naphthalene [155]. Thus, the distortion of the scaffold, conformational flexibility, and deviation from orthogonality of the aromatic core and conjugated segments mean that the use of these structures as models for interactions between π systems is not straightforward.

The 4,5-disubstituted xanthene core has also provided a versatile scaffold for the preparation of stacked oligothiophenes by Swager [156]; perylenediimides by Wasielewski [157] (Fig. 44c) and oligomers and polymers with thiophene [158, 159], carbazole [160, 161], phenylene [162], ferrocene [163], or 9,10-disubstituted anthracene [164] units by Chujo. Swager made use of this scaffold in an exploration of how the formation of intramolecular radical cation π dimers may form the basis for the design of molecular actuators in which a stimulus (i.e., redox switching in this case) brings about a conformational change that may be harnessed in the form of mechanical motion. The two-electron oxidation of a xanthene-scaffolded quaterthiophene in solution at room temperature affords a UV-vis spectrum that is consistent with the formation a radical cation π dimer (Fig. 45a) [156]. As in the case of the bicycloundecanone-scaffolded oligothiophenes, the differential pulse voltammogram of the xanthene-scaffolded compound clearly shows splitting of the oxidation processes for the first and second oxidation. Similarly, a calix[4]arene bearing oligothiophene arms on rings 1 and 3 (Fig. 45b) displays stepwise oxidation with formation of a dicationic state that has the spectral signatures of a radical cation π

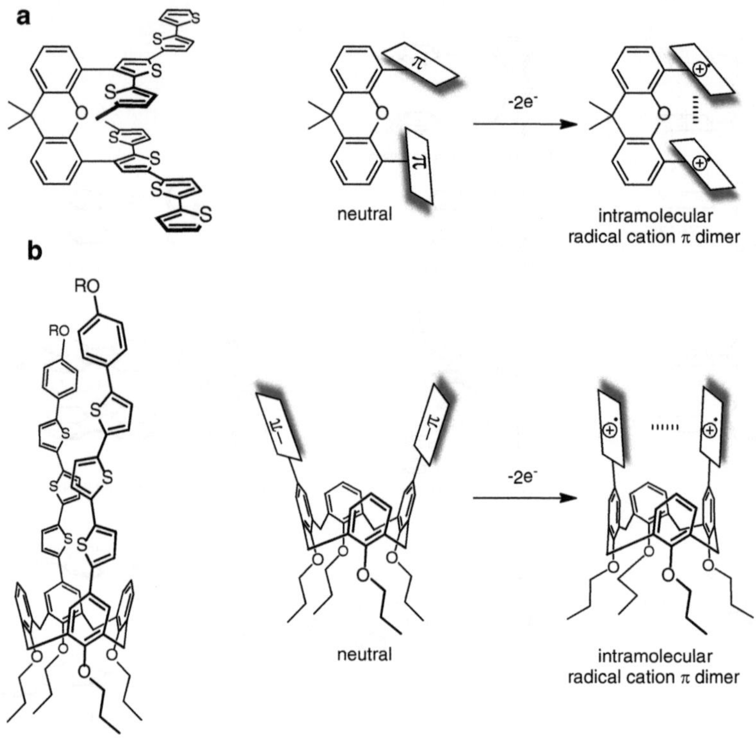

Fig. 45 Proposed components of molecular actuators in which formation of **π** dimers induces a change in conformation

dimer [165]. Similar effects of intersegment interaction are observed with perylenediimide-substituted calixarenes [166], Fig. 44c.

Further insights into the interactions between conjugated segments may also be gained from the spectroscopic analysis of chromophores that are held in well-defined architectures by new stacked scaffolds (e.g., Smith's *m*-terphenyl oxacyclophanes [167, 168] and Swager's benzene-sandwiched phenylene ethynylenes [169]) or by their incorporation into larger macrocycles (e.g., shape-persistent conjugated macrocycles [170, 171]).

Stacked Fused Arenes

While differences in the spectroscopic behavior of stacked conjugated oligomers and their linear unstacked analogs may be attributed to the effect of interactions between π systems, they may also arise from differences in the conformation of the conjugated segments in the two molecular arrangements. For example, the close packing of the conjugated arms of the ***pg*-CP[PV$_n$]$_2$** may enforce a largely planar conformation, whereas the conjugated arms of the pseudopara and pseudoortho analogs are likely to possess more conformational flexibility in solution. Thus,

Table 10 Arene-fused [2.2]paracyclophanes and unstacked fused arene analogs

Syn and *anti* [2.2](4,7)benzothiadiazolophanes [172]

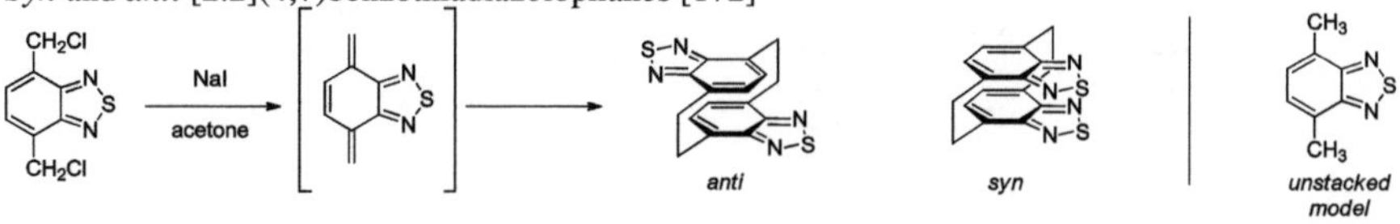

	anti	syn	unstacked model
λ_{ab}[a]	356, 311, 233 nm	349, 310, 236 nm	311, 305, 222 nm
λ_{em}[a]	478 nm	478 nm	402 nm
φ[a]	0.028	0.002	0.005
$E_{½}$ (red)[b]	−2.05 V	−1.92, −1.34 V	−1.98 V

Anti [2.2](1,4)biphenylenophane [86]

	anti	unstacked model
λ_{ab}	257 nm	252, ~270 nm
λ_{em}	537 nm	458, 485, 518 nm
$E_{½}$ (ox) [c]	+1.37, 1.55 V	+1.59 V

Anti [2.2](1,4)-naphthalenophane and *anti* [2.2]isoindolinophanes and dihydro[c]furanophane ([173] and [175], respectively)

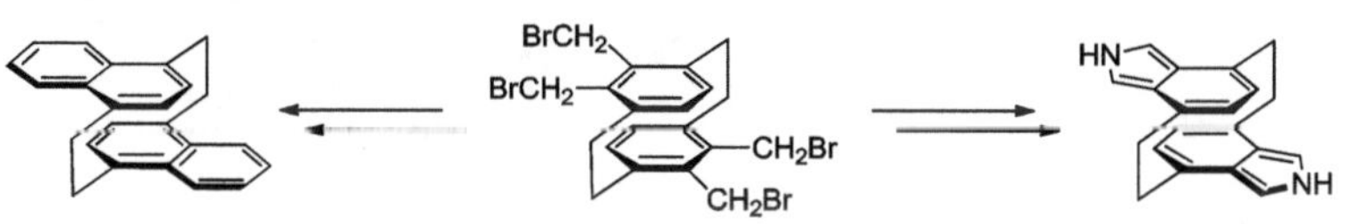

Example of a [2.2](1,4)-anthracenophane [145]

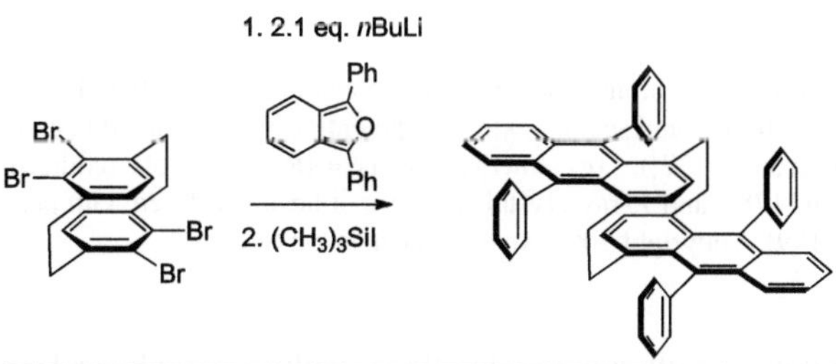

[a]EtOH
[b]CH_2Cl_2/0.1 M n-Bu_4NPF_6, potentials versus Fc/Fc^+
[c]CH_2Cl_2/0.1 M n-Bu_4NClO_4, potentials versus Ag/AgCl

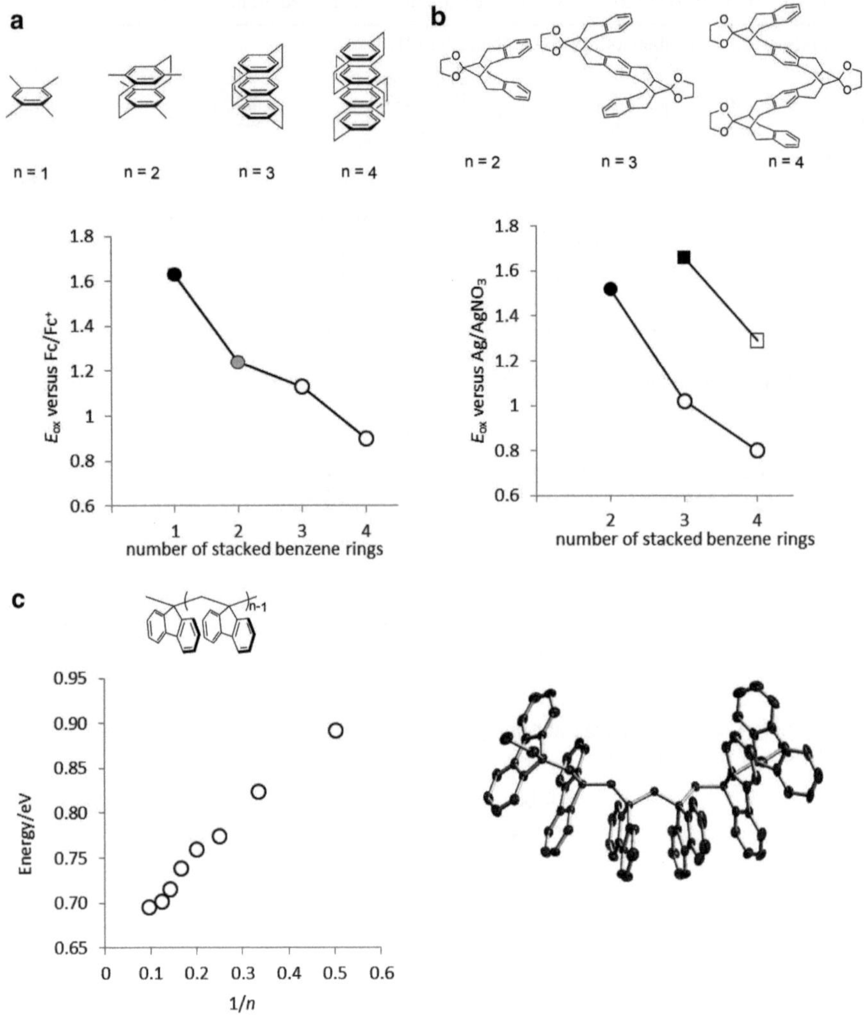

Fig. 46 The effect of multilayer stacking of arenes. (**a**) Oxidation potentials of multitiered [2.2] paracyclophanes (data from [176]); (**b**) first and second oxidation potentials of multitiered benzo-fused bicycloundecanes (data from [177]); (**c**) energy of near-IR band of oxidized oligo(dibenzofulvene)s (data from [178]) and X-ray crystal structure of hexamer. Used with permission from Nakano and Yade [179]. Copyright 2003 American Chemical Society

construction of stacked fused arenes, in which planarity is assured in both stacked and unstacked analogs, might provide better mimics of the conformation and local environment of the backbone of conjugated oligomers and polymers in semiconducting thin films. Examples of fused arenes held in a π-stacked arrangement by a [2.2]paracyclophane scaffold include [2.2](4,7)-benzothiadiazolophane [172], [2.2]

(1,4)-biphenylenophane [86], [2.2](1,4)-naphthalenophane [173, 174], [2.2](4,7)-isoindolinophane [175], and [2.2](1,4)-anthracenophane, shown in Table 10.

Multitiered Architectures

Further insights into the effect of holding conjugated units in a stacked architecture might come from the analysis of series of multilayered compounds. For example, whereas electrochemical oxidation of benzene affords a reactive radical cation that undergoes polymerization, multitiered [2.2]paracyclophanes undergo reversible or quasireversible oxidation (Fig. 46) [176]. The two, three, and four-layered paracyclophanes undergo oxidation at progressively lower potentials, consistent with stabilization of the cationic species by virtue of delocalization over the stacked architecture (Fig. 46a). A similar decrease in the oxidation potential of stacks of benzene rings is observed in Mataka's multitiered bicycloundecanes shown in Fig. 46b [177]. Finally, the effect of π-stacking on the spectroscopy of cationic species is demonstrated by the trend in the energy of the near-IR charge resonance absorption of oxidized oligo(dibenzofulvenes) (Fig. 46c) [178, 179].

Stacked Conjugated Oligomers in Electronic Devices

The studies described in the previous sections explore the role of interchain interactions on the electronic structure of π systems that are held in close proximity and mimic the packing of chains in films of semiconducting conjugated oligomers and polymers. However, they do not directly address the impact of π stacking on charge transport. The study of stacked conjugated oligomers in electronic devices has received little attention and is potentially a rich area for investigation. The covalent

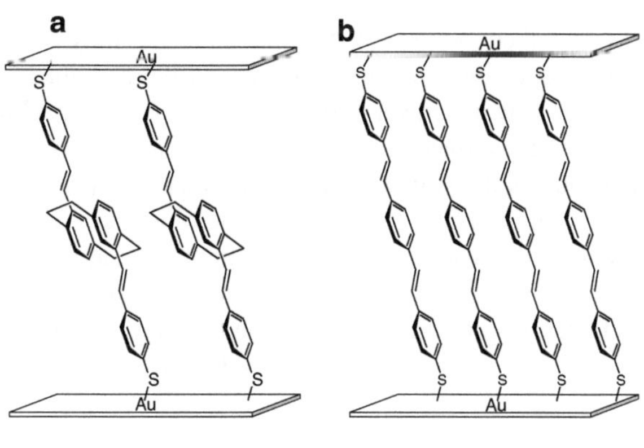

Fig. 47 Metal–molecule–metal molecular junctions constructed. (**a**) A thiol-derivatized pp-**CP[PV$_2$]$_2$**; (**b**) fully conjugated oligophenylene vinylene derivative

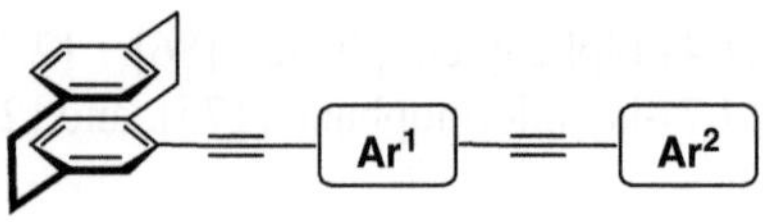

Fig. 48 General structure of arylene ethynylene-substituted [2.2] paracyclophane

attachment of conjugated segments to a scaffold to provide some pre-organization, when combined with methods to optimize molecular packing, may provide new approaches to the creation of materials with enhanced properties.

The potential for strong through space coupling of the π systems of [2.2]paracyclophane to promote charge transport is apparent from a study of the conductivity of a molecular monolayer of a ***pp*-CP[PV$_2$]$_2$** derivative [180–182]. Comparison of *I–V* curves for metal–molecule–metal junctions containing either a thiol substituted derivative of ***pp*-CP[PV$_2$]$_2$** or an unstacked oligophenylene vinylene analog (Fig. 47a, b, respectively) indicates similar conductivity on a per molecule basis (albeit that the molecular packing density of the linear oligomer is approximately twice that of the cyclophane derivative). Thus, the interaction between conjugated segments, in the absence of a fully conjugated pathway, provides a mechanism for charge transport through the junction. While this result clearly indicates that a fully conjugated pathway between the electrodes is not necessary for charge transport, it should be noted that intermolecular chain–chain interactions are present in both cases and presumably have a role to play even in the behavior of the device that incorporates the linear unstacked phenylene vinylene.

Marrocchi and co-workers at the University of Perugia have reported initial experiments to incorporate [2.2]paracyclophane-containing conjugated oligomers in thin film devices (Fig. 48). Their work includes the development of methods to fabricate and pattern thin films and the demonstration of photocurrents from devices that contain oligo(arylene ethynylene)s, shown below [183–185].

Conclusion

As described in the preceding sections, the study of molecules in which conjugated oligomeric segments are held in well-defined stacked arrangements continues to provide new insights into the effect of molecular packing on the electronic structure of conjugated materials. The electronic structure of stacked π-conjugated oligomers depends on the structure of the oligomer (i.e., phenylene vinylene, phenylene ethynylene, thiophene), the length of the oligomer, the position of the interacting segments within each conjugated segment (e.g., at the terminus or interior), the length over which the segments interact, and the number of interacting segments that are held in a stacked arrangement. The presence of a low-energy excited state that leads to excimer-like emission is favored by interaction between central rings within short oligomers (i.e., **X-CP[PV$_3$]$_2$** versus ***pp*-CP[PV$_3$]$_2$**) and a greater length over which the segments interact (e.g., ***pg*-CP[PE$_3$]$_2$** and ***st*-BU[PE$_3$]$_2$** compared to ***pp*-CP[PE$_3$]$_2$**). The stacking of multiple conjugated segments atop one another results in further stabilization of the excited state (e.g., **poly-*pg*-CP[PE$_3$]**). The design of additional scaffolds to allow for control over the packing of π systems, further extension of the length of the segments, and the creation of well-defined, monodisperse, multitier structures will provide new opportunities to explore the evolution of the electronic structure upon proceeding from individual molecules to 3D conjugated materials.

Acknowledgement I wish to thank former students and post-doc Drs. Subodh Jagtap, Glen Brizius, Kurt Knoblock, Bradley Carson, and Fouad Sahli, whose work is described above, for all of their contributions. In addition, I thank Professor Jean-Luc Brédas and Drs. Sukrit Mukhopadhyay and Veaceslav Coropceanu for their computational studies and many hours of discussion. Our work on stacked conjugated oligomers was supported by the National Science Foundation.

References

1. Horowitz G (1946) J Mater Res 2004:19
2. Katz H, Bao Z (2000) J Phys Chem B 104:671
3. Dimitrakopoulos C, Malenfant P (2002) Adv Mater 14:99
4. Desiraju GR, Gavezzotti A (1989) Chem Commun 621
5. Anthony JE, Eaton DL, Parkin SR (2002) Org Lett 4:15
6. Anthony JE (2006) Chem Rev 106:5028
7. MacGillivray LR (2004) Cryst Eng Comm 6:77
8. Horowitz G, Bachet B, Yassar A, Lang P, Demanze F, Fave J.-L, Garnier F (1995) Chem Mater 7:1337
9. Hotta S, Waragai K (1993) Adv Mater 5:896
10. Prosa TJ, Winokur MJ, McCullough RD (1996) Macromolecules 29:3654
11. DeLongchamp DM, Kline RJ, Lin EK, Fischer DA, Richter LJ, Lucas LA, Heeney M, McCulloch I, Northrup JE (2007) Adv Mater 19:833
12. Sirringhaus H, Brown PJ, Friend RH, Nielsen MM, Bechgaard K, Langeveld-Voss BMW, Spiering AJH, Janssen RAJ, Meijer EW, Herwig P, de Leeuw DM (1999) Nature 401:695
13. Kline RJ, McGehee MD, Toney MF (2006) Nat Mater 5:222
14. Fichou D (ed) (1999) Handbook of oligo- and polythiophenes. Wiley-VCH, Weinheim
15. Granier T, Thomas EL, Gagnon DR, Karasz FE, Lenz RW (2003) J Polym Sci B Polym Phys 24:2793
16. Bunz UHF, Enkelmann V, Kloppenburg L, Jones D, Shimizu KD, Claridge JB, zur Loye H-C, Lieser G (1999) Chem Mater 11:1416
17. Tol AJW (1996) Chem Phys 208:73
18. van Haare JAEH, Havinga EE, van Dongen JLJ, Janssen RAJ, Cornil J, Brédas J-L (1998) Chem Eur J 4:1509
19. Apperloo JJ, Janssen RAJ, Malenfant PRL, Groenendaal L, Fréchet JMJ (2000) J Am Chem Soc 122:7042
20. Gao Y, Liu C-G, Jiang Y-S (2002) J Phys Chem A 106:5380
21. Geskin VM, Bredas J-L (2003) ChemPhysChem 4:498
22. Yu Y, Gunic E, Miller LL (1995) Chem Mater 7:255
23. Smie A, Heinze J (1997) Angew Chem Int Ed 36:363

24. Tschuncky P, Heinz J, Smie A, Engelmann G, Kossmehl G (1997) J Electoanal Chem 433:223
25. Heinze J, Tschuncky P, Smie A (1998) J Solid State Electrochem 2:102
26. Engelmann G, Kossmehl G, Heinze J, Tschuncky P, Jugelt W, Welzel H-P (1998) J Chem Soc Perkin Trans 2:169
27. Heinze J, John H, Dietrich M, Tschuncky P (2001) Synth Metals 119:49
28. Miller LL, Mann KR (1996) Acc Chem Res 29:417
29. Zinger B, Mann KR, Hill MG, Miller LL (1992) Chem Mater 4:1113
30. Hill MG, Mann KR, Miller LL, Penneau JF, Zinger B (1992) Chem Mater 4:1106
31. Yu Y, Gunic E, Zinger B, Miller LL (1996) J Am Chem Soc 118:1013
32. Graf DD, Duan RG, Campbell JP, Miller LL, Mann KR (1997) J Am Chem Soc 119:5888
33. Graf DD, Campbell JP, Miller LL, Mann KR (1996) J Am Chem Soc 118:5480
34. Baüerle P, Segelbacher U, Maier A, Mehring M (1993) J Am Chem Soc 115:10217
35. Baüerle P, Segelbacher U, Gaudl KU, Huttenlocher D, Mehring M (1993) Angew Chem Int Ed 32:76
36. Segelbacher U, Sariciftci NS, Grupp A, Baüerle P, Mehring M (1993) Synth Metals 57:4728
37. Tour JM, Wu RL (1992) Macromolecules 25:1901
38. Guay J, Kasai P, Diaz A, Wu R, Tour JM, Dao LH (1992) Chem Mater 4:1097
39. Yassar A, Delabouglise D, Hmyene M, Nessak B, Horowitz G, Garnier F (1992) Adv Mater 4:490
40. Hoeve WT, Wynberg H, Havinga EE, Meijer EW (1991) J Am Chem Soc 113:5887
41. Zotti G, Berlin A, Pagani G, Schiavon G, Zecchin S (1994) Adv Mater 6:231
42. Zotti G, Schiavon G, Berlin A, Pagani G (1993) Chem Mater 5:620
43. Hotta A, Waragai K (1993) J Phys Chem 29:7427
44. Tanaka K, Matsuura Y, Oshima Y, Yamabe T, Hotta S (1994) Synth Metals 66:295
45. Levillain E, Roncali J (1999) J Am Chem Soc 121:8760
46. Apperloo JJ, Raimundo J-M, Frère P, Roncali J, Janssen RA (2000) J Chem Eur J 6:1698
47. Videlot C, Ackermann J, Blanchard P, Raimundo J-M, Frère P, Allain M, de Bettignies R, Levillain E, Roncali J (2003) Adv Mater 15:306
48. Knoblock KM (2006) PhD thesis, Georgia Institute of Technology
49. Nessakh B, Horowitz G, Garnier F, Deloffre F, Srivastava P, Yassar A (1995) J Electroanal Chem 399:97
50. Yamazaki D, Nishinaga T, Tanino N, Komatsu K (2006) J Am Chem Soc 128:14470
51. Ferrón CC, Ruiz Delgado MC, Hernández V, López Navarrete JT, Vercelli B, Zotti G, Capdevila Cortada M, Novoa JJ, Niu W, Hed M, Hartl F (2011) Chem Commun 47:12622
52. Hong Y, Miller LL (1995) Chem Mater 7:1999
53. Lee D, Swager TM (2003) J Am Chem Soc 125:6870
54. Sugiyasu K, Honsho Y, Harrison RM, Sato A, Yasuda T, Seki S, Takeuchi M (2010) J Am Chem Soc 132:14754
55. Frampton MJ, Anderson HL (2007) Angew Chem Int Ed 46:1028
56. Cacialli F, Wilson JS, Michels JJ, Daniel C, Silva C, Friend RH, Severin N, Samorı´ P, Rabe JP, O'Connell MJ, Taylor PN, Anderson HL (2002) Nat Mater 1:160
57. Zhu SS, Caroll PJ, Swager TM (1996) J Am Chem Soc 118:8713
58. Zhu SS, Swager TM (1997) J Am Chem Soc 119:12568
59. Scudder PH, Boekelheide V, Cornutt D, Hopf H (1981) Spectrochim Acta Part A 37A:425
60. Gantzel PK, Trueblood KN (1965) Acta Cryst 18:958
61. Dodziuk H, Szymański S, Jaźwiński J, Marchwiany ME, Hopf H (2010) J Phys Chem A 114:10467
62. Jones PG, Hopf H, Pechlivanidis Z, Boese R (1994) Zeitschrift für Kristallographie 209:673
63. Vorontsova NV, Rozenberg VI, Sergeeva EV, Vorontsov EV, Starikova ZA, Lyssenko KA, Hopf H (2008) Chem Eur J 14:4600
64. Keehn PM, Rosenfeld SM (eds) (1983) Cyclophanes, vol I and II. Academic, New York
65. Vogtle F (1993) Cyclophane chemistry. Wiley-VCH, Chichester
66. Gleiter R, Hopf H (eds) (2004) Modern cyclophane chemistry. Weinheim, Wiley-VCH
67. David ORP (2012) Tetrahedron 68:8977

68. Marrocchi A, Tomasi I, Vaccaro L (2012) Isr J Chem 52:41
69. Reich HJ, Cram DJ (1969) J Am Chem Soc 91:3527
70. Yeh YL, Gorham WF (1969) J Org Chem 34:2366
71. Reich HJ, Cram DJ (1969) J Am Chem Soc 91:3517
72. Pye PJ, Rossen K, Reamer RA, Tsou NN, Volante RP, Reider PJ (1997) J Am Chem Soc 119:6207
73. Braddock DC, Ahmad SM, Douglas G (2004) Tetrahedron Lett 45:6583
74. Izuoka A, Murata S, Sugawara T, Iwamura H (1987) J Am Chem Soc 109:2631
75. El Shaieb K, Narayanan V, Hopf H, Dix I, Fischer A, Jones PG, Ernst L, Ibrom K (2003) Eur J Org Chem 567
76. Psiorz M, Schmid R (1987) Chem Ber 120:1825
77. Zitt H, Dix I, Hopf H, Jones PG (2001) Eur J Org Chem 2298
78. Nikanorov VA, Kharitonov VG, Yatsenko EV, Krut'ko DP, Galakhov MV, Yakushin SO, Mikul'shina VV, Rozenberg VI, Guryshev VN, Yur'ev VP, Reutov OA (1992) *Izv Akad Nauk Ser Khim* 1837
79. König B, Knieriem B, de Meijere A (1993) Chem Ber 126:1643
80. Winberg HE, Fawcett FS, Mochel WE, Theobald CW (1960) J Am Chem Soc 82:1428
81. Winberg HE, Fawcett FS (1973) Org Synth Coll Vol V. Wiley, New York, 883
82. Chow H-F, Low K-H, Wong KY (2005) Synlett 14:2130
83. de la Escosura A, Martínez-Díaz MV, Thordarson P, Rowan EA, Nolte RJM, Torres T (2003) J Am Chem Soc 125:12300
84. Tanaka T, Ozawa T, Inagaki A, Akita M (2007) Dalton Trans 928
85. Morisaki Y, Chujo Y (2002) Polymer Bull 49:209
86. Leung M-K, Balaji Viswanath M, Chou P-Y, Pu S-C, Lin H-C, Jin B-Y (2005) J Org Chem 70:3560
87. Jagtap S, Collard DM (2012) Polym Chem 3:463
88. Hopf H, Raulfs F-W (1986) Tetrahedron 42:1655
89. Dix I, Hopf H, Satyanarayana TBN, Ernst L (2010) Beilstein J Org Chem 6:932
90. Hopf H, Raulfs F-W (1985) Isr J Chem 25:210
91. Hopf H, Kleinschroth J, Böhm I (1981) Org Synth 60:41
92. Hopf H, Raev V, Jones PG, Beilstein J (2011) Org Chem 7:658
93. Bier AK, Bognitzki M, Schmidt A, Greiner A (2012) Macromolecules 45:633
94. Bondarenko L, Dix I, Hinrichs H, Hopf H (2004) Synthesis 16:2751
95. Bondarenko L, Hentschel S, Greiving H, Grunenberg J, Hopf H, Dix I, Jones PG, Ernst L (2007) Chem Eur J 13:3950
96. Boydston AJ, Bondarenko L, Dix I, Weakley TJR, Hopf H, Haley MM (2001) Angew Chem Int Ed 40:2986
97. Bartholemew GP, Bazan GC (2001) Acc Chem Res 34:30
98. Bazan GC (2007) J Org Chem 72:8615
99. Tretiak S, Saxena A, Martin RL, Bishop AR (2000) J Phys Chem B 104:7029
100. Gierschner J, Oelkrug D (2004) In: Nalwa HS (ed) Encyclopedia of nanoscience and nanotechnology, vol 8. American Scientific Publishers, Stevenson Ranch, pp 219–238
101. Oldham WJ, Miao Y-J, Lachicotte RJ, Bazan GC (1998) J Am Chem Soc 120:419
102. Bazan GC, Oldham WJ, Lachicotte RJ, Tretiak S, Chernyak V, Mukamel S (1998) J Am Chem Soc 120:9188
103. Taticchi A, Marrocchi A, Minuti L, Landi S, Gacs-Baitz E (2006) Heterocycles 68:1249
104. Minuti L, Taticchi A, Marrocchi A, Landi S, Gacs-Baitz E (2005) Tetrahedron Lett 46:5735
105. Wang S, Bazan GC, Treriak S, Mukamel S (2000) J Am Chem Soc 122:1289
106. Shabtai E, Rabinovitz M, König B, Knieriem B, de Meijere A (1996) J Chem Soc Perkin Trans 2:2589
107. Bartholemew GP, Rumi M, Pond SJK, Perry JW, Tretiak S, Bazan GC (2004) J Am Chem Soc 126:11529
108. Moran AM, Bartholomew GP, Bazan GC, Kelley AM (2002) J Phys Chem A 106:4928
109. Bartholomew GP, Bazan GC (2002) J Am Chem Soc 124:5183

110. Hong JW, Gaylord BS, Bazan GC (2002) J Am Chem Soc 124:11868
111. Hong JW, Woo HY, Liu B, Bazan GC (2005) J Am Chem Soc 127:7435
112. Woo HY, Hong JW, Liu B, Mikhailovsky A, Korystov D, Bazan GC (2005) J Am Chem Soc 127:820
113. Woo HY, Liu B, Kohler B, Korystov D, Mikhailovsky A, Bazan GC (2005) J Am Chem Soc 127:14721
114. Morisaki Y, Ueno S, Saeki A, Asano A, Seki S, Chujo Y (2012) Chem Eur J 2:4216
115. Morisaki Y, Chujo Y (2005) Bull Chem Soc Jpn 78:288
116. Amthor S, Lambert C (2006) J Phys Chem 110:1177
117. Amthor S, Lambert C (2006) J Phys Chem 110:3495
118. Amthor S, Lambert C (2006) J Phys Chem 110:5204
119. Wolfrum S, Pinzon JR, Molina-Ontoria A, Gouloumis A, Martin N, Echegoyen L, Guldi DM (2011) Chem Commun 2270
120. Molina-Ontoria A, Wielopolski M, Gebhardt J, Gouloumis A, Clark T, Guldi DM, Martín N (2011) J Am Chem Soc 113:2370
121. Mücke P, Winter RF, Novak I, Kowalski K (2012) J Organomet Chem 717:14
122. Ball PJ, Rarog Shtoyko T, Krause Bauer JA, Oldham WJ, Connick WB (2004) Inorg Chem 43:622
123. Boydston AJ, Bondarenko L, Dix I, Weakley TJR, Hopf H, Haley MM (2001) Angew Chem Int Ed 44:2986
124. Hinrichs H, Boydston AJ, Jones PG, Hess K, Herges R, Haley MM, Hopf H (2006) Chem Eur J 12:7103
125. Mukhopadhyay S, Jagtap SP, Coropceanu V, Brédas J-L, Collard DM (2012) Angew Chem 51:11629
126. Greiving H, Hopf H, Jones PG, Bubenitschek P, Desvergne JP, Bouas-Laurent H (1994) Chem Commun 1075
127. Morisaki Y, Wada N, Arita M, Chujo Y (2009) Polym Bull 62:305
128. Kaikawa T, Takimiya K, Aso Y, Otsubo T (2000) Org Lett 2:4197
129. Sakai T, Satou T, Kaikawa T, Takimiya K, Otsubo T, Aso Y (2005) J Am Chem Soc 127:8082
130. Satou T, Sakai T, Kaikawa T, Takimiya K, Otsubo T, Aso Y (2004) Org Lett 6:997
131. Guyard L, Dinh An MN, Audebert P (2001) Adv Mater 13:133
132. Salhi F, Lee B, Metz C, Bottomley LA, Collard DM (2002) Org Lett 4:3195
133. Salhi F, Collard DM (2003) Adv Mater 15:81
134. Guyard L, Dumas C, Miomandre F, Pansu R, Renault-Méallet R, Audebert P (2003) New J Chem 27:1000
135. Tseng J-C, Huang S-L, Lin CL, Lin H-C, Jin B-Y, Chen C-Y, Yu J-K, Chou P-T, Luh T-Y (2003) Org Lett 5:4381
136. Lin H-C, Lin W-Y, Bai H-T, Chen C-H, Jin B-Y, Luh T-Y (2007) Angew Chem Int Ed 119:915
137. Luh T-Y, Lin H-C, Chou C-M (2008) Pure Appl Chem 80:475
138. Lin H-C, Hsu J-H, Lee C-K, Tai OY-H, Wang C-H, Chou C-M, Chen K-Y, Wu Y-L, Luh T-Y (2009) Chem Eur J 15:13201
139. Colpa JP, Hausser KH, Schweitzer D (1978) Chem Phys 29:187
140. Anger I, Sandros K, Sundahl M, Wennerström O (1993) J Phys Chem 97:1920
141. Gierschner J, Mack H-G, Oelkrug D, Waldnerm I, Rau H (2004) J Phys Chem A 108:257
142. Schweitzer D, Hausser KH, Haenel M (1978) Chem Phys 29:181
143. Haenel M, Staab HA (1973) Chem Ber 106:2190
144. Becker B, Bohnen A, Ehrenfreund M, Wohlfarth W, Sakata Y, Huber W, Müllen K (1991) J Am Chem Soc 113:1121
145. König B, Knieriem B, Rauch K, de Meijere A (1993) Chem Ber 126:2531
146. Mataka S, Takahashi K, Mimura T, Hirota T, Takuma K, Kobayashi H, Tashiro M (1985) Chem Commun 973
147. Mataka S, Takahashi K, Mimura T, Hirota T, Takuma K, Kobayashi H, Tashiro M, Imada K, Kuniyoshi M (1987) J Org Chem 52:2653

148. Mataka S, Takahashi K, Hirota T, Takuma K, Kobayashi H, Tashiro M (1986) J Org Chem 51:4618
149. Knoblock KM, Silvestri CJ, Collard DM (2006) J Am Chem Soc 128:13680
150. Jagtap S, Mukhopadhyay S, Coropceanu V, Brizius GL, Brédas J-L, Collard DM (2012) J Am Chem Soc 134:7176
151. Armitt DJ, Crisp GT (2006) J Org Chem 71:3417
152. Armitt DJ, Crisp GT (2006) Tetrahedron 62:1485
153. Crisp GT, Turner PD (2000) Tetrahedron 56:8335
154. House HO, Koepsell DG, Campbell WJ (1972) J Org Chem 37:1003
155. Carson B (2010) PhD thesis, Georgia Institute of Technology
156. Takita R, Song C, Swager TM (2008) Org Lett 10:5003
157. Giaimo JM, Lockard JV, Sinks LE, Scott AM, Wilson TM, Wasielewski MR (2008) J Phys Chem A 112:2322
158. Morisaki Y, Fernandes JA, Chujo Y (2010) Macromol Chem Phys 211:2407
159. Morisaki Y, Fernandes JA, Chujo Y (2009) Macromol Rapid Commun 30:2107
160. Fernandes JA, Morisaki Y, Chujo Y (2010) Polym Bull 65:465
161. Morisaki Y, Fernandes JA, Wada N, Chujo Y (2009) J Polym Sci A Polym Chem 47:4279
162. Morisaki Y, Imoto H, Miyake J, Chujo Y (2009) Macromol Rapid Commun 30:1094
163. Morisaki Y, Murakami T, Chujo Y (2009) J Inorg Organomet Polym Mater 19:104
164. Morisaki Y, Sawamura T, Murakami T, Chujo Y (2010) Organic Lett 12:3188
165. Song C, Swager TM (2008) Org Lett 10:3575
166. Hippius C, Stokkum IHM, Zangrando E, Williams RM, Wykes M, Beljonne D, Würthner F (2008) J Phys Chem C 112:14626
167. Mangalum A, Morgan BP, Hanley JM, Jecen KM, McGill CJ, Robertson GA, Smith RC (2010) Chem Commun 46:5136
168. Smith RC (2009) Macromol Rapid Commun 30:206
169. McNeil AJ, Müller P, Whitten JE, Swager TM (2006) J Am Chem Soc 128:12426
170. Moore JS (1997) Acc Chem Res 30:402
171. Becker K, Fritzche M, Höger S, Lupton JM (2008) J Phys Chem B 112:4849
172. Watanabe M, Goto K, Fujitsuka M, Tojo S, Majima T, Shinmyozu T (2010) Bull Chem Soc Jpn 83:1155
173. Kleinschroth J, Hopf H (1978) Tetrahedron Lett 11:969
174. Cram DJ, Dalton CK, Knox GR (1963) J Am Chem Soc 85:1088
175. Mourad AFE, Nour-el-Din AM, Mekhamer RA (1986) J Chem Engin Data 31:367
176. Ohya-Nishiguchi H, Terahara A, Hirota N, Sakata Y, Misumi S (1982) Bull Chem Soc Jpn 55:1782
177. Mataka S, Shigaki K, Sawada T, Mitoma Y, Taniguchi M, Thiemann T, Ohga K, Egashira N (1998) Angew Chem Int Ed 37:2532
178. Nakano T, Yade T (2008) Chem Lett 37:258
179. Nakano T, Yade T (2003) J Am Chem Soc 125:15474
180. Seferos DS, Trammell SA, Bazan GC, Kushmerick JG (2005) Proc Natl Acad Sci U S A 102:8821
181. Seferos DS, Szuchmacher Blum A, Kushmerick JG, Bazan GC (2006) J Am Chem Soc 128:11260
182. Seferos DS, Banach DA, Alcantar NA, Israelachvili JN, Bazan GC (2004) J Org Chem 69:1110
183. Valentini L, Mengoni F, Meloni F, Marrocchi A, Seri M, Taticchi A, Kenny JM (2008) Thin Solid Films 516:7193
184. Valentini L, Mengoni F, Taticchi A, Marrocchi A, Kenny JM (2006) J Mater Chem 16:1622
185. Valentini L, Mengoni F, Kenny JM, Marrocchi A, Taticchi A (2007) Small 3:1200

π-Stacking on Density Functional Theory: A Review

Takao Tsuneda and Tetsuya Taketsugu

Introduction

Van der Waals Interactions in DFT

In line with increasing use of density functional theory (DFT) in quantum chemistry, it is presently employed in more than 80 % of van der Waals calculations. Since most van der Waals calculations target at large-scale systems such as biomolecules and nanomaterials, it is natural to use DFT having features of both high speed and high accuracy. Nevertheless, it has been reported that DFT provides poor van der Waals bonds for many years [1]. For example, until recently, no exchange-correlation functional gives meaningful potential energy curves for the van der Waals bonds of rare gas dimers in Kohn–Sham calculations [1]. The main cause for the poor DFT results of van der Waals bonds is the neglect of van der Waals interactions in conventional exchange-correlation functionals [2].

By definition, van der Waals interaction is the collective term of dipole–dipole, dipole–induced dipole, and dispersion interactions [3]. The dipole–dipole interaction is the electrostatic interaction between permanent dipoles in polar systems. For the interactions between systems *A* and *B*, the corresponding potential is given classically as

$$V_{\mu-\mu}(r) = -\frac{\mu_A \mu_B}{R_{AB}^{\ 3}} \tag{1}$$

T. Tsuneda (✉)
Fuel Cell Nanomaterials Center, University of Yamanashi, Kofu 400-0021, Japan
e-mail: ttsuneda@yamanashi.ac.jp

T. Taketsugu
Department of Chemistry, Faculty of Science, Hokkaido University,
Sapporo 060-0810, Japan

T. Nakano (ed.), *π-Stacked Polymers and Molecules: Theory, Synthesis, and Properties*,
DOI 10.1007/978-4-431-54129-5_5, © Springer Japan 2014

where μ_X is the permanent dipole of system X and R_{AB} is the distance between systems A and B. Note that atomic units ($\hbar = e = m_e = 1/(4\pi\,\varepsilon_0) = 1$, energies are in hartree, and distances are in bohr) are used in all equations of this review. In the Kohn–Sham equation [4], which is usually the main equation of DFT, this interaction is contained as a part of Coulomb interactions. The dipole–induced dipole interaction is the interaction between polar and nonpolar systems. Assuming the permanent dipole moment of polar system A as μ_A and the polarizability, the linear response for electric field producing induced dipole moment, of nonpolar system B as α_B, the classical potential of this interaction is given as [3]

$$V_{\mu-\alpha}(r) = -\frac{\mu_A^2 \alpha_B}{R_{AB}^6} \tag{2}$$

Since this interaction is very weak, it causes the low solubility of polar molecules in nonpolar solvents. This interaction is also incorporated in Kohn–Sham SCF calculations. The dispersion interaction is a universal interaction, which acts even between the bodies of neither charge nor multipole moment. As a classical expression, the potential function between two heterogeneous bodies, which London developed using perturbation theory, is used [5],

$$V_{disp}^{London}(r) = -\frac{3}{2}\frac{\alpha_A \alpha_B}{R_{AB}^6}\frac{I_A I_B}{I_A + I_B} \tag{3}$$

where I_X is the ionization potential of partial system X. This dispersion interaction is interpreted as the interaction between an instantaneous dipole moment by the fluctuation of electron distribution and an induced dipole moment by the electric field formed of the instantaneous dipole moment. That is, two spatially separated electron distributions fluctuate around these equilibrium distributions by electron correlation to produce interactions between two bodies. Therefore, dispersion interaction is a pure electron correlation between two bodies, which cannot be incorporated by one-body mean-field approximation, and it is long-range correlation explicitly acting between distant electrons. This dispersion interaction is the only van der Waals interaction that is not incorporated in Kohn–Sham calculations using conventional correlation functionals.

Although dispersion interaction should be included in correlation functionals, it has usually been not taken into consideration in conventional correlation functionals. Conventional generalized-gradient-approximation (GGA) correlation functionals are classified into two types: density gradient expansion-type and Colle–Salvetti-type correlation functionals. In density gradient expansion-type correlation functionals, local density approximation (LDA) correlation functionals are corrected using the functionals of density gradient. For example, Perdew–Wang 1991 (PW91) [6] and Perder–Burke–Ernzerhof (PBE) [7] correlation functionals are involved in these functionals. However, since electron density and its gradient are essentially one-electron functions, these functionals cannot give dispersion interactions, which are explicit long-range electron–electron interactions as shown in Eq. (3).

Colle–Salvetti-type correlation functionals [8] are derived from correlation wavefunctions, in which uncorrelated wavefunctions are multiplied by the factor of correlation holes. Lee–Yang–Parr (LYP) [9] and one-parameter progressive (OP) correlation functionals [10] are included in these functionals. Since these functionals contain only short-range correlation resulting from correlation holes, long-range correlations including dispersion interactions are neglected in these functionals. Naturally, Kohn–Sham calculations using these GGA correlation functionals have almost always failed with no dispersion interaction to give van der Waals bonds even qualitatively. It is, therefore, reasonable to consider that dispersion interaction should be explicitly supplemented in conventional correlation functionals.

π-Stacking Interactions

π-Stacking interaction is one of the dispersion interactions and often significantly contributes to the structures of large-scale systems. The name "π-stacking" comes from the structures of π-stacked systems, in which π orbitals seem to overlap. However, it is theoretically revealed that π orbitals do not play a special role in π-stacking interactions [11], as implicated by experimental phase-change studies. That is, π-stacking interactions are nothing more than usual dispersion interactions. π-Stacking interactions have also been interpreted by various models: e.g. solvophobic (entropic interaction), donor–acceptor (charge transfer interaction), and atomic charge (electrostatic interaction between uneven charge distributions) models [12]. However, these classical models have been disconfirmed both experimentally and theoretically [12]. Hunter suggested a model, in which the repulsive electrostatic interactions of quadrupole moments are balanced with attractive dispersion interactions [13]. However, as clearly shown in section "Long-Range Exchange Interactions", long-range exchange interactions are actual repulsions against the attractions of long-range dispersion correlations in π-stacking bonds [14]. In most cases, π-stacking interactions indicate the noncovalent interactions between aromatic rings. π-Stacking interactions have three types of conformations: parallel, T-shaped, and parallel-displaced configurations (Fig. 1). As anticipated from the above discussion, similar potential energy curves are given for the bond dissociations of parallel and T-shaped configurations, although parallel configurations tend to give shorter bond distances with smaller binding energies than those of T-shaped configurations [14]. Therefore, most theoretical studies have assumed that π-stacking interactions are accurately reproduced by appropriate methods giving correct dispersion interactions.

So far, it has been experimentally reported that π-stacking interactions play important roles in various large systems: e.g., carbon nanomaterials (stackings on fullerene, graphene, and nanotube available in hydrogen storage materials [15], drug-delivery vehicles [16] and biochemical sensors [17, 18]), biomolecules (DNA stackings [19, 20] and protein foldings [21, 22]), and supramolecules (molecular recognitions [23, 24], crystal packings [25, 26], and host–guest interactions [27, 28]). Therefore, past

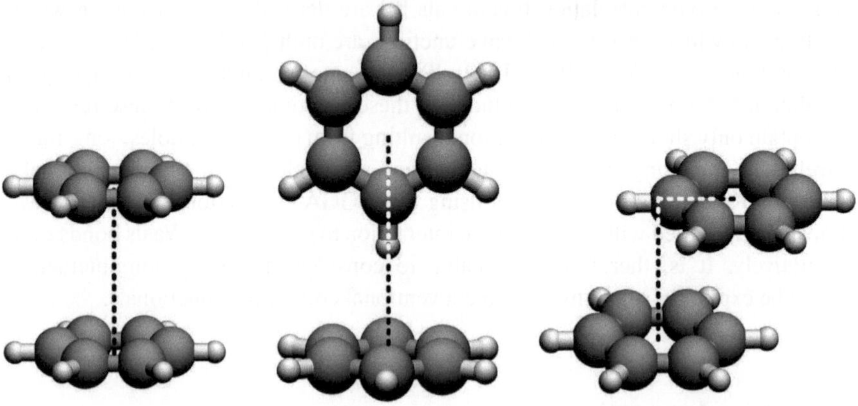

Fig. 1 Three conformation types of benzene dimer

theoretical investigations have also mainly approached the π-stacking natures of these large systems. Carbon nanomaterials having polycyclic structures use π-stackings to solvate in organic sovents. In DNA, π-stacking interactions form stacking bonds between the bases to build up the double-helical structure [29]. Protein foldings are also critically affected by the π-stacking of aromatic rings in proteins. For supramolecules, π-stackings are usually used to self-assemble π-conjugated molecules [30]. On the other hand, π-stacking interactions hardly contribute to chemical reactions. For the substitution reaction of benzene dimer shown in Fig. 2, Wheeler and Houk suggested that even for the reaction of π-stacked benzene derivatives, the substituent group (X) directly interacts with the aromatic ring to proceed the reaction with no contribution of the π-stacking [31]. This is analogous to Diels–Alder reactions, in which dispersion interactions between π orbitals hardly affect the reaction enthalpies [32]. However, it is also found that dispersion interactions clearly decrease reaction barriers of these Diels–Alder reactions [32]. We, therefore, suppose that π-stackings may contribute to reactivities in reactions of π-stacked systems, although they may not directly affect reaction mechanisms.

As described above, π-stackings essentially consist of dispersion interactions. Dispersion effects should, therefore, be explicitly taken into consideration to calculate π-stackings. However, many conventional DFT calculations have used exchange-correlation functionals containing no explicit dispersion interaction such as pure LDA and GGA functionals and consequently have provided poor results in π-stacking calculations. To open the door to quantitative discussions of van der Waals bonds including π-stackings, a wide variety of dispersion corrections have been developed in the field of DFT. In this review, we will briefly explain major dispersion corrections for DFT exchange-correlation functionals on their basic concepts in section "Dispersion Corrections in DFT". We will then list past DFT

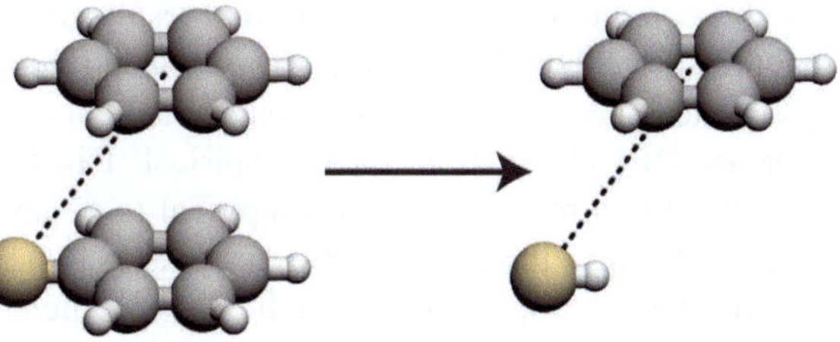

Fig. 2 The displacement reaction of benzene dimer

applications to π-stacking systems with mentioning their limitations in section "DFT Calculations of π-Stacking Systems".

Dispersion Corrections in DFT

So far, various types of dispersion corrections have been suggested. These dispersion corrections are generally classified to five types: classical dispersion corrections, combinations with perturbation theories, linear-response theories, van der Waals (dispersion) functionals, and semiempirical dispersion-corrected functionals. Besides these dispersion corrections, long-range exchange interactions and correlation functionals are also significant in calculating van der Waals bonds. In this section, these dispersion corrections and the effects of long-range exchange interactions and correlation functionals are briefly reviewed.

Classical Dispersion Corrections

The simplest dispersion correction may be the empirical correction for the Kohn–Sham energy using the London's classical interatomic dispersion energy,

$$E_{disp}^{London} = -\sum_{A>B} \frac{C_6^{AB}}{R_{AB}^6} f_{damp}(R_{AB}) \quad (4)$$

where A and B are usually the labels of atoms and C_6^{AB} is the interatomic dispersion coefficient parameterized, and f_{damp} is a damping function for cutting off unnecessary short-range interactions. The feature of this correction method is much less computational time required in dispersion calculations than those of other dispersion corrections. Since this classical correction also accurately reproduces experimental results with well-calibrated dispersion coefficients, it is the most widely used dispersion correction in classical molecular dynamics (MD) simulations. It is, however, difficult to apply this correction to the calculations of new uninformed systems due to the empirically parameterized dispersion coefficient and it has been reported that this correction gives much different results dependent on exchange-correlation functionals combined. Therefore, the right-hand side of Eq. (4) is usually multiplied by adjusted parameters dependent on exchange-correlation functional combined as seen in DFT-D (section "Semiempirical Dispersion-Corrected Functionals"). However, these methods are too empirical to be appropriate for a wide variety of systems. To solve this problem, Becke developed the exchange-hole dipole moment (XDM) method [33, 34], in which the C_6^{AB} coefficient in Eq. (4) is calculated by

$$C_6^{AB} = \frac{\langle d_x^2 \rangle_A \langle d_x^2 \rangle_B \alpha_A \alpha_B}{\langle d_x^2 \rangle_A \alpha_B + \langle d_x^2 \rangle_B \alpha_A} \tag{5}$$

$$\langle d_x^2 \rangle_A = \sum_\sigma \int d^3 r \rho_\sigma(r) d_{x\sigma}^2(r) \tag{6}$$

and

$$d_{x\sigma}(r) = \left\{ \frac{1}{\rho_\sigma(r)} \sum_{ij} \left[\int d^3 r' r' \phi_{i\sigma}(r') \phi_{j\sigma}(r') \right] \phi_{i\sigma}(r) \phi_{j\sigma}(r) \right\} - r \tag{7}$$

Although this method is also empirical, it has a physical meaning at least in the dispersion coefficient and gives more accurate dispersion interactions than those of the London's classical dispersion energy.

Combination with Perturbation Theories

Perturbation theories such as the second-order Møller–Plesset perturbation (MP2) method [35] have been appreciated as *ab initio* wavefunction theories reproducing dispersion interactions with relatively short computational time. Therefore, dispersion interactions can be incorporated in the Kohn–Sham method by combining with the perturbation theories in principle. One of the methods based on this concept is DFT symmetry-adapted perturbation theory (DFT-SAPT), which uses Kohn–Sham orbitals to calculate the perturbation energies [36]. In contrast to *ab initio* SAPT calculating both inter- and intra-molecular electron correlations, only intermolecular electron correlations are calculated as a dispersion correction for the Kohn–Sham method in DFT-SAPT. Consequently, this drastically reduces the computational cost to tenths or hundredths of an *ab initio* SAPT one with similar accuracies. Although DFT-SAPT is a promising dispersion calculation method for clearly separated systems, it cannot reproduce intramolecular dispersion interactions. Moreover, despite the drastically reduced computational time, DFT-SAPT calculations need much more computational time than those of Kohn–Sham calculations even for the lowest-order DFT-SAPT2. Therefore, these are consequently applicable to systems containing only up to several dozens of atoms with current commercial computer performance. As another method combining DFT with perturbation theories, there are double-hybrid functionals, which mix perturbation energies in correlation functionals at a constant rate [37]. That is, these functionals extend hybrid functionals by mixing correlation functionals with MP2 perturbation energies such as

$$E_{xc} = (1-a_x)E_x + a_x E_{HF} + (1-a_c)E_c + a_c E_{MP2} \tag{8}$$

where E_{MP2} is MP2 electron correlation energy. B2PLYP functional [38] is one of the double-hybrid functionals. Although this method gives middle-range electron correlations, it is not appropriate for dispersion calculations because of the incomplete dispersion interactions incorporated.

Linear-Response Theories

By using linear-response theories, dispersion interactions can be calculated directly in the framework of the Kohn–Sham method. Adiabatic connection/fluctuation-dissipation theorem (AC/FDT) method is a linear-response theory for calculating dispersion interactions based on the Kohn–Sham method [39]. In this AC/FDT method, electron correlation is calculated as the energy response quantity for the spontaneous fluctuations of electronic motions coming from the perturbation of interelectronic interactions as

$$E_c^{AC/FDT} = -\int_0^1 d\lambda \iint d^3\mathbf{r} d^3\mathbf{r}' \frac{1}{|\mathbf{r}-\mathbf{r}'|} \times \left[\frac{1}{2\pi} \int_0^\infty du \left\{ \chi_\lambda(\mathbf{r},\mathbf{r}',iu) - \chi_0(\mathbf{r},\mathbf{r}',iu) \right\} \right], \tag{9}$$

where $\mathbf{r}$ and $\mathbf{r}'$ are the position vectors of electrons. In this equation, χ_λ and χ_0 are density response functions for interacting and independent electrons, respectively, and these are obtained by solving the Dyson equation as

$$\begin{aligned} \chi_\lambda(\mathbf{r},\mathbf{r}',\omega) &= \chi_0(\mathbf{r},\mathbf{r}',\omega) + \iint d^3\mathbf{r}_1 d^3\mathbf{r}_2 \chi_0(\mathbf{r},\mathbf{r}_1,\omega) \\ &\quad \times \left\{ \frac{\lambda}{|\mathbf{r}_1-\mathbf{r}_2|} + f_{xc}^{\lambda}(\mathbf{r}_1,\mathbf{r}_2,\omega) \right\} \chi_\lambda(\mathbf{r}_2,\mathbf{r}',\omega), \end{aligned} \tag{10}$$

where f_{xc}^{λ} is the exchange-correlation integral kernel for interacting systems. Since the electron correlation in Eq. (9) contains dispersion interactions as the long-range correlation, the long-range part of this correlation energy is often used as a dispersion correction, which is called "RPAx" [40]. Analogously to the time-dependent response Kohn–Sham (TDKS) method, the correlation energy is calculated by solving the TDKS matrix equation. Although this AC/FDT method is obviously the most superior dispersion correction from a physical point of view, this method requires an enormous amount of computational time, which is more than thousand times the time needed in Kohn–Sham calculations, unless drastic approximations would be adopted.

Van der Waals (Dispersion) Functionals

Van der Waals (dispersion) functionals have been developed to reduce the enormous computational time required in the AC/FDT method with keeping the accuracies and to be easy-to-use like the London's classical potential. Lundqvist and coworkers proposed a dispersion functional called Andersson–Langreth–Lundqvist (ALL) functional by using a local density approximation for the electron density response function of the AC/FDT method [41],

$$E_{\rm disp}^{\rm ALL}\left[\rho\right] = -\frac{6}{4\pi^{3/2}}\int_{V_1} d^3\mathbf{r}_1 \int_{V_2} d^3\mathbf{r}_2 \frac{\sqrt{\rho(\mathbf{r}_1)}\sqrt{\rho(\mathbf{r}_1)}}{\sqrt{\rho(\mathbf{r}_1)}+\sqrt{\rho(\mathbf{r}_1)}}\frac{1}{r_{12}^6}. \tag{11}$$

As almost the same functional, Dobson and coworkers independently suggested a dispersion functional based on local density approximation in the same year [42]. Note that explicit numerical two-electron integrals, which usually take long computational time to be calculated, are included in the formulations of these functionals. However, in actual calculations, the computational time is usually less than that of the Kohn–Sham calculation, because spatial regions of small momentum variations and core regions can be neglected in the integral calculations. On the other hand, these functionals are applicable only to two-body systems having well-separated electron distributions and need damping functions $f_{\rm damp}$ shown in Eq. (4) for short-range electron–electron distances. To solve this problem, many researchers have attempted to develop dispersion functionals applicable to the regions of overlapped electron distributions. Lundqvist and coworkers also proposed a dispersion functional available for such electronic regions [43]. This functional has a complicated form using $\phi(\mathbf{r}_1, \mathbf{r}_2)$ function containing the spatial coordinates of two electrons and the electron density and its gradient at these coordinates,

$$E_{\rm disp}^{\rm DRSLL}\left[\rho\right] = \int d^3\mathbf{r}_1 \int d^3\mathbf{r}_2 \rho(\mathbf{r}_1)\phi(\mathbf{r}_1,\mathbf{r}_2)\rho(\mathbf{r}_2), \tag{12}$$

$$\phi(\mathbf{r}_1,\mathbf{r}_2) = \frac{2}{\pi^2}\int_0^\infty daa^2 \int dbb^2 W(a,b) T\left(v_1(a), v_1(b), v_2(a), v_2(b)\right), \tag{13}$$

$$T(w,x,y,z) = \frac{1}{2}\left[\frac{1}{w+x}+\frac{1}{y+x}\right]\left[\frac{1}{(w+y)(x+z)}+\frac{1}{(w+z)(y+x)}\right], \tag{14}$$

and

$$\begin{aligned} W(a,b) = \frac{2}{a^3b^2}\Big[&(3-a^2)b\cos b\sin a + (3-b^2)a\cos a\sin b \\ &+(a^2+b^2-3)\sin a\sin b - 3ab\cos a\cos b\Big], \end{aligned} \tag{15}$$

where

$$v_i(y) = y^2/2\left(1-\exp(-4\pi y^2/9d_i^2)\right), \tag{16}$$

and

$$d_i = r_{12}q_0(\mathbf{r}_i). \tag{17}$$

In Eq. (17), q_0 is given using the Fermi momentum, $k_F=(3\pi^2\rho)^{1/3}$, as

$$q_0(\mathbf{r}) = k_F(\mathbf{r})\left[1+0.09434\left(\frac{\rho(\mathbf{r})}{2k_F(\mathbf{r})\rho(\mathbf{r})}\right)^2\right]. \tag{18}$$

This functional requires no damping function, because it naturally approaches zero for short electron–electron distance. Therefore, this functional can reproduce intramolecular dispersion interactions. This dispersion functional is used in, e.g., vdW-DF dispersion correction [43] combining with revPBE or another GGA functional. As other interesting dispersion functionals, Vydrov–van Voorhis 2009 (VV09) functional [44], which uses the dielectric model smoothing the real-space cutoff, and local response dispersion (LRD) functional [45], which combines the local response approximation functional of Dobson and coworkers [42] with the real-space cutoff [44], are also suggested. Combining with long-range corrected DFT, this LRD functional has succeeded to reproduce various kinds of reaction energy diagrams and photochemistries, which have been poorly given by conventional DFTs, as mentioned in section "Long-Range Exchange Interactions".

Semiempirical Dispersion-Corrected Functionals

Semiempirical dispersion-corrected functionals intend to reproduce accurate properties including van der Waals interaction energies by using dispersion terms with many adjusted semiempirical parameters like CHARMM [46] and Amber [47] molecular force fields in classical MDs of biological systems. DFT-D functionals such as BLYP-D, B3LYP-D, and B97-D functionals [48] and Mx-series functionals such as M05-2x and M06-2x functionals [49] are included in these semiempirical dispersion-corrected functionals. In DFT-D functionals, there are three versions, DFT-D1, -D2, and -D3, based on the level of dispersion corrections. For deep understanding of dispersion corrections, it is interesting to show the highest-level DFT-D3 functional [50],

$$E^{\text{DFT-D3}} = E^{\text{KS}} - E_{\text{disp}}^{(2)} - E_{\text{disp}}^{(3)} \tag{19}$$

$$E_{\text{disp}}^{(2)} = -\sum_{A>B}\sum_{n=6,8,10,\cdots} s_n \frac{C_n^{\text{AB}}}{R_{\text{AB}}{}^n} f_{\text{damp}}^n(R_{\text{AB}}), \tag{20}$$

$$E_{\mathrm{disp}}^{(3)} = -\sum_{\mathrm{A>B>C}} \frac{C_9^{\mathrm{ABC}}\left(3\cos\theta_a\cos\theta_b\cos\theta_c + 1\right)}{\left(R_{\mathrm{AB}}R_{\mathrm{BC}}R_{\mathrm{CA}}\right)^3} f_{\mathrm{damp}}^{(3)}(\bar{R}_{\mathrm{ABC}}), \tag{21}$$

where A, B, and C are atomic labels, θ_a, θ_b, and θ_c are the internal angles of ABC triangle, and $\bar{R}_{\mathrm{ABC}}$ is geometrically averaged radii. The damping functions are given by

$$f_{\mathrm{damp}}^{n}(R_{\mathrm{AB}}) = \frac{1}{1 + 6\left(R_{\mathrm{AB}} / (s_{r,n} R_0^{\mathrm{AB}})\right)^{-\alpha_n}}, \tag{22}$$

and

$$f_{\mathrm{damp}}^{(3)}(\bar{R}_{\mathrm{ABC}}) = \frac{1}{1 + 6\left(\bar{R}_{\mathrm{ABC}} / (4R_0^{\mathrm{ABC}}/3)\right)^{-16}}, \tag{23}$$

All the remaining parameters are semiempirical: R_0^{AB} and R_0^{ABC} are cutoff radii adjusted to each atomic pairwise and three-body. Coefficients s_n ($n = 8, 10, \cdots$) are fitted in benchmark calculations depending on functionals combined, while s_6 is one or an adjusted value less than 1. For dispersion coefficients, C_n^{AB} and C_9^{ABC}, time-dependent DFT (TDDFT) and recursion relations are used to determine the values for each atomic pairwise and three-body. The lowest-order C_6^{AB} is expressed in the Casimer–Polder formula [51],

$$C_6^{\mathrm{AB}} = \frac{3}{\pi}\int_0^{\infty} d\omega\, \alpha^{\mathrm{A}}(i\omega)\alpha^{\mathrm{B}}(i\omega), \tag{24}$$

where $\alpha^{\mathrm{A}}(i\,\omega)$ is the averaged dipole polarizability of atom A at imaginary frequency. Other C_n^{AB} coefficients are derived by the recursion relations [52] as

$$C_8^{\mathrm{AB}} = 3C_6^{\mathrm{AB}}\sqrt{Q_{\mathrm{A}}Q_{\mathrm{B}}}, \tag{25}$$

$$C_{10}^{\mathrm{AB}} = \frac{49}{40}\frac{(C_8^{\mathrm{AB}})^2}{C_6^{\mathrm{AB}}}, \tag{26}$$

and

$$C_{n+4}^{\mathrm{AB}} = C_{n-2}^{\mathrm{AB}}, \left(\frac{C_{n+2}^{\mathrm{AB}}}{C_n^{\mathrm{AB}}}\right)^3 \tag{27}$$

with

$$Q_{\mathrm{A}} = \sqrt{Z_{\mathrm{A}}}\frac{\langle r^4\rangle_{\mathrm{A}}}{\langle r^2\rangle_{\mathrm{A}}}, \tag{28}$$

where Z_A is the nuclear charge of atom A and $\langle r^4 \rangle_A$ and $\langle r^2 \rangle_A$ are the expectation values derived from the electron density of atom A. The remaining triple–dipole C_9^{ABC} coefficient is given by

$$C_9^{ABC} = \frac{3}{\pi} \int_0^{\infty} d\omega \alpha^A(i\omega)\alpha^B(i\omega)\alpha^C(i\omega) \approx -\sqrt{C_6^{AB} C_6^{BC} C_6^{CA}}. \qquad (29)$$

These coefficients are determined by TDDFT calculation for each atomic pairwise and three-body. Since these values are fixed after one TDDFT calculation, these dispersion calculations are not the bottleneck of DFT-D3 calculations. For DFT-D3 functionals, BLYP-D3 and B2PLYP-D3 functionals are suggested [50]. In Mx-series and other semiempirical dispersion-corrected functionals, dispersion interactions are incorporated in the similar manner, although only the C_6^{AB} term is usually contained. This type of dispersion corrections is obviously efficient, because dispersion interactions are easily incorporated with high accuracy only by using functionals. However, it has been reported that the calculated results depend on the parameters used and the R^{-6} decay of dispersion interaction cannot be reproduced.

Long-Range Exchange Interactions

Although only dispersion corrections have so far been presented in this section, we should notice that repulsions balanced with dispersion attractions are equivalently significant in dispersion calculations. As seen in the Lennard–Jones potential (section "Biomolecules: DNA and Proteins"), the repulsions have been interpreted to come from long-range exchange interactions. Nevertheless, long-range exchange interactions have been neglected in conventional exchange functionals similarly to dispersion interactions in conventional correlation functionals. Figure 3a displays the dissociation potential energy curves of Ar dimer, which are calculated using various exchange functionals [1]. It is well known that the bond of Ar dimer consists only of dispersion interactions. This figure clearly indicates that the dissociation potentials of van der Waals bonds are significantly affected by exchange functionals used: LDA, Becke 1988 GGA (B88), and Perdew–Burke–Ernzerhof GGA (PBE) exchange functionals give qualitatively different potential energy curves, despite these functionals have been combined with dispersion corrections mentioned above. In contrast, the figure shows that long-range correction clearly makes these potential energy curves close to each other. Therefore, this figure reveals that the different potential energy curves of GGA functionals mainly come from the lack of long-range exchange interactions.

Tsuneda and coworkers suggested the long-range correction, in which exchange interactions are divided into short- and long-range parts, and then the short-range part of a general exchange functional is combined with the long-range part of the Hartree–Fock (HF) exchange integral [53]. In this correction, the two-electron operator $1/r_{12}$ is divided by the error function as

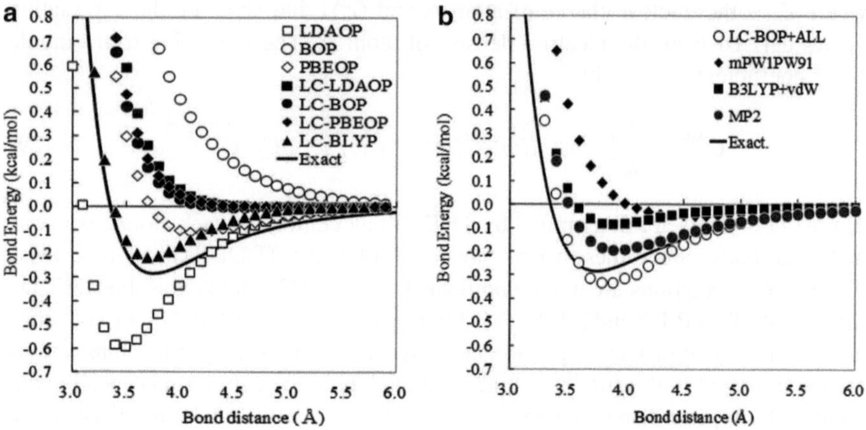

Fig. 3 Dissociation potential energy curves of argon dimer calculated with (**a**) various pure and long-range corrected (LC) functionals and (**b**) LC functionals combined with ALL dispersion functional. The curves of CCSD(T) and other dispersion-corrected DFT are also shown for comparison

$$\frac{1}{r_{12}} = \frac{1 - \mathrm{erf}(\mu r_{12})}{r_{12}} + \frac{\mathrm{erf}(\mu r_{12})}{r_{12}}. \tag{30}$$

Since usual exchange functionals, however, have no corresponding density matrices, the short-range part of exchange functionals are derived assuming that all the difference in exchange functionals is reflected to the momentum k_σ as

$$E_{\mathrm{x}}^{\mathrm{LC(sr)}} = -\frac{1}{2}\sum_\sigma \int d^3\mathbf{r}\rho_\sigma^{4/3} K_\sigma \left\{1 - \frac{8}{3}a_\sigma \left[\sqrt{\pi}\,\mathrm{erf}\left(\frac{1}{2a_\sigma}\right) + 2a_\sigma\left(b_\sigma - c_\sigma\right)\right]\right\}, \tag{31}$$

where a_σ, b_σ, and c_σ are given as

$$a_\sigma = \frac{\mu}{2k_\sigma} = \frac{\mu K_\sigma^{1/2}}{6\sqrt{\pi}\rho_\sigma^{1/3}}, \tag{32}$$

$$b_\sigma = \exp\left(-\frac{1}{4a_\sigma^2}\right) - 1, \tag{33}$$

and

$$c_\sigma = 2a_\sigma^2 b_\sigma + \frac{1}{2}. \tag{34}$$

A dimensionless coefficient K_σ is defined in usual GGA exchange functional form as

$$E_x = -\frac{1}{2}\sum_\sigma \int d^3\mathbf{r}\rho_\sigma^{4/3} K_\sigma. \tag{35}$$

The significant feature of this correction is to use the following momentum k_σ, which becomes the Fermi momentum when combined with the LDA exchange functional,

$$k_\sigma = \left(\frac{9\pi}{K_\sigma}\right)^{1/2} \rho_\sigma^{1/3}. \tag{36}$$

This momentum makes long-range corrected (LC) LDA functional identical to the form that Savin previously proposed as a long-range correction for LDA exchange functional, which is called range-separation hybrid (RSH) functional [54]. For the long-range part of the HF exchange integral, the two-electron operator is simply replaced with the error function such as

$$E_{\mathrm{x}}^{\mathrm{LC(lr)}} = -\frac{1}{2}\sum_\sigma \sum_i^n \sum_j^n \iint d^3\mathbf{r}_1 d^3\mathbf{r}_2 \phi_{i\sigma}^*(\mathbf{r}_1)\phi_{j\sigma}^*(\mathbf{r}_1)\frac{\mathrm{erf}(\mu r_{12})}{r_{12}}\phi_{i\sigma}(\mathbf{r}_2)\phi_{j\sigma}(\mathbf{r}_2). \tag{37}$$

The only parameter μ is determined for each exchange functional: e.g., for B88 [55] and PBE [7] exchange functionals, $\mu = 0.33$ in response property calculations [56] and 0.47 in ground-state property calculations [57] are adopted. The computational time needed in LC-DFT calculations is almost the same as that in B3LYP calculations. Nevertheless, it has been reported that long-range correction solves or improves various problems in conventional DFT calculations: e.g., charge transfers, Rydberg excitations, and oscillator strengths in TDDFT [56], optical response properties of long-chain molecules [58] and diradicals [59], and orbital energies of valence molecular orbitals [60]. Due to its high applicabilities, many types of LC functionals have been developed: e.g., CAM-B3LYP [61], LC-ωPBE [62], and ωB97X [63] functionals. By using this long-range correction, long-range exchange interactions are naturally incorporated in exchange functionals.

Figure 3b illustrates the potential energy curves of LC-DFT combining with a dispersion functional (ALL functional in section "Van der Waals (Dispersion) Functionals"). As shown in the figure, LC functionals give very accurate potential energy curves, which are close to *ab initio* coupled-cluster singles, doubles, and perturbative triples (CCSD(T)) one [64], in contrast to those of a dispersion-corrected semiempirical functional (mPW1PW91) and a classical dispersion-corrected one (B3LYP+vdW). In addition, it was suggested that long-range exchange interactions play a determinant role in van der Waals bond angles [65]. Besides ALL dispersion functional, LC functionals have been combined with various dispersion corrections: e.g., LRD dispersion functional (section "Van der Waals (Dispersion) Functionals") [45], AC/FDT dispersion energy (section "Linear-Response Theories") [40], semiempirical functionals (section "Semiempirical Dispersion-Corrected Functionals") such as ωB97X-D [66], and perturbation energies (section "Combination with Perturbation Theories") such as ωB97X-2 [67]. Actually, LC functionals can be easily combined with any dispersion correction.

Effects of Correlation Functionals

In DFT calculations of dispersion interactions, the choice of correlation functionals is also important. Figure 3a also compares the dissociation potential energy curves of Ar dimer for using one-parameter progressive (OP) [10] and Lee–Yang–Parr (LYP) [9] correlation functionals, which are both Colle–Salvetti-type correlation functionals [8]. The Colle–Salvetti-type correlation functionals are derived from correlated wavefunctions, in which uncorrelated wavefunctions such as the HF wavefunction are multiplied by a correlation factor (Jastrow factor) satisfying the correlation cusp condition for short-range interelectrons. Therefore, these functionals essentially contain only short-range electron correlations and neglect long-range electron correlations including dispersion interactions. Nevertheless, artificial van der Waals bonds are given using the LYP correlation functional with no dispersion interaction. This is attributed to the inappropriate behavior of this functional. To give appropriate weak bonds like van der Waals bonds, correlation functionals have to obey the high-density-gradient low-density (HDGLG) limit condition for correlation energy [68, 69],

$$\lim_{x\to\infty} \rho^{-1}\overline{E}_{c} = 0, \tag{38}$$

where $x = |\nabla\rho|/\rho^{4/3}$. However, the LYP functional does not satisfy this condition differently from the OP functional [1]. The Kohn–Sham method using a correlation functional violating this condition usually overstabilizes van der Waals bonds. Therefore, we should carefully select a correlation functional satisfying the HDGLG limit condition in DFT calculations of dispersion interactions.

DFT Calculations of π-Stacking Systems

In this section, past DFT studies on π-stacking interactions are briefly reviewed. Except for small benchmark molecules, the most frequently raised π-stacked systems are carbon nanomaterials including nanotube and graphene, biomolecules including DNA and proteins, and supramolecules. Summarizing previous dispersion-corrected DFT calculations of these π-stacked systems, we will anticipate the future trend of DFT studies on π-stacking interactions.

Small π-Stacked Systems

The applicabilities of dispersion corrections have so far been examined by calculating small weakly bonded systems including π-stacked systems. In particular, benzene and naphthalene dimers have been used as test targets of dispersion corrections

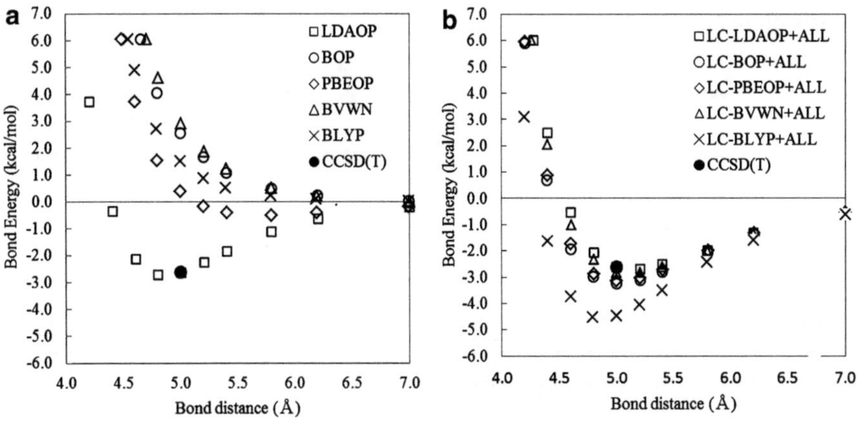

Fig. 4 Dissociation potential energy curves of benzene dimer for the parallel conformation calculated using (**a**) various pure and long-range corrected (LC) functionals and (**b**) LC functionals combined with ALL dispersion functional. The curves of CCSD(T) and other dispersion-corrected DFT are also shown for comparison

for π-stacking interactions [14, 70, 71], because these dimers are the simplest systems for discussing three conformations shown in Fig. 1. For clarifying the reproducibilities of π-stacking interactions, it is one of the best strategies to calculate the dissociation potential energy curves of these dimers. Figure 4 illustrates the dissociation potential energy curves of benzene dimer for the parallel conformation calculated with (a) pure LDA and GGA functionals and LC functionals and (b) LC functionals with ALL dispersion functional. As shown in Fig. 4a, traditional DFTs using pure LDA and GGA functionals give much different dissociation potential energy curves for benzene dimer [14] similarly to the curves of rare-gas dimers [1]. This difference is also attributed to the lack of long-range exchange interactions in exchange functionals (section "Long-Range Exchange Interactions") and the poor HDGLG limit of some correlation functionals (section "Effects of Correlation Functionals") [14]. Figure 4b indicates that all LC exchange functionals with OP correlation functional, which gives the correct HDGLG limit, produces very accurate potential energy curves close to CCSD(T) one by combining with a dispersion functional. Since similar curves are obtained for naphthalene dimer [14], this indicates that besides dispersion corrections, long-range exchange interactions should be included with an appropriate correlation functional satisfying the HDGLG limit condition in DFT calculations of dispersion interactions. Nonetheless, further quantitative validation is required, because some other dispersion-corrected DFTs also give correct potential energy curves for benzene dimer with no long-range correction [70, 72].

As a benchmark set for the quantitative validation of dispersion corrections, Hobza and coworkers suggested the S22 set [73], which contains 22 dimers including benzene dimer and other π-stacked systems (Fig. 5). This S22 benchmark set provides interaction energies of hydrogen-bonded, dispersion-bonded, and mixed

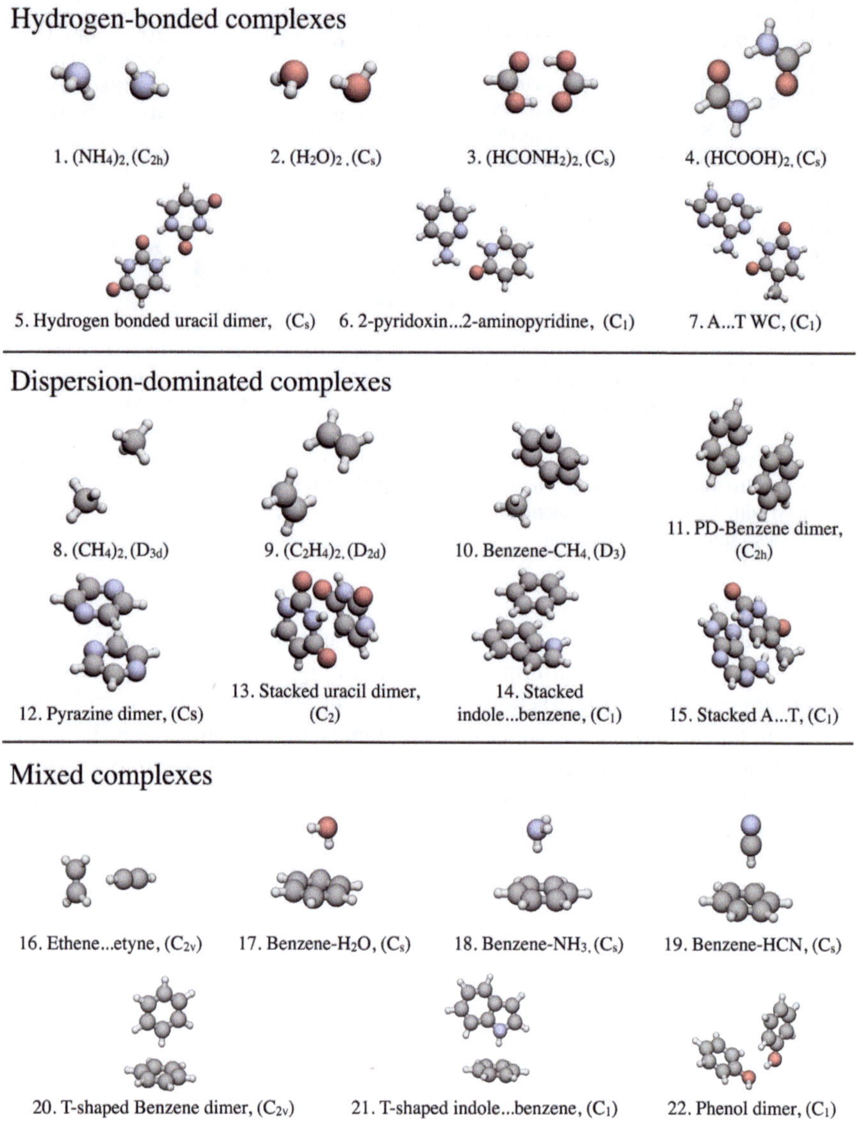

Fig. 5 The S22 benchmark set of weakly bonded systems

complexes, which are the calculated results of the CCSD(T) method at the complete basis set (CBS) limit [74]. Due to its convenience, this benchmark set has been used not only in testing the accuracies of dispersion corrections but also in determining the adjustable parameters of semiempirical dispersion-corrected functionals. Table 1 displays the mean absolute deviations (MAD) of various dispersion-corrected DFT calculations for the S22 benchmark set in the ascending order. This table clearly shows that independent of dispersion corrections combined, LC functionals give better accuracies than those of semiempirical functionals and vdW

Table 1 Mean absolute deviations (MAD) of DFT calculations using various types of dispersion corrections for the S22 benchmark set in ascending order (kcal/mol). The MAD of the MP2 method at complete basis set (CBS) limit is also displayed for comparison. For each method, the section giving an explanation in this chapter is appended

Method	Type of correction(s)	Section(s)	MAD	Reference
ωB97X-D	LC + semiempirical	"Semiempirical Dispersion-Corrected Functionals" and "Long-Range Exchange Interactions"	0.22	[66]
BLYP-D3	Semiempirical	"Semiempirical Dispersion-Corrected Functionals"	0.23	[50]
ωB97X-2	LC + perturbation	"Combination with Perturbation Theories" and "Long-Range Exchange Interactions"	0.26	[67]
LC-BOP+LRD	LC + vdW functional	"Van der Waals (Dispersion) Functionals" and "Long-Range Exchange Interactions"	0.27	[45]
B2PLYP-D3	Semiempirical + perturbation	"Combination with Perturbation Theories" and "Semiempirical Dispersion-Corrected Functionals"	0.29	[50]
RSH+RPAx	LC + AC/FDT	"Linear-Response Theories" and Long-Range Exchange Interactions	0.32	[40]
M06-2x	Semiempirical	"Semiempirical Dispersion-Corrected Functionals"	0.44	[75]
BLYP-D	Semiempirical	"Semiempirical Dispersion-Corrected Functionals"	0.55	[48]
B97-D	Semiempirical	"Semiempirical Dispersion-Corrected Functionals"	0.61	[48]
B3LYP-D	Semiempirical	"Semiempirical Dispersion-Corrected Functionals"	0.70	[48]
MP2/CBS	Perturbation	"Combination with Perturbation Theories"	0.78	[50]
HF+VV09	vdW functional	"Van der Waals (Dispersion) Functionals"	0.89	[76]
M05-2x	Semiempirical	"Semiempirical Dispersion-Corrected Functionals"	0.90	[75]
vdW-DF(rPW86)	vdW functional	"Van der Waals (Dispersion) Functionals"	1.03	[76]
rPW86+VV09	vdW functional	"Van der Waals (Dispersion) Functionals"	1.20	[76]
vdW-DF(revPBE)	vdW functional	"Van der Waals (Dispersion) Functionals"	1.44	[76]
vdW-DF(HF)	vdW functional	"Van der Waals (Dispersion) Functionals"	2.80	[76]

functionals. In contrast, relatively poor results are given by vdW-DF functionals, which combine pure functionals such as revPBE or the HF exchange integral with a vdW functional. Therefore, this table obviously supports the above suggestion that long-range exchange interactions should be included to reproduce van der Waals bonds. It is interesting to note that some dispersion-corrected DFTs give more accurate results than those of the MP2 method at the CBS limit. In this connection, it is reported that DFT-SAPT (section "Combination with Perturbation Theories") gives close accuracies at the CBS limit to those of M06-2x and BLYP-D for the S22 set, although it is not listed in this table due to the unrecording of MAD [74]. These accuracies may be taken as poorer-than-expected, because perturbation theories have been believed to give highly accurate dispersion interactions. However, these results are reasonable because the poor interaction energies in MP2 calculations have been found for several π-stacked dimers having near-degenerate orbitals that usually cause instability in perturbation calculations [65].

Carbon Materials: Nanotube and Graphene

Besides benzene and naphthalene dimers, various carbon materials have also attracted attentions in the context of π-stacking interactions. The most frequently targeted systems are nanotube and graphene for their electronic, adsorptive, and reactive properties. So far, many DFT studies have been reported especially for single-walled carbon nanotube (SWCNT) and graphene in the fields of both quantum chemistry and solid state physics.

For SWCNT, some studies focus on the π-stacking-mediated solubility in organic solvent (Fig. 6). SWCNT is a promising multifunctional material to be an efficient gas storage element, a sensor, and a drug-delivery agent. However, it is well known in the field of carbon materials that SWCNT is hardly dissolved in any solvent. This may impose restrictions on industrial or medical use. For solving this problem, many DFT calculations have been performed to suggest organic molecules solvating SWCNT by π-stacking interactions. The π-stacking on SWCNT has been examined with various model systems of SWCNT for the adsorptions of various organic molecules: e.g., planar organic molecules [77], liquid crystal molecules [78], and benzene derivatives [79]. SWCNT has also attracted attentions as gas storage materials [15]. Therefore, DFT studies have also focused on the adsorptions of various small molecules [80]: e.g., hydrogen [81] and methane [82] molecules, on SWCNT. These studies have reported that conventional DFTs provide reasonable interaction energies. However, further investigations are required for approaching to the adsorptive properties of SWCNT, because most conventional calculations have used unreliable dispersion-uncorrected functionals such as pure LDA and GGA functionals for evaluating the dispersion interactions. The adsorptive property of SWCNT is, therefore, still a challenge in DFT calculations.

Graphene is also one of the most frequently studied π-stacking systems in carbon materials because of its high carrier mobility, high thermal and electronic conductivity, and strong material behavior. For graphene, the adsorptive property has been

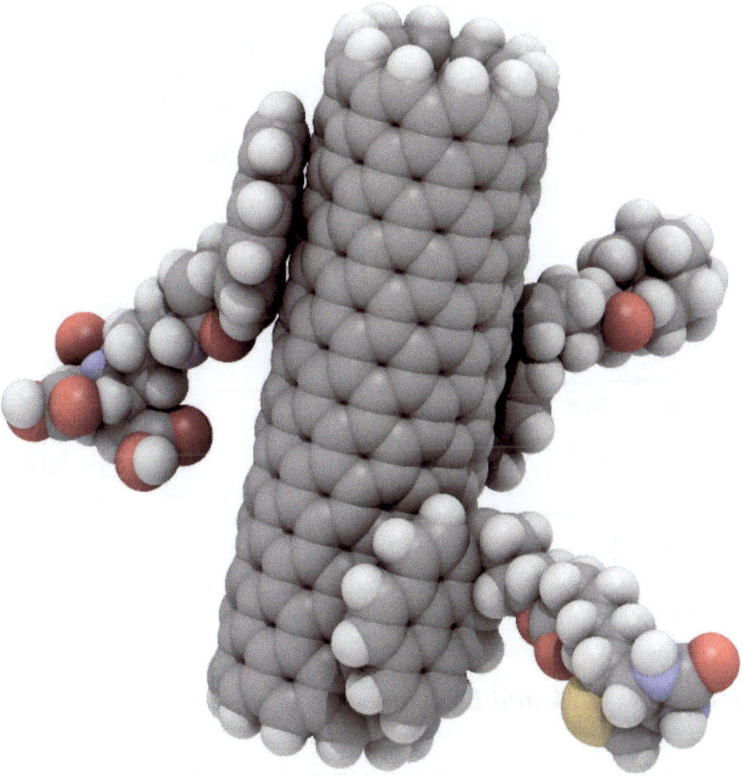

Fig. 6 An example of the π-stacking of single-walled carbon nanotube

explored in the π-stacking calculations of its model systems such as coronene [83, 84]. Dispersion-corrected DFTs have provided accurate π-stacking interaction energies between these graphene models, if appropriate dispersion corrections are incorporated. Therefore, graphene models are usually used as trial systems for examining dispersion corrections. These calculations have consequently revealed the adsorptive properties of planar polyaromatic systems. For example, π-stacking interaction energies of these systems are found to increase as a function of the number of hydrogen atoms at the edge [84]. Graphene is also theoretically investigated for the adsorption of biomolecules such as DNA in the context of drug delivery, molecule immobilization, and biosensor [85, 86]. As a result, it is suggested that the nucleobases of DNA form weak hydrogen-bond-like interactions with graphene between the pyramidal amino groups of the bases and the π-center of graphene [86]. Graphene is also well known as a semiconductor, which has zero band gap in the electronic state. In association with this unique electronic state, band gap calculations are also performed for the π-stacking of aromatic systems on graphene to develop high performance nanodevices [87]. Since graphene is attractive for its unique electronic properties, we expect that the π-stacking of graphene would be much more investigated from the viewpoint of electronic structure.

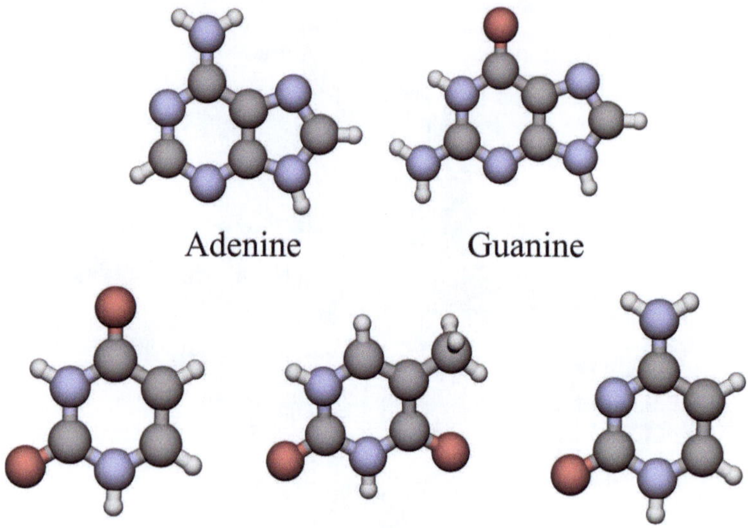

Fig. 7 DNA nucleobases

Biomolecules: DNA and Proteins

π-Stackings have also been attracted attentions in the field of biological science. Besides π-stacking effects on the structures of DNA and proteins, there are various DFT studies on π-stackings of biomolecules: e.g., DNA-nanotube [88] and DNA-graphene [86] stackings. However, most DFT studies have drawn attentions to the base stackings of DNA and protein foldings in this field.

The base stackings in DNA have been studied using the hydrogen-terminated models of five nucleobases (Fig. 7), i.e. guanine, cytosine, adenine, thymine, and uracil, in DFT calculations. For the base stackings, there is even a benchmark set in JSCH-2005 database [73], which contains hydrogen-bonded DNA base pairs, stacked base pairs, and amino acid pairs besides S22 set mentioned above, for examining the applicabilities of dispersion corrections. However, this benchmark set has been calculated only with DFT-D (section "Semiempirical Dispersion-Corrected Functionals") [89] in dispersion-corrected DFTs, while this set is frequently calculated using semiempirical molecular orbital theories combined with semiempirical dispersion corrections. This is due to the computational cost needed in the calculation of this set containing 143 complexes. Most dispersion-corrected DFT calculations of DNA focus only on the stackings of five nucleobases. So far, DFT-D [38], vdW-DF (section "Van der Waals (Dispersion) Functionals") [90], and DFT-SAPT (section "Combination with Perturbation Theories") [91] have been applied to the base-stacking calculations and consequently have provided accurate interaction energies. For the stacked bases, excited state calculations have also been performed by TDDFT using LC functionals (section "Long-Range Exchange

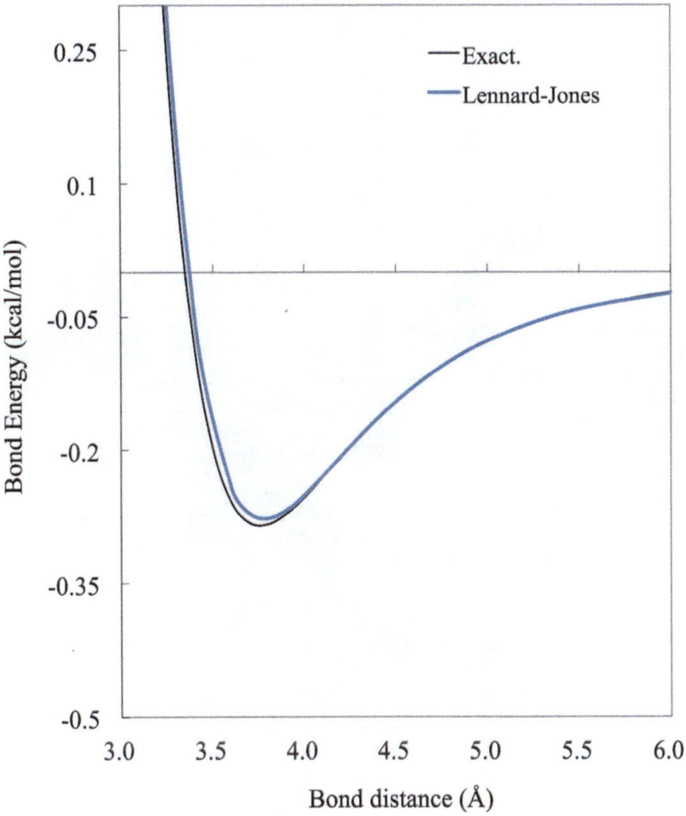

Fig. 8 The Lennard–Jones potential and empirical potential of argon dimer

Interactions") [92, 93]. We see that dispersion-corrected DFT studies of DNA would shift attentions from base-stacking structures to electronic properties.

In most dispersion calculations of proteins, folding structures have been discussed with no particular attention to π-stackings. These folding structure calculations are carried out by classical MDs using empirical potentials such as CHARMM [46] and Amber [47]. In these empirical potentials, dispersion interactions are incorporated by London's classical dispersion energy in Eq. (4) multiplied by an empirical parameter in the Lennard–Jones potential (Fig. 8). Note that exchange interaction is also incorporated as the repulsion term in this potential as supported by the above discussion. Most dispersion-corrected DFT studies of proteins have used hydrogen-terminated small amino acid residues to discuss π-stackings in proteins [49], because DFT calculations are too time consuming to determine the folding structures of proteins. In practice, since protein structures are essentially not so sensitive to the quality of calculation methods, classical potential calculations may be the best way to obtain the structures. However, proteins would be further investigated for the electronic structures in near future at the request of drug discovery field. We, therefore, suppose that dispersion-corrected DFT becomes more important even in the research field of proteins.

Fig. 9 Buckycatcher

Supramolecules

Recently, supramolecules become one of the hottest π-stacked systems in DFT calculations. In supramolecules, molecules self-assemble with weak bonds such as hydrogen and van der Waals bonds. We should notice that optimized structures do not always provide enough information to design new materials. Material design often requires the detailed information on electronic properties, which are not given in classical MD calculations. Dispersion-corrected DFTs are best suited to obtaining this information. Fullerene is one of the most frequently used constituent materials in π-stacking supramolecules. In particular, C_{60} fullerene has been studied for its π-stacking with polycyclic organic molecules to produce supramolecules such as buckycatcher (Fig. 9) [94, 95]. Self-assembled nanowires are also frequently raised systems in DFT calculations. For example, the nanowires of azaadamantanetrione [96] and cyclic oligothiophene [97] have been calculated. Moreover, metal complexes self-assembling to huge structures are noteworthy. That is, many metal complexes coordinating aromatic ring molecules are π-stacked to form huge systems. Dispersion-corrected DFTs are promising even in determining the structures of these metal complexes, because it is hard to optimize the geometries of metal-contained systems in classical MD calculations. Actually, several calculations are recently performed [98]. DFT calculations have also been used for other self-assembled systems to analyze their electronic states. Since most of these calculations have been carried out in recent 5 years, the use of dispersion-corrected DFTs may keep increasing in the calculations of supramolecules.

Conclusions

π-Stacking is a dispersion interaction between cyclic compounds such as aromatic rings. Since π-stacking often determines the structures of significant large systems such as carbon nanomaterials, biomolecules, and supramolecules, it currently draws attentions in a wide variety of scientific and industrial fields. Although the name "π-stacking" implies the contribution of π orbitals, it is confirmed that π orbitals play no particular role in π-stackings. Namely, π-stacking is nothing but usual dispersion interaction. Dispersion interaction is the only van der Waals interaction that the Kohn–Sham SCF calculation using conventional exchange-correlation functionals cannot reproduce. Therefore, an explicit dispersion correction is required for functionals to investigate π-stackings on the Kohn–Sham SCF calculations.

Various dispersion corrections have, so far, been developed in the field of DFT. Fortunately, these corrections are generally classified into only five categories: classical dispersion corrections, combinations with perturbation theories, linear-response theories, van der Waals (dispersion) functionals, and semiempirical dispersion-corrected functionals. Naturally, these corrections have both advantages and disadvantages as explained in the text. However, due to practical reasons such as computational time and user-friendliness, only van der Waals and semiempirical dispersion-corrected functionals seem to be frequently used in current DFT calculations. What should not be overlooked is the repulsion, which balances with the dispersion attraction in van der Waals bonds. In the Lennard–Jones potential, this repulsion appears as the R^{-12} term, which is interpreted to come from long-range exchange interaction. Note that this long-range exchange interaction is also deficient in conventional functionals.

The significance of the long-range exchange interaction is also numerically shown in the potential energy curves of small π-stacked benzene and naphthalene dimers and the dispersion interaction energies of the S22 benchmark set. As a result of these calculations, long-range corrected (LC) DFTs, in which long-range exchange interactions are complemented to exchange functionals, give the best performance in all dispersion-corrected DFTs. Nevertheless, LC-DFTs have not been applied to significant π-stacked systems that various dispersion-corrected DFTs have been examined: nanotube, graphene, DNA, protein, and supramolecule. However, the use of other dispersion-corrected DFTs is also not so common in the calculations of these systems. Since this is mainly due to the high computational cost required in the calculations, we expect that dispersion-corrected DFTs become one of the main tools for investigating these systems theoretically.

Acknowledgements This research was supported by the Japanese Ministry of Education, Culture, Sports, Science and Technology (MEXT) (Grant: 23225001 and 24350005).

References

1. Kamiya M, Tsuneda T, Hirao K (2002) J Chem Phys 117:6010
2. Tsuneda T, Sato T (2009) Butsuri 64:291
3. Israelachvili JN (1992) Intermolecular and surface forces. Academic Press, London
4. Kohn W, Sham LJ (1965) Phys Rev A 140:1133
5. London FW (1930) Z Phys 63:245
6. Perdew JP, Wang Y (1992) Phys Rev B 45:13244
7. Perdew JP, Burke K, Ernzerhof M (1996) Phys Rev Lett 77:3865
8. Colle R, Salvetti O (1975) Theor Chim Acta 37:329
9. Lee C, Yang W, Parr RG (1988) Phys Rev B 37:785
10. Tsuneda T, Suzumura T, Hirao K (1999) J Chem Phys 110:10664
11. Grimme S (2008) Angew Chem Int Ed 47:3430
12. Hunter CA, Sanders JKM (1990) J Am Chem Soc 112:5525
13. Hunter CA (1993) Angew Chem 105:1653
14. Sato T, Tsuneda T, Hirao K (2005) J Chem Phys 123:104307
15. Dillon AC, Heben MJ (2001) Appl Phys A 72:133
16. Dresselhaus MS, Dresselhaus G, Avouris P (2000) Carbon nanotubes: synthesis, structure, properties and applications. Springer, Berlin
17. Yang W, Ratinac KR, Ringer SP, Thordarson P, Gooding JJ, Braet F (2010) Angew Chem Int Ed 49:2114
18. Shao Y, Wang J, Wu H, Liu J, Aksay IA, Lin Y (2010) Electroanalysis 22:1027
19. Lerman LS (1961) J Mol Biol 3:18
20. Zimm BH (1960) J Chem Phys 33:1349
21. Burley SK, Patsko GA (1985) Science 229:23
22. Hunter CA, Singh J, Thornton JM (1991) J Mol Biol 218:837
23. Quiocho FA, Vyas NK (1984) Nature (London) 310:381
24. Vyas NK, Vyas MN, Quiocho FA (1987) Nature (London) 327:635
25. Vyas NK, Vyas MN, Quiocho FA (1988) Science 242:1290
26. Jorgensen WL, Severance DL (1990) J Am Chem Soc 112:4768
27. Hunter CA (1994) Chem Soc Rev 23:101
28. Kryger G, Silman I, Sussman JL (1998) J Physiol (Paris) 92:191
29. Cerny J, Kabelac M, Hobza P (2008) J Am Chem Soc 130:16055
30. Hoeben FJM, Jonkheijm P, Meijer EW, Schenning PHJ (2005) Chem Rev 105:1491
31. Wheeler SE, Houk KN (2008) J Am Chem Soc 130:10854
32. Singh RK, Tsuneda T (2013) J Comput Chem 34:379
33. Becke AD, Johnson ER (2005a) J Chem Phys 123:154101
34. Becke AD, Johnson ER (2005b) J Chem Phys 125:154105
35. McWeeny R (1992) Methods of molecular quantum mechanics, 2nd edn. Academic Press, San Diego
36. Williams HL, Chabalowski CF (2001) J Phys Chem A 105:646
37. Schwabe T, Grimme S (2007) Phys Chem Chem Phys 9:3397
38. Grimme S (2006) J Chem Phys 124:034108

39. Langreth DC, Perdew JP (1975) Solid State Comm 17:1425
40. Zhu W, Toulouse J, Savin A, Angyan JG (2009) J Chem Phys 131:174105
41. Andersson Y, Langreth DC, Lundqvist BI (1996) Phys Rev Lett 76:102
42. Dobson JF, Dinte BP (1996) Phys Rev Lett 76:1780
43. Dion M, Rydberg H, Schröder E, Langreth DC, Lundqvist BI (2004) Phys Rev Lett 92:246401
44. Vydrov OA, van Voorhis T (2009) J Chem Phys 130:104105
45. Sato T, Nakai H (2009) J Chem Phys 131:224104
46. Brooks RE, Bruccoleri BR, Olafson BD, States DJ, Swaminathan S, Karplus M (1983) J Comput Chem 4:187
47. Pearlman DA, Case DA, Caldwell JW, Ross WS, Cheatham TEI, DeBolt S, Ferguson D, Seibel G, Kollman P (1995) Comput Phys Comm 91:1
48. Antony J, Grimme S (2006) Phys Chem Chem Phys 8:5287
49. Zhao Y, Truhlar DG (2008a) Theor Chem Acc 120:215
50. Grimme S, Antony J, Ehrlich S, Krieg H (2010) J Chem Phys 132:154104
51. Casimir H, Polder D (1948) Phys Rev 73:360
52. Starkschall G, Gordon R (1972) J Chem Phys 56:2801
53. Iikura H, Tsuneda T, Yanai T, Hirao K (2001) J Chem Phys 115:3540
54. Savin A (1996) In: Seminario JJ (ed) Recent developments and applications of modern density functional theory. Elsevier, Amsterdam
55. Becke AD (1988) Phys Rev A 38:3098
56. Tawada Y, Tsuneda T, Yanagisawa S, Yanai T, Hirao K (2004) J Chem Phys 120:8425
57. Song J-W, Tsuneda T, Sato T, Hirao K (2010) Org Lett 12:1440
58. Kamiya M, Sekino H, Tsuneda T, Hirao K (2005) J Chem Phys 122:234111(1)
59. Kishi R, Bonness S, Yoneda K, Takahashi H, Nakano M, Botek E, Champagne B, Kubo T, Kamada K, Ohta K, et al (2010) J Chem Phys 132:094107
60. Tsuneda T, Song J-W, Suzuki S, Hirao K (2010) J Chem Phys 133:174101
61. Yanai T, Tew DP, Handy NC (2004) Chem Phys Lett 91:51
62. Vydrov OA, Heyd J, Krukau A, Scuseria GE (2006) J Chem Phys 125:074106(1)
63. Chai J-D, Head-Gordon M (2008a) J Chem Phys 128:084106(1)
64. Giese TJ, Audette VM, York DM (2003) J Chem Phys 119:2618
65. Sato T, Tsuneda T, Hirao K (2007) J Chem Phys 126:234114
66. Chai J-D, Head-Gordon M (2008b) Phys Chem Chem Phys 10:6615
67. Chai J-D, Head-Gordon M (2009) J Chem Phys 131:174105
68. Ma SK, Brueckner KA (1968) Phys Rev 165:18
69. Dreizler RM, Gross EKU (1990) Density-functional theory an approach to the quantum many-body problem. Springer, Berlin
70. Meijer EJ, Sprik M (1996) J Chem Phys 105:8684
71. Tsuzuki S, Lüthi H (2001) J Chem Phys 114:3949
72. Ehrlich S, Moellmann J, Grimme S (2012) Acc Chem Res. doi:10.1021/ar3000844
73. Jurecka P, Sponer J, Cerny J, Hobza P (2006) Phys Chem Chem Phys 8:1985
74. Riley KE, Pitonak M, Jurecka P, Hobza P (2010) Chem Rev 110:5023
75. Pernal K, Podeszwa R, Patkowski K, Szalewicz K (2009) Phys Rev Lett 103:263201
76. Vydrov OA, Van Voorhis T (2010) J Chem Phys 132:164113
77. Tournas F, Latil S, Heggie MI, Charlier J-C (2005) Phys Rev B 72:075431
78. Park KA, Lee SM, Lee SH, Lee YH (2007) J Phys Chem C 111:1620
79. Woods LM, Badescu SC, Reinecke TL (2007) Phys Rev B 75:155415
80. Zhao J, Buldum A, Han J, Lu JP (2002) Nanotechnologoy 13:195
81. Tada K, Furuya S, Watanabe K (2001) Phys Rev B 63:155405
82. Grimme M, Steinmetz S, Korth M (2007) J Org Chem 72:2118
83. Mackie ID, DiLabio GA (2008) J Phys Chem A 112:10968
84. Björk J, Hanke F, Palma C-A, Samori P, Cecchini M, Persson M (2010) J Phys Chem Lett 1:3407
85. Gowtham S, Scheicher RH, Ahuja R, Pandey R, Karna SP (2007) Phys Rev B 76:033401
86. Panigrahi S, Bhattacharya A, Banerjee S, Bhattacharrya D (2012) J Phys Chem C 116:4374
87. Zhang Z, Huang H, Yang X, Zang L (2011) J Phys Chem Lett 2:2897

88. Gao H, Kong Y (2004) Annu Rev Mater Res 34:123
89. Morgado C, Vincent MA, Hillier IH, Shan X (2007) Phys Chem Chem Phys 9:448
90. Cooper VR, Thonhauser T, Puzder A, Schröder E, Lundqvist BI, Langreth DC (2007) J Am Chem Soc 130:1304
91. Hesselmann A, Jansen G, Schütz M (2006) J Am Chem Soc 128:11730
92. Lange AW, Rohrdanz MA, Herbert JM (2008) J Phys Chem B Lett 112:6304
93. Santoro F, Barone V, Improta R (2009) J Am Chem Soc 131:15232
94. Chakrabarti S, Ruud K (2009) J Phys Chem A 113:5485
95. Zhao Y, Truhlar DG (2008b) Phys Chem Chem Phys 10:2813
96. Sumpter BG, Meunier V, Valeev EF, Lampkins AJ, Li H, Castellano RK (2007) J Phys Chem C 111:18912
97. Wong BM, Ye SH (2011) Phys Rev B 84:075115
98. Choudhury SR, Gamez P, Robertazzi A, Chen C-Y, Lee HM, Mukhopadhyay S (2008) Cryst Growth Des 8:3773

MIX
Papier aus verantwortungsvollen Quellen
Paper from responsible sources
FSC® C105338

If you have any concerns about our products,
you can contact us on
ProductSafety@springernature.com

In case Publisher is established outside the EU,
the EU authorized representative is:
Springer Nature Customer Service Center GmbH
Europaplatz 3, 69115 Heidelberg, Germany

Printed by Libri Plureos GmbH
in Hamburg, Germany